The Use of Pulsating Stars in Fundamental Problems of Astronomy

International Astronomical Union
Union Astronomique Internationale

The following Colloquia of the International Astronomical Union are published for the Union by Cambridge University Press.

82. Cepheids. *Edited by Barry F. Madore.* 0 521 30091 6. 1985.

91. History of Oriental Astronomy. *Edited by G. Swarup, A. K. Bag and K. S. Shukla.* 0 521 34659 2. 1987.

92. Physics of Be Stars. *Edited by A. Slettebak and T. P. Snow.* 0 521 33078 5. 1987.

101. Supernova Remnants and the Interstellar Medium. *Edited by R. S. Roger and T. L. Landecker.* 0 521 35062 X. 1988

106. Evolution of Peculiar Red Giant Stars. *Edited by Hollis Johnson and Ben Zuckerman.* 0 521 36617 8. 1989.

111. The Use of Pulsating Stars in Fundamental Problems of Astronomy. *Edited by Edward G. Schmidt.* 0 521 37023 X. 1989.

International Astronomical Union
Union Astronomique Internationale

The Use of Pulsating Stars in Fundamental Problems of Astronomy

The proceedings of International Astronomical Union 111th *colloquium*

Edited by

EDWARD G. SCHMIDT

CAMBRIDGE UNIVERSITY PRESS

Cambridge

New York   Port Chester

Melbourne   Sydney

Published by the Press Syndicate of the University of Cambridge
The Pitt Building, Trumpington Street, Cambridge CB2 1RP
40 West 20th Street, New York, NY 10011, USA
10 Stamford Road, Oakleigh, Melbourne 3166, Australia

© Cambridge University Press 1989

First published 1989

Printed in Great Britain at the University Press, Cambridge

British Library cataloguing in publication data available

Library of Congress cataloguing in publication data available

ISBN 0 521 37023 X

CONTENTS

1. Stellar Pulsation and Evolution

The Masses and Pulsation Modes of Classical Cepheids — Arthur N. Cox — 1

The Evolution of Stars of Medium Mass — Cesare Chiosi — 19

Progress Toward an Improved Equation of State and Opacity for Stellar Envelopes — Dimitri Mihalas — 59

Stellar Mass Loss and Pulsation — L. A. Willson — 63

2. Morphology and History of the Galaxy

Structure and Evolution of the Milky Way Galaxy — Gerard Gilmore & Rosemary F. G. Wyse — 83

Horizontal Branch Evolution — R. T. Rood & D. A. Crocker — 103

The Absolute Magnitudes of RR Lyrae Stars and the Age of the Galaxy — Allan Sandage — 121

The Globular Cluster ω Centauri and its RR Lyrae Variables — R. J. Dickens — 141

3. Comparison of the Milky Way with Other Galaxies

Variable Stars and the Cosmic Distance Scale — Jeremy Mould — 169

Cepheids in Local Group Galaxies — Edward G. Schmidt — 177

The Baade–Wesselink Technique — Thomas J. Moffett — 191

Mira Variables, Stellar Evolution and Galactic Structure — M. W. Feast — 205

Anomalous Cepheids and Population II Blue Stragglers — James M. Nemec — 215

4. Abstracts of Poster Papers

Type I Intermittent Chaos in Hydrodynamic Pulsation Models — Toshiki Aikawa — 247

Double Mode Pulsating Stars and Opacity Changes — G. K. Andreasen — 248

The Pulsation of Some δ Scuti Stars with Unusual Light Curves — E. Antonello & E. Poretti — 249

Problems with the Baade–Wesselink Method — E. Böhm-Vitense, P. Garnavich, M. Lawler, J. Mena-Werth, S. Morgan, E. Peterson & S. Temple — 250

Multiple Close Frequencies of the δ Scuti Star θ^2 Tau: The Second Multisite Campaign — M. Breger, R. Garrido, Huang Lin, Jiang Shi-yang, Guo Zi-he, M. Frueh & M. Paparo — 251

Cepheid Evolution with Pulsationally-Driven Mass Loss — Wendee M. Brunish & Lee Anne Willson — 252

The Distance and Age of the Globular Cluster M5 — Bruce W. Carney, David W. Latham, Rodney V. Jones & Judy A. Beck — 253

Discontinuity Modes in Polytropes — Bradley W. Carroll — 254

The Luminosity–Metallicity Relation for RR Lyrae Stars and Its Implications for the Astronomical Distance Scale — G. Clementinni & C. Cacciari — 255

A Possible Pulsation Mechanism for B Stars — Arthur N. Cox — 256

Solar-Like Oscillations in Late Spectral Class Stars 257
Arthur N. Cox & Julius H. Cahn
Oscillation Frequencies of Solar Models 258
Arthur N. Cox, Joyce A. Guzik & Russell B. Kidman
The Effects of Time-Dependent Convection on White Dwarf Radial Pulsations 259
Arthur N. Cox & Sumner G. Starrfield
S Sge: A Cepheid Triple System 260
N. R. Evans, M. H. Slovak and D. L. Welch
The Blue Edge of the Helium Star Instability Strip 261
Yu. A. Fadeyev
Envelope Distention and Mass Loss in W Vir Pulsating Variables 262
Yu. A. Fadeyev & H. Muthsam
Forthcoming Cepheid Database 263
J. Donald Fernie & Nancy Remage Evans
Hydrodynamic Models of Low-Mass Pulsating Supergiants with Radiation Transfer 264
A. B. Fokin
A Study of Period Doubling in a One-Zone Pulsating Stellar Model 265
A. B. Foken
The Period–Radius Relation from 101 Cepheid Radii 266
W. P. Gieren, T. G. Barnes III & T. J. Moffett
Irregularity Interpreted as Low Dimension Chaos for Convective Models of W Vir Variables 267
S. Ami Glasner & J. Robert Buchler
Period Doubling in Variable Stars: A Tentative Interpretation of Observed Light Curves of Variable White Dwarfs and Mira Stars 268
M. J. Goupil, A. Baglin & M. Auvergne
Star Formation History of the Large Magellanic Cloud and Asymptotic Giant Branch Evolution Obtained From a Study of the Long Period Variables 269
Shaun M. G. Hughes & P. R. Wood
On the Irregular Light Variation of the RV Tau Star R Sct 270
Z. Kollath & G. Kovacs
Is Delta Scuti Seismology Possible? 271
G. Kovacs
The Luminosities of 13 Field RR Lyrae Stars: The Correlation with Metallicities 272
T. Liu & K. A. Janes
Globular Cluster Distances from the RR Lyrae Log (Period) – Infrared Magnitude Relation 273
A. J. Longmore, R. Dixon, I. Skillen, R. F. Jameson & J. A. Fernley
RR Lyrae Stars and the Sandage Period-Shift Effect Examined Using IR-Derived Temperatures 274
A. J. Longmore, R. Dixon, I. Skillen, R. F. Jameson & J. A. Fernley
Secular Changes in the Light Curve of the Short-Period Cepheid EU Tau 275
J. M. Matthews & W. P. Gieren
Preliminary Results of a World-Wide Photoelectric Campaign on δ Scuti 276
B. McNamara, S. Baggett, J. Pena, K. Thompson, G. Moore, L. Mantegazza, K. Sekiguchi & M. Candy
Observations of Variability in the Radial Velocity of α Boo 277
W. J. Merline
The Effects of Metal Enhancement on the Period Ratio of Double Mode Cepheids 278
Siobahn Morgan & Arthur N. Cox
The Importance of 3:1 Resonances in Stellar Pulsations 279
Pawel Moskalik & J. Robert Buchler
Statistics of Pulsating Variables 280
M. F. Novikova & Yu. A. Fadeyev
Nonlinear RR Lyrae Models and Double Mode Pulsation 281
Dale A. Ostlie
Pulsations of Eötvös Spheres 282
W. Dean Pesnell

Contents

A Search for RR Lyraes in Wide Binary Systems   283
Charles F. Prosser

The Binary Cepheid S Sagittae Revisited   284
Mark H. Slovak, Thomas G. Barnes III, Nancy R. Evans, Douglas L. Welch & Thomas J. Moffett

CCD Observations of Variable Stars in Globular Clusters   285
H. A. Smith, J. R. Kuhn & J. Curtis

The Chromosphere of β Cassiopiae   286
Terry J. Teays, Edward G. Schmidt, Massimo Fracassini & Laura E. Pasinetti Fracassini

The Unusual Period Distribution of the RR Lyrae Variables in NGC 5897   287
Amelia Wehlau

Pulsation of α Cir (HD 128898)   288
Werner W. Weiss & Hartmut Schneider

THE MASSES AND PULSATION MODES OF CLASSICAL CEPHEIDS

Arthur N. Cox
Theoretical Division, Los Alamos National Laboratory

Abstract. Recent general observations pertaining to the masses and pulsation modes of classical Cepheids are reviewed. Certain special ones of these variables that display unusual behavior such as long term amplitude variations, overtone pulsations, and double-mode behavior, including one Cepheid in the first and second overtones as well as those in the fundamental and first overtone, are discussed in some detail. I suggest that the amplitude varying supergiant Cepheid HR 7308, which seems to be pulsating in the second radial overtone, is alternating between states where there is enough helium to barely give kappa and gamma effect driving, and where the helium has settled too deep for this driving. Reestablishment of the helium may be due to rapid levitation of the CNONe elements which cause convection and which then dredge-up the helium again to suppress the convection and also to drive pulsations. The use of Fourier fitting to the light and velocity curves have revealed some features that are interesting and are discussed, such as the pulsation mode discrimination and the derivation of accurate Wesselink radii. The seven methods of determining the masses of these stars are presented with a short critique of each. Evolution and pulsation masses depend on uncertain observed luminosities, and these have been revised recently. Evolution masses also depend on the blue looping yellow giant evolution tracks that may need revision because of convective core overshooting and possible opacity, Z, or Y abundance increases. The most discrepant masses, however, are those that depend on period ratios - the "bump" and "beat" masses. The possibility that the Cepheid internal structure can be modified enough by doubling the material opacity seems unlikely. It is suggested, though, that perhaps the period ratio mass anomaly solution can be found in CNONe element enhancement by radiation absorption levitation that would give higher opacities by abundance effects instead of any revisions in the opacity calculations. Some details for this mass anomaly "solution" are presented.

INTRODUCTION

A review of the status of research in classical Cepheids is an enormous task. Observational work continues unabated, it seems, and theoretical investigations still center on several unsolved problems. To limit this review properly, I will concentrate only on the aspects of Cepheid research that have to do with stellar structure, evolution, and pulsation, leaving the important features of the period-luminosity-color relations and the distance scale to other galaxies to the other reviewers. Unfortunately, I cannot review the results being presented at this conference, and I am sure that some of the things I say will be shown to be incorrect. Further some of my questions may be already answered.

Cox (1980) in his review at Los Alamos, suggested three problems that he considered at the frontier of the theoretical research on Cepheids. These were the full amplitude behavior of Cepheid pulsations as determined from nonlinear calculations, convection, and the masses of the Cepheids. While I will categorize things a bit differently here, the problems that he discussed remain with us today. Mostly, I will review the problem of the masses, attempting to put some of the recent controversies at least into focus.

One can tell that the Cepheid masses are always a topic of interest, because at the last pulsation meeting in Los Alamos, Norman Simon (1987) reviewed the status of these masses. In fact, his entire review centered about the mass anomalies, especially those for the shorter period ones that display period ratios by the light and velocity curve bumps or directly by having two pulsation modes going simultaneously.

Let me close this introduction by noting that Cepheids are everywhere. The recent loss of one of them about 10,000 years ago (plus 170,000 years) by becoming supernova 1987A (Woosley and Phillips, 1988) reminds us of how important they are to studies of stellar evolution.

GENERAL OBSERVATIONAL DATA

The extensive photometric data of Moffett and Barnes (1987 and others) have been a major advance in the knowledge of Cepheid pulsations. The use of these photometric (B-V) and (V-R) data and simultaneous radial velocity measurements published by Barnes, Moffett and Slovak (1987) will hopefully allow them to derive Baade-Wesselink radii for as many as 112 Cepheids. Results to date for 63 Cepheids give radii that are consistent with masses obtained by conventional methods, but accuracies as high as needed to influence research on Cepheid mass discrepancies are not available. To get radii with uncertainties of only a few percent, and masses with uncertainties of less than 10 percent, new data will have to be obtained.

Observations of Cepheid mass loss have been made by Deasy and Butler (1986), McAlary and Welch (1986), and Deasy (1988). They have used IRAS data together with their spectra to show that the mass loss rate rarely exceeds 10^{-8} solar masses per year. With that rate, it is not likely that the evolution in blue loops in the Hertzsprung-Russell (H-R) diagram will be affected much, but work by Brunish and Willson (1987) on evolution with even larger pulsation induced mass loss is continuing, as is presented at this conference. Deasy notes that the mass loss rates for non-variable supergiants are of the same order of magnitude as for the Cepheids. We must be careful not to confuse the observations for the low mass and gravity type II Cepheids, which show considerable mass loss, with those for the classical Cepheids that have apparently low mass loss rates. We also need to realize that close binary stars frequently have large mass loss rates, which are induced by the near companion, and they are not typical of a general process.

These mass loss rates are not that much different from those scaled from the observed solar rates by accounting for the larger Cepheid surface area. Mass loss rates of 10^{-10} solar masses per year, if the Cepheid wind is helium deficient as for the sun, are just those proposed by Cox, Michaud, and Hodson (1978) for considerable surface helium enhancement. Winds 100 times stronger, if they do not destroy any hydrogen-helium fractionation, can give all the surface helium enrichment desired for low mass Cepheids without affecting their evolution.

I would like to comment on another general observation, that of the orbit of U Aql (Welch, et al. 1987). This Cepheid, like several others now known from IUE spectra (Böhm-Vitense and Proffitt, 1985), has an early type companion. With reasonable guesses for the companion the mass of the U Aql Cepheid can be estimated to have an upper limit of between 6.4 and 8.8 solar masses. This star and others with approximate masses such as SU Cyg (Evans and Bolton, 1987) and S Mus and V636 Sco (Böhm-Vitense, 1986) as well as the approximate radii determined by many over the years shows that the masses of Cepheids are not as low as the two solar masses that are needed to explain the double-mode Cepheid period ratios.

Special Cepheids: RU Cam, HR 7308 (V473 Lyr), Polaris, η Aql

What has happened to those two intriguing Cepheids RU Cam (Demers and Fernie, 1966) and HR 7308 (Breger, 1981)? These two have displayed an amplitude change over a few years that is not predicted by pulsation theory. Could it be that somehow the helium kappa and gamma pulsation driving decreased by some internal composition change? Could the helium settle out fast enough, as it does for some δ Scuti variables, to make them pulsationally stable? Under such low helium conditions the CNONe elements might be levitated to large local abundances and even produce a convection zone. And then could convective dredgeup mixing reestablish the helium abundance, destroy the convection zone, and produce enough helium driving to once again barely overbalance the ever-present radiative damping?

I note that RU Cam is actually a type II Cepheid with probably a low mass like 0.6 solar mass. Its gravity must be very low. Also the variable HR 7308, F6 I-IIb supergiant, has a very low gravity. Settling of helium out of the pulsation driving region is quite reasonable for these two stars. Our recent calculations and those of Michaud, Vauclair and Vauclair (1983) show that helium depletion is a real possibility. Since HR 7308 is pulsating in the second overtone, which often, and in this case, is the most driven mode, the variable may be starving for helium. This is especially so because the star is actually very near the instability strip red edge at about 6100K (Burki, Mayor, and Benz, 1982) where usually only the fundamental mode is observed.

HR 7308 has an amplitude behavior that has been varying with a period of 1210 days (Breger, 1981) to maybe as long as 1400 days (Burki, et al. 1986). The varying shape of the light curve with amplitude should produce data for detailed theoretical studies of the light curve skewness.

As Simon (1987) has suggested, the Cepheids are indeed a homogeneous class with no large variations of mass or luminosity from star to star, but these two quantities are global, evolutionary ones. But pulsation depends on the surface layer compositions, which could easily vary somewhat from case to case, giving a range of pulsation behavior.

The Cepheid Polaris is interesting. Its period has been increasing for 40 years, and its amplitude is decreasing. I believe that the suggestion by Arellano Ferro (1983) that it is evolving to the red out of the instability strip is correct. However, it could be losing its pulsation region helium so that its driving is decreasing. At the mass of near six solar masses and luminosity of about a thousand solar luminosities, we apparently have a good marker of the pulsational instability strip red edge (maybe 5800K). It would be interesting to find other quite red Cepheids with period increases so that they might mark the instability strip red edge at other luminosities. T Mon with a 27 day period might be a good example at a much higher luminosity and cooler surface temperature.

The case of η Aql is another one with the pulsation period increasing, and details have recently been given by Jacobsen and Wallerstein (1987). A summary of period changes was given by Fernie (1984), showing that there was not any large discrepancy between theory and observation. Fernie has pointed out that fast evolving stars with large period changes would not be easily observed, and slow evolution at blue loop tips could give very small period changes even with standard evolution tracks. He believes that slow period changes do not necessarily indicate mass loss than can trap a Cepheid's blueward evolution in the instability strip.

<u>Pulsation modes:</u> <u>SU Cas</u>, <u>HD 144972</u>, <u>DT Cyg</u>, <u>EU Tau</u>, <u>EW Sct</u>, <u>CO Aur</u>
The case of SU Cas is a favorite one for me because there seemed to be a continual argument about its pulsation mode. In my Cepheid mass review (Cox, 1979), I concluded that it is likely to be in the fundamental mode, and Turner, et al. (1985) as well as Turner and

Evans (1984) have given observational support by estimating its
luminosity. However, Gieren (1976) has attracted the attention of most
observers by advocating first overtone pulsation, and the low amplitude
sinusoidal light and velocity curves (and the ratio of their first two
Fourier amplitude coefficients, Simon and Lee, 1981) show that also.
This is now confirmed by IUE observations that see the spectrum of both
the Cepheid and the blue companion. The luminosity of SU Cas is then
1530 times that for the sun, at least 50 percent larger than the value I
used in 1979. Such a bright and big variable then must be in the
overtone to match its observed very short period (Simon and Aikawa,
1986).

Except for SU Cas I have thought that all galactic Cepheids were
fundamental mode pulsators, but I do have to realize that the recent
data for HD 144972 and DT Cyg point to first overtone pulsation (Moffett
and Barnes, 1986). At least they have nearly sinusoidal light curves.
EU Tau now has joined this list, with a suggestion by Burki (1985) that
it is even in the second overtone. Gieren (1985) derives a radius much
smaller so that it even could be a fundamental mode pulsator, but Simon
and Lee (1981) from Fourier analysis of the light curve suggest that EU
Tau is in the first overtone. Presently there is a discussion by Gieren
and Matthews (1987) and Fernie (1987b) about whether there is a second
very low amplitude higher order mode present.

There is an expectation that if the Cepheid is hot enough, it will
pulsate in the first overtone, just as in the case for the RR Lyrae
Bailey c-type variables. Since they have small amplitudes usually,
their light curves are nearly sinusoidal. These are the s-Cepheids, but
just because the light curves have that shape certainly does not insure
that the mode is an overtone. Recently Fernie and Chan (1986) have
shown that fundamental mode "bump" Cepheids often have almost sinusoidal
light curve shapes, and that is also seen in the ratio of the first two
Fourier terms for these stars. Near ten days, several fundamental mode
Cepheids have more sinusoidal light curves than the overtone pulsators
do as displayed by Simon and Moffett (1985). Overtone pulsators are
quite rare in our galaxy and apparently common in the lower metallicity
Magellanic Clouds. In their place we have the double-mode Cepheids
(Stobie 1977).

Before discussing these double-mode Cepheids, I must comment on the
suggestion by Böhm-Vitense (1988) that most short period Cepheids are
overtone pulsators. The use of Wesselink radii is very unreliable, as
Simon (1988, preprint), recently, and Cox (1979), long ago have shown.
Gieren (1982 and 1986b) has derived quite small radii that imply these
small masses. The even smaller masses, if the overtone mode is assumed,
are much less than those that can barely evolve onto the blue loops in
the H-R diagram where the Cepheids are observed. Since masses greater
than the four solar masses minimum are actually observed for the binary
SU Cyg (3.85 day pulsation period) and implied from the more extensive
Wesselink radii obtained by Moffett and Barnes (1987), it seems that the
Gieren Wesselink radii are not reliable for mass determinations. Burki
(1985) has given Wesselink radii that show, if anything, that masses are

anomalously low for the longest periods, not the shortest ones. Appeal to stellar models for the anomalously low "beat" masses, and to hydrodynamic calculations by Christy (1968) and Stobie (1969) and others for the low "bump" Cepheid masses is not warranted because of the many theoretical uncertainties in the compositions and opacities of Cepheid envelopes.

The list of galactic double-mode classical Cepheids is growing slowly with EW Sct being added to the eleven that I spoke of at Boulder in 1982 (Antonello et al. 1987). More are sure to be discovered. CO Aur seems to be in the first and second overtones (Mantegazza, 1983 and most recently Babel and Burki, 1987). I suppose that theoretically these modes may be acceptable if they are indeed at the very blue edge of the instability strip. From the blue edges given by Cox and Hodson (1978), CO Aur seems to be just at a possible transition line between first and second overtone pulsation.

It may be good news that Andreasen (1987) has started to find double-mode Cepheids in the Large Magellanic Cloud. One can hope that DV 14 and HV 2345 are confirmed. The period ratio near the galactic value of 0.70 will tell us whether the envelope structure of the short period Cepheids depends on the interior metallicity.

It seems that the high metallicity of our galaxy favors double-mode behavior instead of the very common first and maybe even second overtone pure mode pulsation in the Magellanic Clouds. This should be a clue in trying to understand both the anomalous period ratios for the double-mode Cepheids and how the double-mode behavior arises. Unfortunately, the high galactic metallicity possibly could promote both more helium deficient mass loss to give helium enhanced surface layers (Cox, Hodson, and King, 1979) or be more effective in producing a higher opacity in the pulsating regions (Simon, 1982). We need further clues from observations to understand double-mode Cepheids.

The finding of double-mode Cepheids at a period ratio of 0.70 in the low metallicity Clouds may actually be unwelcome, because it has been hoped for a long time that if the period ratio can be explained, the cause of the double-mode behavior (maybe related to the metallicity?) then would follow naturally. Ostlie in a poster paper at this conference alleges that time dependent convection is an essential ingredient in producing double-mode behavior, and how that relates to metallicity is not clear.

Fourier Fitting
The Barnes and Moffett light and velocity curves have been useful for discussion of the general properties of galactic Cepheids. These data have been used with the Fourier fitting methods by Simon and Moffett (1985) to give us an understanding of how the light curves evolve with changing period. These fits have been useful for the determination of Baade-Wesselink radii, as recently discussed by Simon (1988, preprint), as well as for mode identifications and even variable star classification as for XZ Cet (Teays and Simon, 1985).

SEVEN TYPES OF CEPHEID MASSES

The most basic type of mass that any star can exhibit is that determined from the solution of an orbit for a binary or multiple star system. It seems that there is only one case available to us that has a direct bearing on current mass anomalies. That is the case of SU Cyg, that has been investigated relentlessly by Nancy Evans for almost 10 years. With a luminosity of 1530 solar luminosities, evolution tracks (Z=0.02) would predict that the mass for this luminosity would be about 5.5 solar masses. Using evolution tracks that allow for convective core overshooting would reduce this mass to as low as 4.3 solar mass. The orbital solution (Evans and Bolton, 1987) gives 6.3 solar mass with a lower limit of 5.9, verifying, for this single case, that masses of Cepheids with known luminosities can be approximately determined from standard evolutionary tracks in the H-R diagram. At least a very low mass like 2 solar masses that double-mode Cepheids suggest from their period ratios are definitely ruled out. To compensate for the low evolution mass at the observed luminosity, one could use a larger Z approximately equal to 0.03 that would change the 5.5 solar masses to the observed 6.3 solar masses. Such a Z would not be unexpected for the recently born Cepheids.

We should note that the much brighter overshooting evolution tracks give rather low masses, and the degree of overshooting of the convective core may be too large in those evolution tracks that are becoming available from Chiosi. If anything, the SU Cyg case seems to show that the evolution tracks should be fainter, for a given mass, than the standard Becker, Iben, and Tuggle (BIT, 1977) tracks. SU Cyg may rule out the large overshooting that has recently been popular.

Böhm-Vitense (1986) has given masses for two other Cepheids based on orbital information, but these cases depend on the assumed masses of the blue companions. Therefore their accuracy is not very good. Even so, the masses of V636 Sco and S Mus are near 5 solar masses, rather low, but not in bad conflict with evolution masses for them at respectively, 6.80 and 9.66 days period.

Böhm-Vitense (1985) and Evans and Arellano Ferro (1987) have considered the cases for Cepheids with blue companions. The relative luminosity can be determined for these binaries from IUE spectra, and with the assumption of the blue companion luminosity (from its surface effective temperature and an assumed zero age main sequence), the Cepheid luminosity can be determined. Use of stellar evolution tracks then can allow identification of the mass that goes with this Cepheid luminosity. The sometimes quite low Cepheid luminosity implies a low Cepheid mass. However, Feast and Walker (1987) have pointed out that if the blue companion has evolved to a higher luminosity near the main sequence, the resulting Cepheid luminosity and mass may be considerably larger than assumed.

Standard evolutionary tracks have been studied to see if newer ideas about overshooting or increased opacities change their positions in the H-R diagram. Becker and Cox (1982) have showed that an increase of about 0.7 magnitude in luminosity for a fixed mass (9 solar mass) is certainly appropriate to allow for convective core overshooting. This overshooting was also considered by Huang and Weigert (1983) with similar results. Becker (1985) in another investigation demonstrated that a change in opacity in the temperature range between 100,000K and 500,000K did not have any important effect on the tracks. For an known luminosity of a Cepheid from the relative luminosity discussed above or from a known distance, an evolution mass can be determined by finding the mass of the evolution track that has that luminosity.

Schmidt (1984) has studied the Cepheids in eight galactic clusters and has found that their luminosities are perhaps 0.5 magnitude less luminous than believed when I wrote the Cepheid mass reviews in 1979 and 1980 (Cox, 1980). Others like Turner (1986), studying NGC 6087 and Gieren (1986a), using surface brightness and radial velocity measurements for 30 galactic Cepheids, feel that Schmidt is correct in reducing the Cepheid luminosities, but by a slightly smaller amount than 0.5 magnitude. It does seem that these newer luminosities are being accepted, and that causes severe problems for Cepheid masses.

As we reported at Toronto (Kidman and Cox, 1985), we need to be careful about such low luminosities and masses. It takes a mass of 4 or 5 solar masses merely to evolve into the pulsation instability strip, and that result has been verified by several evolution track studies. Using the standard BIT evolution tracks, the low Schmidt luminosities frequently give masses smaller than this lower limit. The way out of this dilemma seems to be to have tracks that are less bright (larger Z) for masses around 5 solar masses. Another intriguing way is to have surfaces enhanced by helium that will give smaller radii for these models and bluer evolution into the instability strip at slightly lower masses. This effect is limited, however, because the blue looping is caused by the structure of the deep composition gradient set up by prior evolution, and not much influenced by composition variations. Cepheids could have somewhat lower masses, and many problems such as the discrepancy between the observed and theoretical ratios of the numbers of long and short period Cepheids discussed by BIT would be alleviated.

Another way of determining the mass of a Cepheid is to use the observed period and an approximate effective temperature. Four equations involving these observed quantities: the definition of the effective temperature, the evolution mass-luminosity relation, the period-mean density relation, and a theoretical fit for the pulsation constant Q, can be used to derive four values: the mass, radius, luminosity, and pulsation constant. These "theoretical masses" (Cox, 1979) differ little from the evolution masses, determined from the observed luminosity only, even though they do not require a known luminosity. They are similar because of the strong influence of the mass-luminosity relation obtained from theoretical evolution tracks.

The classical pulsation mass first discussed by Cogan (1970) is most often computed for variable stars. It needs the luminosity, effective temperature, and the pulsation period with the mode assumed. These masses have been moved around as the luminosity (and effective temperature) scale of the Cepheids have changed. With the increase of distances and luminosities by 0.26 magnitude, when the Hyades distance was increased, the masses increased accordingly. Then with the Schmidt observation that the distances of galactic clusters containing Cepheids needed to be reduced from those luminosities by 0.5 magnitude, the masses have decreased by a factor of two. Even though evolution and theoretical masses may have to be reduced by modified overshooting evolution tracks, the lower pulsation masses are now appreciably out of line now with evolution masses.

Observers have long wanted to settle this mass question by measuring the Baade-Wesselink radii that can be inserted into the period-mean density relation with assumed pulsation mode to quite directly give the Cepheid mass. The problem was, and still is, that the radius, which needs to be taken to the third power in the mass determination, is not accurate enough. If Fourier fitting and other special techniques can give radii to the expected accuracy of a few percent, as Simon (1988, preprint) suggests for a Cepheid like U Sgr, then Wesselink masses of great value can be made available.

It does seem currently, that these masses are not significantly discrepant with respect to the earlier mentioned dynamical, evolution, and theoretical Cepheid masses, but the accuracy available is often very poor. For example, Gieren (1986b) presents radii and masses for 30 galactic Cepheids using the surface brightness method of Barnes and Evans with simultaneous BVRI photometric and radial velocity measurements. His small radii often give small masses compared to the theoretical masses (Cox, 1979). A possible fix for this problem is to have the theoretical evolution tracks at a higher luminosity for a given mass than those given by BIT, but that is the opposite direction than needed to fix the SU Cyg mass problem. My opinion is that we cannot rely yet on the Wesselink masses for Cepheids.

Carson and Stothers (1988) now claim that masses based on the phase of the bump in the light and velocity curves for Cepheids with periods between about 5 and 21 days are close to all those given by dynamical and evolution methods. Long ago Simon and Schmidt (1976) demonstrated that the bump was a manifestation of the resonance between the fundamental and second overtone modes which appears when the ratio of these periods lies between 0.47 and 0.53. Using the Carson opacities in nonlinear one-dimensional calculations, Carson and Stothers can produce the velocity curve bumps at a mass centered at 6 solar masses. This "bump" mass is only 15 percent smaller than given by Cox in 1979 for evolution, theoretical or pulsation masses for periods near 10 days. This is an unexpected result, because previous authors (Fricke, Stobie, and Strittmatter, 1972, and many others) got "bump" masses near 4 solar masses, just over half the then conventional masses for the 10 day period. The "bump" masses have been anomalously low since Christy (1968) and Stobie (1969) first noticed this effect in the nonlinear calculations for classical Cepheids.

A comparison of the Los Alamos Opacity Library opacities for the Carson 312 mixture (Y=0.25, Z=0.02) shows that the reason Carson and Stothers can get near conventional masses in their nonlinear calculations is that the Carson opacities are just about as much larger than the Los Alamos ones in the 150,000K to 800,000K temperature range as Simon (1982) first suggested to alleviate the mass anomaly. The larger Carson opacities reduce the second overtone to fundamental mode period ratio for six solar mass Cepheids to near resonance.

Figure 1 compares the Carson and Los Alamos opacities, just as previously presented by Cox and Kidman (1984). The line labeled 4 at 10^{-5} g·cm^{-3} indicates that the Carson opacities are approximately twice the Los Alamos opacities in the interesting temperature and density region. Since the Carson opacities in this temperature region have been shown to be incorrect (Carson et al. 1984), and opacities in this region

Figure 1. The logarithm of the ratio of the Carson 312 opacities to the Los Alamos Opacity Library opacities are plotted versus temperature for log density (g·cm^{-3}) = -8, -7, -6, and -5. The CNO bump is seen at the highest temperatures.

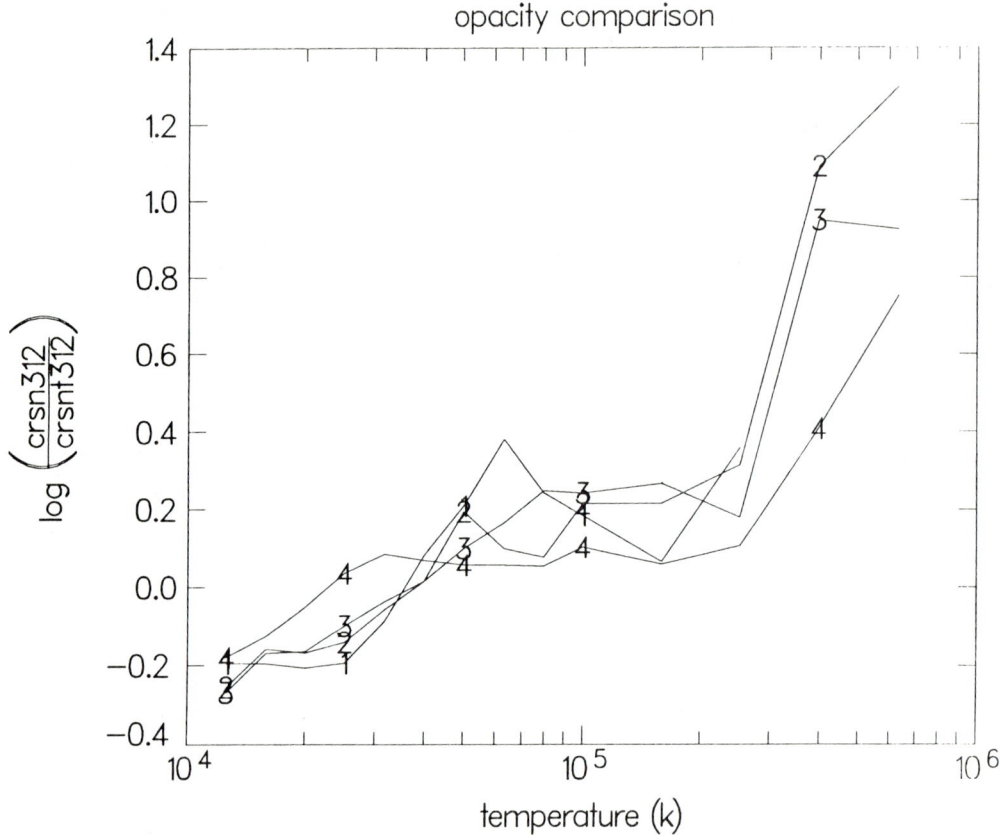

are important to the "bump" Cepheid structure, we cannot accept this "reconciliation" between theory and observations that they propose.

I cannot support the Carson and Stothers statement that use of the Carson opacities makes no difference; I find that it indeed does. In linear theory studies, 10 day, 7 solar mass models (5420K, 1.536×10^{37} erg·sec^{-1}) have a second overtone to fundamental mode period ratio of 0.52 using the Carson opacities and 0.55 with the King4a mixture of Cox and Tabor (1976). This reduction of the period ratio is just about what one would expect for the larger Carson opacities at the sensitive temperatures.

Another problem that Carson and Stothers have is that the amplitudes of the velocity curves, just as for earlier work for the RR Lyrae variables, are too large, probably because of their different helium opacities and their larger derivative with respect to temperature. This may explain why theoretical bumps are seen between 5 and 21 days whereas the observed light and velocity curves have these bumps only between 7 and 12 days. I suggest that the large helium opacities drive these nonlinear models to such a high amplitude that the bumps occur way outside the observed resonance region.

I also want to note that Fernie and Chan (1986) have discovered that in this second overtone to fundamental mode resonance region, all Cepheids seem to pulsate with about the same moderate amplitude, midway between that for many of both shorter and longer period. Apparently the resonance locks-in a ten percent radius amplitude, almost sinusoidal shape behavior that is not displayed for Cepheids at other periods where a wide range of amplitudes is observed.

I must, however, point out that Cogan, Cox and King (1980) were not able to make sense of Cepheid amplitudes. My best opinion is that the Cepheids have a small range of helium composition in the driving region that produces a range of amplitudes for most classical Cepheids. However, when there is the near two to one resonance, the driving becomes limited regardless of the helium abundance.

I very much like the recent paper by Andreasen and Petersen (1988) and another by Andreasen (preprint) who attempt to solve the "bump" and "beat" Cepheid low mass anomaly by increasing the opacity of the stellar material. This is the idea of Simon (1982), and it has had some notoriety because earlier ideas about this "bump" mass and the double-mode Cepheid mass anomalies possibly are not so desirable. The goal is to get the two classes of period ratio Cepheids ("bump" and "beat") to have an envelope structure such that the period ratios decrease for a given mass from those using conventional models.

Currently the popular idea is that the opacity in the temperature range of 150,000K to 800,000K needs to be increased by approximately a factor of 2.5 to change the surface layer structure of Cepheids so that the observed period ratios of the dozen double-mode classical Cepheids can be matched. This idea has been taken up by Andreasen and extended to other yellow giant classes such as the "bump" Cepheids, the RR Lyrae and δ Scuti variables. This adjustment has been successful in changing the second overtone to fundamental mode period ratio resonance to near 10 days as observed for the central period for the "bump" Cepheids. It also has increased the mass of the RR Lyrae variables to as much as 0.75 solar mass so that, with the asymptotic giant branch evolution, some additional mass loss can occur to produce the common form of white dwarf at 0.6 solar mass. Finally, the δ Scuti variable period ratios, for those dozen or so that show two radial periods, are predicted closer to those observed.

At Los Alamos, we have had great difficulty in accepting that the opacities in the required temperature range are wrong by a factor of more than two. Magee, Merts, and Huebner (1984) have investigated this problem in detail and find no opacity problem. I must note that the recent Iglesias, Rogers, and Wilson (1987) result that for iron, the same-shell (n=3 to n=3) transitions are larger than expected, giving a big opacity increase for iron at 20ev (232,000K) is not likely relevant to the stellar composition of Cepheids. The number fraction of iron in this composition is 3.7×10^{-5}, and investigations by myself have shown that iron only rarely has any influence on mixture opacities.

But even if the iron proves to be an important component of the opacity at the temperature of kT=20 ev, the Livermore opacity work has shown that at 60 ev, the Los Alamos values are unchanged. Thus the desire to have opacities more than doubled over the entire temperature range of 150,000K to 800,000K does not yet seem in hand. For classical Cepheids, the opacities do not have be increased at these highest temperatures, however. It is only the δ Scuti variable period ratios that dictate opacity increases at 800,000K as Andreasen at this conference demonstrates.

We now propose, however, that instead of an increase in the opacity, the mass anomalies can be solved by an increase in the heavy element abundances in a thin layer hidden below the surface. This then increases the opacities as desired, but without any large error in the atomic physics.

In a poster paper Siobahn Morgan, working with me this summer at Los Alamos, shows that with an effective Z of 0.30, period ratios of double-mode Cepheids can be reduced to the observed level. This indeed is an increase in the opacity by the amount suggested by Simon and by Andreasen. We are now running the Iben evolution program with element diffusion and element levitation to see if such abundance increases actually can be predicted.

We need to be careful about a great increase in the heavy elements because an inverted mu gradient will be Rayleigh-Taylor unstable to downward mixing. Figure 2 shows the Z variation throughout the envelope. This structure is formally stable against downward mixing because the radiative gradient there is so subadiabatic. Any mixing, which will actually be rather rapid, will have to be countered by an effective and rapid element levitation. This large Z means that actually only 4 percent in the particles are CNONe elements. This last number is important, because Michaud privately to me has noted that it is difficult to get number fraction enhancements of more than a few percent when all the photons are being absorbed by the levitating elements.

This opacity increase causes a thin convection zone to appear in just the region where the increased opacity is desired. While unrelated to this problem, we have suggested that the high luminosity (high photon flux) B stars may also have a thin layer of enhanced Z elements that

Figure 2. The Z mass fraction abundance is plotted versus the mass into a 5 solar mass model for a double-mode Cepheid. Going inward from the surface, the Z abundance starts up at a mass fraction of $-\log(1-q)$ of 4.92 at a temperature of 61,000K. The plateau is reached at 4.89 and 63,000K. The inverted mu gradient begins at $-\log(1-q)$ of 3.30 and ends at 2.07 with temperatures, respectively, of 200,000K and 420,000K. The hydrogen convection zone lies between $-\log(1-q)$ of 6.68 and 6.14 with temperatures of 6000K and 25,800K, and the helium convection zone is between 5.77 and 5.29 with temperatures of 34,000K and 49,000K. A thin convection zone also exists on the plateau between 4.88 and 4.70 with temperatures of 64,800K and 75,000K.

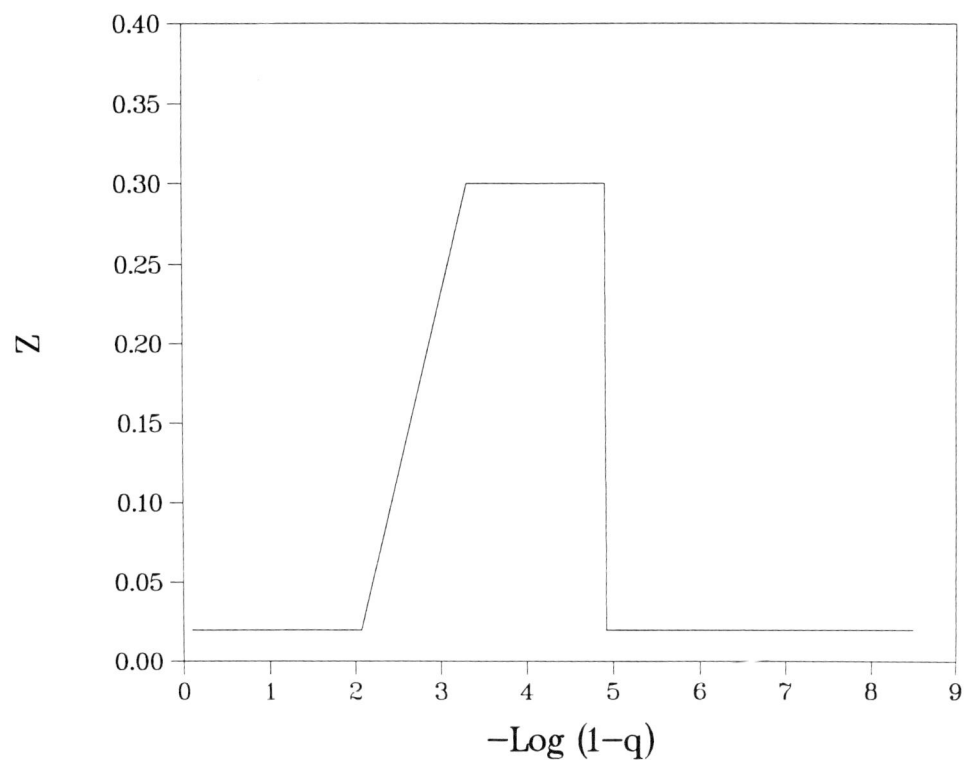

produces a convection zone. For the B stars, it just may be that this slowly reacting convection is the cause of the β Cephei variable star pulsations that for at least a quarter of a century has remained unexplained.

Cox, Deupree, King and Hodson (1977) suggested long ago an enhanced helium layer at the surface down to a temperature of 250,000K that would make the model appear to the fundamental and first overtone modes to be less concentrated. Thus these modes would have their periods increased by over 5 percent with the fundamental feeling the decreased concentration more than the first overtone. Then the period ratio could be reduced to the observed values with conventional masses. Because of the Rayleigh-Taylor instability of such a composition inversion, this idea has never been generally accepted. Now, however, we are again investigating whether it may be possible that the gradients are shallow enough to be stable in the presence of the very subadiabatic radiative gradient.

Stothers (1979) suggested strong magnetic fields that could deconcentrate the models by having this additional pressure contribution in these same surface layers. Again the idea has not obtained acceptance, because such strong fields are not thought to be possible at a mass fraction depth of 0.001 into the star.

FUTURE TASKS

For the future, it seems to me that there are several observational and theoretical tasks that are appropriate for understanding the pulsations of Cepheids and the structure of the instability strip in the H-R diagram. We are interested in the masses, radii, luminosities, compositions, and pulsation periods and modes.

Observers are indeed busy trying to identify cases where binary stars can measure stellar masses. Is it possible that the eclipsing binary BM Cas (Fernie, 1987a) will prove to be the Rosetta Stone after all?

Baade-Wesselink radii should be available with only a few percent uncertainty, but this status has not yet been achieved for any Cepheid. Investigations now being pursued may make this accuracy a reality with very accurate photometry in many colors including the infrared. Hopefully then the radius and mass problems will be alleviated at least for a few special Cepheids. It would seem appropriate that SU Cyg with its known dynamical mass be included in these Baade-Wesselink radius studies.

The more traditional ways of getting Cepheid masses using observed luminosities needs to be refined. This means that there is a need for an accurate luminosity scale; current uncertainties are greater than 0.1 magnitude. This implies an error of at least 15 percent in the mass, and that is just marginally acceptable for understanding Cepheids.

Double-mode Cepheids are useful for understanding the structure of the instability strip. The placement of the double-mode Cepheids at the transition line between the fundamental and overtone modes (Barrell, 1981) is valuable for guiding theoretical analyses, and with CO Aur at 6700K, it seems that this Cepheid is at the transition line between the first and second overtones. Is it certain that the overtone pulsators, SU Cas, HD144972, DT Cyg, and EU Tau are pure overtone pulsators? If so, as we now believe, the first overtone region must have a significant temperature width in the H-R diagram.

Searches for Magellanic Cloud double-mode Cepheids, and determination of their period ratios would give indications of how the period ratio depends on metallicity if there is any dependence at all. Observational information on what causes the low period ratio is badly needed.

A final need from observers is a clue to the pulsation amplitude changes for HR 7308. My proposed disappearance and reappearance of helium in the 40,000K temperature levels in the star might result in small surface composition changes in levitated elements like CNONe.

An important theoretical task is the calculation of evolution tracks for cases with convective core overshooting and for cases of possible CNONe element levitation with helium being dragged up also. The large SU Cyg mass seems to indicate that we do not need core overshooting, but a decrease in the track luminosities for a given mass such as for a larger Z.

For understanding the double-mode problem, more linear theory models are needed to obtain periods in unconventional composition cases. Clearly the opacity uncertainty needs to be settled.

Full amplitude nonlinear calculations are needed for a variety of cases with varying compositions (probably only helium), so that the large range of amplitudes observed can be understood. And these especially need to be done near the second overtone to fundamental mode resonance to explain the uniformity of light and velocity curves there.

Simple applications of time-dependent convection need to be made. The question is whether this feature is an essential one for obtaining double-mode behavior.

As a final important task for the theoreticians, a demonstration with nonlinear methods, of the double-mode behavior for classical Cepheids is most important. Hopefully, one of the period ratio anomaly solutions, such as surface enhanced helium, increased envelope opacities, or augmented Z in the deeper pulsating layers, will also allow simultaneously a demonstration of double-mode Cepheids.

ACKNOWLEDGEMENTS

I would like to thank Siobahn M. Morgan for her help in the preparation of this review, including finding almost 200 published papers and discussing the manuscript.

REFERENCES

Andreasen, G.K. (1987). Astron. & Astrophys.,186,159.
Andreasen, G.K. & Petersen, J.O. (1988). Astron. & Astrophys.,92, L4.
Antonello, E., Mantegazza, L. & Poretti, E. (1987). Lecture Notes in Physics,274, 191.
Arellano Ferro, A. (1983). Ap.J.,274, 755.
Babel, J. & Burki, G. (1987). Astron. & Astrophys.,181, 34.
Barnes, T.G., Moffett, T.J., & Slovak M.H. (1987). Ap.J.Suppl.,65, 307.
Barrell, S.L. (1981). M.N.R.A.S.,196, 357.
Becker, S.A. (1985). In Cepheids: Theory and Observations, ed. B.F. Madore, p. 104. Cambridge: Cambridge University Press.
Becker, S.A. & Cox, A.N. (1982). Ap.J.,260, 707.
Becker, S.A., Iben, I. & Tuggle, R.S. (BIT) (1977). Ap.J.,218, 633.
Böhm-Vitense, E. (1985). Ap.J.,296, 169.
Böhm-Vitense, E. (1986). Ap.J.,303, 262.
Böhm-Vitense, E. (1988). Ap.J.Lett.,324, L27.
Böhm-Vitense, E. & Proffitt, C. (1985). Ap.J.,296, 175.
Breger, M. (1981). Ap.J.,249, 666.
Brunish, W.M. & Willson, L.A. (1987). Lecture Notes in Physics,274, 27.
Burki, G. (1985). In Cepheids: Theory and Observations, ed. B.F. Madore, p. 34. Cambridge: Cambridge University Press.
Burki, G., Mayor, M. & Benz, W. (1982). Astron. & Astrophys.,109, 258.
Burki, G., Schmidt, E.G., Arellano Ferro, A., Fernie, J.D., Sasselov, D., Simon, N.R., Percy, J.R., & Szabados, L. (1986). Astron. & Astrophys.,168, 139.
Carson, T.R., Huebner, W.F., Magee, N.H., & Merts, A.L. (1984). Ap.J.,283, 466.
Carson, T.R. & Stothers, R.B. (1988). Ap.J.,328, 196.
Christy, R.F. (1968). Quart.J.R.A.S.,9, 13.
Cogan, B.C. (1970). Ap.J.,162, 139.
Cogan, B.C., Cox, A.N., & King, D.S. (1980). Space Sci. Rev.,27, 419.
Cox, A.N. (1979). Ap.J.,229, 212.
Cox, A.N. (1980). Ann. Rev. Astron. Astrophys.,18, 15.
Cox, A.N., Deupree, R.G., King, D.S., & Hodson, S.W. (1977). Ap.J.Lett.,214, L127.
Cox, A.N. & Hodson, S.W. (1978). In The H-R Diagram, IAU Symposium 80, eds. A.G.D. Philip & D.S. Hayes, p. 237. Dordrecht: Reidel.
Cox, A.N., Hodson, S.W. & King, D.S. (1979). Ap.J.Lett.,230, L109.
Cox, A.N. & Kidman, R.B. (1984). In Observational Tests of the Stellar Evolution Theory, IAU Symposium 105, eds. A. Maeder & A. Renzini. Dordrecht: Reidel.
Cox, A.N., Michaud, G., & Hodson, S.W. (1978). Ap.J.,222, 621.
Cox, A.N. & Tabor, J.E. (1976). Ap.J.Suppl.,31, 271.
Cox, J.P. (1980). Space Sci. Rev.,27, 389.

Deasy, H.P. (1988). M.N.R.A.S., 231, 673.
Deasy, H. & Butler, C.J. (1986). Nature, 320, 726.
Demers, S. & Fernie, J.D. (1966). Ap.J., 144, 437.
Evans, N.R. & Arellano Ferro, A. (1987). Lecture Notes in Physics, 274, 183.
Evans, N.R. & Bolton, C.T. (1987). Lecture Notes in Physics, 274, 163.
Feast, M.J. & Walker, A.R. (1987). Ann. Rev. Astron. & Astrophys., 25, 345.
Fernie, J.D. (1984). In Observational Tests of the Stellar Evolution Theory, eds A. Maeder & A. Renzini, p. 441. Reidel: Dordrecht.
Fernie, J.D. (1987a). Lecture Notes in Physics, 274, 167.
Fernie, J.D. (1987b). P.A.S.P., 99, 1093.
Fernie, J.D. & Chan, S.J. (1986). Ap.J., 303, 766.
Fricke, K., Stobie, R.S., & Strittmatter, P.A. (1972). Ap.J., 171, 593.
Gieren, W. (1976). Astron. & Astrophys., 47, 211.
Gieren, W. (1982). Ap.J., 260, 208.
Gieren, W. (1985). In Cepheids: Theory and Observations, ed. B.F. Madore, P. 98. Cambridge: Cambridge University Press.
Gieren, W.P. (1986a). Ap.J., 306, 25.
Gieren, W.P. (1986b). M.N.R.A.S., 222, 251.
Gieren, W.P. & Matthews, J.M. (1987). A.J., 94, 431.
Huang, R.Q. & Weigert, A. (1983). Astron. & Astrophys., 127, 309.
Iglesias, C.A., Rogers, F.J., & Wilson, B.G. (1987). Ap.J., 322, L45.
Jacobsen, T.S. & Wallerstein, G. (1987). P.A.S.P., 99, 138.
Kidman, R.B. & Cox, A.N. (1985). In Cepheids: Theory and Observations, ed. B.F. Madore, p. 256. Cambridge: Cambridge University Press.
Magee, N.H., Merts, A.L., & Huebner, W.F. (1984). Ap.J., 283, 264.
Mantegazza, L. (1983). Astron. & Astrophys., 118, 321.
McAlary, C.W. & Welch, D.L. (1986). A.J., 91, 1209.
Michaud, G., Vauclair, G., & Vauclair, S. (1983). Ap.J., 267, 256.
Moffett, T.J. & Barnes, T.G. (1986). M.N.R.A.S., 219, 45p.
Moffett, T.J. & Barnes, T.G. (1987). Ap.J., 323, 280.
Schmidt, E.G. (1984). Ap.J., 285, 501.
Simon, N.R. (1982). Ap.J.Lett., 260, L87.
Simon, N.R. (1987). Lecture Notes in Physics, 274, 148.
Simon, N.R. & Aikawa, T. (1986). B.A.A.S., 17, 894.
Simon, N.R. & Lee, A.S. (1981). Ap.J., 248, 291.
Simon, N.R. & Moffett, T.J. (1985). P.A.S.P., 97, 1078.
Simon, N.R. & Schmidt, E.G. (1976). Ap.J., 205, 162.
Stobie, R.S. (1969). M.N.R.A.S., 144, 485.
Stobie, R.S. (1977). M.N.R.A.S., 180, 631.
Stothers, R. (1979). Ap.J., 229, 1023.
Teays, T.J. & Simon, N.R. (1985). Ap.J., 290, 683.
Turner, D.G. (1986). A.J., 92, 111.
Turner, D.G. & Evans, N.R. (1984). Ap.J., 283, 254.
Turner, D.G., Forbes, D.W., Lyons, R.W., & Havlen, R.J. (1985). In Cepheids: Theory and Observations, ed. B.F. Madore, p. 95. Cambridge: Cambridge University Press.

Welch, D.L., Evans, N.R., Lyons, R.W., Harris, H.C., Barnes, T.G., Slovak, M.H., & Moffett, T.J. (1987). P.A.S.P., 99, 610.
Woosley, S.E. & Phillips, M.M. (1988). Science, 240, 750.

THE EVOLUTION OF STARS OF MEDIUM MASS

Cesare Chiosi
Department of Astronomy
University of Padova
Vicolo dell'Osservatorio 5
35122 Padova, Italy

Abstract. The main properties of the evolution of low, intermediate, and high mass stars are reviewed focusing on a few issues tightly related to the interpretation of Pop I Cepheid stars. After a summary discussion of the physical mechanism responsible for the Cepheid pulsation, the classical results of stellar evolution theory for the main evolutionary phases (main sequence, core He-burning, and later) all over the HR diagram are presented, putting into evidence the various points of disagreement with current observational data. We then review the models incorporating the effect of convective overshoot, and present in some detail a study on the rich, young cluster NGC 1866 in the Large Magellanic Cloud, in which they are compared with the observational data. Arguments are given to favour the adoption of models with convective overshoot instead of the classical ones. The topics of the mass discrepancy and frequency-period distribution of Cepheid stars are then discussed in the light of models with convective overshoot, showing how they can better account for the observational data. Finally, we address the question of the enhancement in the radiative opacity of heavy elements in the middle temperature region (approximately where the ultimate ionization of the CNO group of elements occurs), which has been suggested to explain the mass anomalies and period ratios encountered in double-mode and bump Cepheids. Evolutionary models for intermediate mass stars incorporating the above modification in the radiative opacity are then presented and discussed in some detail.

INTRODUCTION
Becker (1985), in his recent review on the evolution of intermediate mass stars in relation to Cepheids, separates these stars in two loose groups according to their metallicity: the Pop I Cepheids when $Z > 0.005$, and Pop II Cepheids when $Z < 0.005$. Galactic Pop I Cepheids have periods in the range 1-80 days, whereas those of the Large Magellanic Cloud may pulsate with periods up to 200 days. Galactic Cepheids are concentrated toward the galactic plane, and possess effective temperatures from 6600 to 5500 K (spectral type F5 to G5) and visual absolute magnitudes from -1 to -6. These pulsators obey a well defined period-luminosity relationship, and from an evolutionary point of view are commonly thought to arise as the result of normal post main

sequence evolution of intermediate to high mass stars (3 to 10 $M_\odot$) in core He-burning stages. With the possible exception of Pop II stars of relatively high mass (M > 2 $M_\odot$) that may occur in galaxies like LMC and SMC, and which would behave similarly to classical Pop I Cepheids, Pop II Cepheids are a separate class of stars which lie in the HR diagram between RR Lyrae, RV Tauri and LPV stars, obey a different period-luminosity relationship, are broken into three main groups (BL Her stars, W Vir stars, and Anomalous Cepheids), and are generally associated to low mass stars undergoing post core He-burning evolution, although other explanations are possible for the last group.

1. THE CLASSICAL SCENARIO FOR POPULATION I CEPHEIDS

In the following, we will limit ourselves only to the Pop I Cepheids, paying major attention to those aspects tightly related to the evolutionary behaviour of intermediate and high mass stars.

There are two main facts that concur to determine the Cepheid star phenomenon: 1) the instability strip itself (location in the HR diagram, slope, width, etc.); 2) the crossing of the instability strip by a star in the course of its evolution.

In fact, whenever a star crosses the instability strip, it becomes pulsationally unstable according to the well known mechanisms, even if differences, arising from the particular evolutionary stage during which the crossing occurs, are likely to be expected.

1.1 The Cepheid instability strip

Although the physical mechanism responsible for the pulsation of Cepheid stars has been well known for a long time, so that a lengthy discussion is superfluous (see also the recent review by Cox 1985), nonetheless a short summary of the basic ideas may be useful. Given that nuclear reactions play a totally negligible role in the phenomenon of pulsation in most types of variable stars, since pulsations in stellar interiors, where nuclear reactions occur, are of very small amplitude, the source of instability has been identified with the envelope ionization mechanism. The ability of the envelope ionization to drive pulsations resides in the fact that the ionization of abundant elements results in a modulation of the flux variations, and these modulations are such that the regions absorb heat when they are most compressed and lose heat when most expanded. As a consequence of this, maximum pressure in these layers will come after maximum density, thus favouring instability. Finally, the ionization region is located in the external layers of a star where the pulsation amplitudes happen to be high. In contrast, the innermost regions oscillate at small amplitudes and show the opposite tendency, thus damping all incipient pulsations. The effects of modulations of the flux can be understood by means of the work integral, written in the approximate form (see J. P. Cox 1980 for details)

$$W = -L_T \int_0^M (\Gamma_3 - 1)(\delta\rho/\rho) \, d/dm \, (\delta L/L) \, dm \qquad (1)$$

where all variations denote the space parts of the corresponding variables, nuclear energy sources are neglected, L is the interior luminosity at radius r, $(\delta\rho/\rho)$ is assumed real and positive (regions are compressed at minimum stellar radius), and finally the equilibrium value of L is assumed constant with the mass and equal to the total luminosity L_T. Negative and positive values of W indicate stability and instability, respectively. Assuming that the energy is carried only by radiation (a reasonable approximation in most cases), the luminosity variations are given by

$$(\delta L/L) = 4 (\delta r/r) - (\delta\kappa/\kappa) + 4(\delta T/T) + (dr_o/d[\ln T_o]) \, d/dr_o \, (\delta T/T), \qquad (2)$$

where the quantities in brackets have their usual meaning and the subscript o refers to the equilibrium configuration. Using the quasi-adiabatic approximation,

$$(\delta T/T) = (\Gamma_3-1) \, (\delta\rho/\rho), \qquad (3)$$

assuming a Kramers-like law for the opacity $\kappa = \kappa_o \rho^n T^{-s}$, and neglecting the last term in equation (2) we get,

$$(\delta L/L)_a = 4(\delta r/r) + [(s+4)(\Gamma_3-1)-n](\delta\rho/\rho) \qquad (4)$$

together with

$$(\delta L/L) = 4(\delta r/r) - n(\delta\rho/\rho)(s+4)(\delta T/T). \qquad (5)$$

In most physical situations, since $d(\delta L/L)/dm$ turns out to be positive, W is negative, and stability is established. This can be easily understood as follows: at maximum compression, the luminosity variations increase outward, hence more heat flows out from each elementary shell than flowing in from inside and the pulsations are damped. This is a result of the opacity variation, which usually decreases upon compression (see the temperature, density dependence of the Kramers law). The important effect of the ionization of abundant elements like H, He, C, and O is that (Γ_3-1) becomes very small, since most of the work of adiabatic compression goes into ionization rather than into kinetic energy of thermal motion, so that temperature remains about constant upon compression. This can be easily understood considering that ionization represents additional internal degrees of freedom. As a consequence of it, the adiabatic luminosity variations exhibit a marked dip in the ionization zone. The strong outward decrease of $(\delta L/L)_a$ in the inner part of the ionization zone indicates absorption of energy at maximum compression and acts as the driving source of destabilization (see equation (1)). The diminution of (Γ_3-1) and $(\delta L/L)_a$ results in part from the smaller temperature variation in occurrence of ionization. This direct effect of the temperature variations on the luminosity variations is referred to as the γ-mechanism. If the opacity obeys a Kramer-like relation, it follows that the small values of $(\delta T/T)$ in regions of small (Γ_3-1) may cause κ to increase upon compression, whereas normally the opposite occurs. This local increase in opacity contributes still further to the driving (instability). This constitutes the so-called κ-mechanism. Furthermore, upon compression, negative ($\delta r/r$), the total radiating area is reduced, so that more energy is trapped into the region, thus contributing to destabilize it still further. This is referred to as the r-mechanism [first term in equations (4) and/or (5)]. Finally, if the exponent s is large and negative (as may be the case in the H ionization zone), there may be a damming up of radiation upon compression, hence driving, even if Γ_3-1 has the normal value. This fact has been found important in stars like δ Scuti and β Cephei (Stellingwerf 1978, 1979) and it is referred to as the "bump mechanism". Arguments similar to those developed above lead to the conclusion that $(\delta L/L)_a$ must increase rapidly outward in the outer layers of the ionization zone, thus giving the opposite effect. In general, the strong driving of the innermost portions is largely cancelled by the opposite trend in the outermost portions. However, because of nonadiabatic effects, $(\delta L/L)$ and $(\delta L/L)_a$ may be quite different in the very outer layers of a star. To understand the role of the nonadiabatic behaviour of the layers far out in a star, it is convenient to distinguish three separate regions. The first is the inner region in which oscillations are usually nearly adiabatic and any driving is almost exactly cancelled by an equal amount of damping, so that there is no net contribution. The second is the outermost region, which may contain a large fraction of the radius and in which ionization and highly nonadiabatic effects may occur, but whose mass is so small that the perturbations of the radiative flux are effectively "frozen in" so that, owing to the absence of modulation, there is no net driving. The third region, called the transition region, TR, is located in between the first and the second one, and can be roughly defined by the condition that the total internal energy of the layers above it equals the energy radiated by the star in one pulsation period,

$$(c_v \; T^* \; \Delta M^*) = L \; P, \tag{6}$$

where c_v is the specific heat at constant volume, T^* is the mean temperature of the TR, ΔM^* is the mass above it, L is the luminosity, and P the period. The location of the TR is mostly regulated by the equilibrium radius R of the star ($T \propto R^{-0.5}$) and it moves outward in mass as R increases. The main effects of nonadiabaticity in the outer layers are to decouple $(\delta L/L)$ from $(\delta L/L)_a$ in the regions exterior to TR. Density and temperature variations are such that $(\delta L/L)$ is kept about constant. If R is so small that the ionization zones lie above the TR, $(\delta L/L)$ is constant in the region in spite of the pronounced dip in $(\delta L/L)_a$. Therefore $(\delta L/L)$ is positive and increases outward until the TR is reached and then remains constant. Therefore only damping exists in the outer layers and the star is stable. At somewhat larger radii, the TR moves inward and eventually becomes nearly equal to the characteristic temperature of ionization of elements like He. In this case the ionization region and the TR almost coincide. The inner layers of the ionization zone essentially lie in the adiabatic interior, where $(\delta L/L) = (\delta L/L)_a$ and where the outward decrease in (Γ_3-1) induces a strong decrease both in $(\delta L/L)_a$ and $(\delta L/L)$. The outer layers of the ionization region lie essentially in the nonadiabatic domain, where

($\delta L/L$) has become frozen in at the small value it had at the dip. In this case the strong damping in the outer layers is eliminated by nonadiabatic effects, and the strong driving due to the rapid decrease of ($\delta L/L$) in the inner layers is now effective and the star is unstable. It can be easily seen that condition (6) defining the TR implies that the instability region is sharply bounded on the high effective temperature side, and that this boundary is nearly vertical on the HR diagram. The large body of literature on the Cepheid star instability strip has clarified that the He^+ envelope ionization is the driving agent of their pulsation and the detailed calculations confirm the above oversimplified picture (see Cox 1985 for updated references). The envelope ionization mechanism, which so successfully accounts for the blue side of the instability strip, does not account, on the basis of pure radiative transfer, for the return to stability indicated by the red side of the strip. However, as the radius of the star increases, convection sets in in the outer layers and simple minded arguments permit us to see that convection in the envelope actually quenches pulsations. In fact, convective transfer is most efficient when the star is most compressed. Near maximum compression, the rate of energy losses is most effective, yielding an ideal situation for damping and hence stability (Deupree 1977). Unfortunately, the modulation of the flux upon the occurrence of convection is much more complicated than in the radiative case, so that the exact determination of the red boundary of the instability strip is uncertain. This implies that the width of the instability strip is still a matter of debate. Along the blue edge of the instability strip we may express condition (6) in an alternative form

$$L = Q_i^\alpha \ Y^\beta \ Z^\gamma \ M^\delta \ T_{eff}^\delta \qquad (7)$$

where Q_i is the pulsational constant for the mode under consideration, Y and Z are, respectively, the helium and heavy element abundances, and α, β, γ, and δ are appropriate exponents to be determined by detailed pulsation calculations (see Iben & Tuggle 1972a,b, 1975). Relation (7) has a number of important applications. For instance, it embodies the period luminosity relationship. If T_{eff} is expressed as a function of L, and R, and R in turn is expressed as a function of P, M and Q_i are taken from the period-mean density relationship, then at given Y and Z relation (7) becomes a relation among L, M and P and, in turn, assuming a L-M dependence, it yields L = L(P) (see also J.P. Cox 1985). The period is known to increase with the luminosity. Equation (7) is also useful to estimate how the blue edge of the instability strip is affected by variations in the chemical composition. At fixed L, an increase in Y moves the blue edge to higher T_{eff}, whereas an increase in Z moves the blue edge to lower T_{eff}. This arises primarily through the opacity; an increase in Y implies a decrease in the hydrogen content (normally the main source of opacity in the envelope). Since the material in the envelope is more transparent, the He^+ ionization temperature is reached at higher pressure or equivalently for larger values of ΔM^*. Therefore T_{eff} must increase in order to satisfy equation (6). Similarly for variations in Z. Furthermore, the blue edge of higher mode pulsations (smaller Q_i) should in general lie at higher T_{eff} than a lower mode blue edge. This prediction is however not always confirmed by detailed calculations (Iben & Tuggle 1972a,b) since equation (6) is only a necessary, not a sufficient, condition for instability.

1.2 The classical picture for the evolution of intermediate and high mass stars

In the following, we will consider only those stars which, by virtue of the relatively high initial mass and He core mass, avoid the core He-flash and ignite helium in non-degenerate conditions. As is well known the minimum core mass for this to occur is about 0.45 to 0.55 $M_\odot$, with which an initial mass (otherwise known as M_{HeF}) of about 2.2 to 1.8 $M_\odot$ is associated, respectively, according to the initial chemical composition. Stars lighter than M_{HeF} are not of interest here. They in fact either never cross the instability strip (or do it only once on a very short time scale) or, crossing it during the core He-burning phase in the horizontal branch (HB), give rise to the RR Lyrae stars, pulsators whose instability mechanism is similar to that of Cepheids, but whose periods are of the order of a few hours. In turn we distinguish the intermediate mass stars from the massive ones by looking at the stage of carbon ignition in the core. By intermediate mass we mean those stars which, following core He-exhaustion, develop an highly degenerate carbon-oxygen (C-O) core, and as asymptotic giant branch (AGB) stars experience helium shell flashes or thermal pulses. The AGB phase is terminated either by envelope ejection and formation of a white dwarf ($M_{HeF} < M_i < M_W$) or carbon ignition and deflagration in a highly degenerate core once it has grown to the Chandrasekhar limit of 1.4 $M_\odot$. The limit mass M_W is regulated by the efficiency of mass loss by stellar wind during the red giant (RGB) and AGB phases (see Iben & Renzini 1983). This point will be discusssed in more detail below. The minimum mass of the C-O core, below which carbon ignition in non-degenerate condition fails and the above scheme holds is 1.06 $M_\odot$ to which an initial mass from 7 to 9 $M_\odot$ corresponds, depending on the chemical composition. This particular value of the initial mass is known as M_{up}. Finally, massive stars are those that ignite carbon non-violently and through a series of nuclear burnings proceed either to the construction of an iron core and subsequent photodissociation instability with core collapse and supernova explosion ($M_i > M_{mas}$), or following a more complicated scheme undergo core collapse and supernova explosion ($M_{up} < M_i < M_{mas}$). M_{mas} is about 12 $M_\odot$. As a conclusion, the range of initial masses that matters in our discussion is confined between M_{HeF} and an upper value of the mass, called M_{Ceph}, located in the domain of massive stars (approximately 30 $M_\odot$), above which under the action of mass loss by stellar wind (see below) stars never get into the red supergiant area and, therefore, never intersect the instability strip.

i) <u>Core H-burning</u>. When, as a result of the pre-main sequence overall gravitational contraction, temperatures near the centre become sufficiently large that H-burning reactions are initiated, the star settles on the main sequence, where it remains for most (up to 80%) of its entire nuclear burning life. Conversion of hydrogen into helium in the core, causes the increase in the mean molecular weight together with core contraction and heating up determined by the release of gravitational energy. In the mass range we are interested in, the transfer of the liberated energy (nuclear and gravitational) is secured by convection, which, mixing the material over a finite fraction of the stellar mass, on one hand fuels the innermost regions of fresh hydrogen, thus determining the duration of the H-burning phase, on the other hand gives rise to the building up of a helium-rich core. The extension of the convective core is classically determined by the so-called Schwarzschild condition in a local, adiabatic treatment of convection. The mass fraction of the convective core and, in turn, the mass of the He-rich core at the end of the central H-burning, increase with the star mass, whereas the duration of the core H-burning decreases with it. During the core H-burning phase the luminosity and radius of the star increase, whereas the effective temperature decreases. Core H-burning models are expected to describe a band in the HR diagram, whose width is essentially determined by the size of the convective core (among the most massive stars also by the efficiency of mass loss by stellar wind and opacity in the CNO ionization region, see below), to which main sequence stars should correspond. The core H-burning phase of stars of any mass does not deserve a detailed discussion owing to the large body of literature existing on the subject. Suffice it to recall that massive stars may be affected by semiconvective instability and, more important, by mass loss by stellar wind, which in most massive stars may be efficient enough to deeply alter even the simple core H-burning phase (see Chiosi & Maeder 1986).

Semiconvection (thereinafter the H-semiconvection) has long been the characterizing feature of the structure of massive stars evolved at constant mass. In brief, the existence of semiconvection is due to the fact that radiation pressure and electron scattering dominate pressure and opacity in massive stars. As a consequence of the high radiation pressure, the convective core tends to grow and to expand as evolution proceeds, giving rise to a chemical discontinuity at the border of the core, which due to the electron scattering opacity does not allow radiative equilibrium outside the border of the convective core to be established. The difficulty is removed assuming that partial mixing occurs to restore radiative equilibrium. The basic mechanism for the so-called semiconvective mixing is unknown however. Whether or not the matter in this region becomes fully convective during the overall contracting phase terminating the main sequence phase, depends on the choice for the stability criterion, over which there was considerable controversy (see Chiosi & Maeder 1986 and references therein). The occurrence or non-occurrence of a fully convective shell influences where in the HR diagram core He-burning will take place. The advent of models with mass loss by stellar winds has put semiconvection aside till recently, when it has been opened up again in relation to the problem of the progenitor star of the supernova 1987A.

To date the most salient signature of the evolution of massive stars is the occurrence of mass loss by stellar winds (see Chiosi & Maeder 1986). The major differences with respect to the evolution at constant mass are that these stars evolve at lower luminosity, live longer and tend either to cover a wider range of effective temperatures or, on the contrary, to evolve near or to the left of the zero age main sequence as the rate of mass loss increases. Finally, semiconvection is strongly inhibited by mass loss by stellar wind. Fortunately, all this is seen to occur only in stars of very high mass (M_i greater than about 30-40 $M_\odot$), which, as a consequence of the high mass loss, never enter the red supergiant region and therefore are not relevant to our discussion of the Cepheid stars. Massive stars of relatively lower mass (say in the range 10 to 40 $M_\odot$) are only marginally affected by mass loss during the core H-burning phase, and they evolve similarly to constant mass stars. The major uncertainties are in the radiative opacities, and in the correct treatment of convection in central interiors, this latter giving rise to the well known phenomenon of convective overshoot (see below).

ii) <u>Core He-burning</u>. After hydrogen disappears at the centre, the H-exhausted core of the star contracts rapidly and, driven by the energy generated in a thin H-burning shell, the envelope expands. The star becomes a red giant and settles along the Hayashi line, where it undergoes the first dredge up episode, during which external convection may penetrate deep enough to pass the layer at which the chemical structure of the star has been previously altered by nuclear processing and/or internal mixing. In such a case changes in the surface chemical composition are expected. The H-exhausted core continues to grow in mass due to the outward migration of the H-burning shell, to contract and to heat up until He-burning temperatures are reached at the centre (this occurring, as already mentioned, well before electrons become degenerate there and while the star is at the tip of the Hayashi line). The question as to why stars become red giants has been debated for many years without receiving a satisfactory answer. The last studies of this topic are by Renzini (1984a), who identifies in a thermal instability in the envelope, primarily determined by the derivatives of the opacity in the middle temperature region (see also Iben & Renzini 1984), the physical cause for the rapid expansion of the envelope to red giant dimensions. During the ensuing core He-burning phase, hydrogen continues to burn in a shell at about the same rate as it did during the main sequence phase. The rate at which helium is burnt in the convective core determines the rate at which the star evolves. Typically, the lifetime in the core He-burning stage is about 15 to 30% of the main sequence lifetime, being larger in models of smaller mass. Like the main sequence phase, at the end of the core He-burning, which turns helium into carbon, oxygen and traces of heavier species, a C-O core is built up whose dimensions depend on the total mass of the star and once more the kind of physical model adopted to describe central convection and its efficiency. Since the mass of the convective core increases at increasing total mass, so does the mass of the C-O core.

The core He-burning phase of intermediate mass stars, in particular of those toward the lower mass end, and of low mass stars (horizontal branch stages) is known to be affected by the same kind of convective instability (semiconvection) that was encountered by massive stars. The onset of this instability (hereafter the He-semiconvection) is due to the fact that as helium is turned into carbon and oxygen, the carbon-rich mixture inside the core is more opaque than the carbon-poor mixture outside the core. However, He-semiconvection, which requires a particular algorithm to properly fix the boundary of the convective core, alters the structure of intermediate mass stars only very marginally and can be ignored to any practical purpose. This is not the case for HB stars where the models are sensitive to the effects induced by semiconvection. Much more important, on the contrary, is the kind of convective instability met in the models at the very end of the core He-burning, when the central He content has dropped below a few per cent (Castellani et al. 1985). Although the physical nature of this instability, otherwise known as the breathing pulses of convection, is still uncertain, the net result of it is an oscillatory increase in the mass of the convective core, refueling the He-burning reactions, prolonging the core He-burning lifetime, and giving rise to a much larger C-O core. This phenomenon, although deeply affecting the ensuing AGB stages, has little relevance in the context of Cepheid stars.

Since all intermediate mass stars expand to red giant dimensions before igniting helium in the core, they cross the Cepheid instability strip at least once (first crossing). This is also true for most models of massive stars ($10 < M_i < 40\text{-}50\ M_\odot$) under favourable circumstances. However, since the above expansion takes place on the Kelvin-Helmholtz time scale (10^6 to 10^3 yr with increasing mass of the star), the first crossing has little relevance to the discussion of the properties of Cepheids, even though the possibility that some of these are in the first crossing cannot be excluded. Other possible crossings are discussed below.

iii) <u>Evolution in the HR diagram during core He-burning</u>. In intermediate mass stars, slow evolution during core He-burning takes place in two distinct regions, a first near the Hayashi line and a second at high effective temperatures. Stars of the first region are in early stages of core He-burning. Approximately, when the energy released by the burning core (increasing) equals the energy released by the H-burning shell (decreasing), a rapid contraction of the envelope readjusts the outer layers from convective to radiative and the star leaves the red region moving to high effective temperatures, where the major phase of core He-burning occurs. The location of these two regions and the lifetime spent in each region are function of the mass, chemical composition, nuclear reaction rates, the $^{12}C(\alpha,\gamma)^{16}O$ in particular, extension of the convective core, opacity, mass loss along the red giant branch, and other both physical and computational details. For any choice of the composition, as the stellar mass decreases, the location of the blue region, otherwise known as the blue loop, moves toward the Hayashi line, and eventually merges with the red giant region. Thus for an assigned chemical composition, core He-burning occurs in bands, one roughly corresponding to the locus of the Hayashi line or red giant stars, and another that breaks off the red giant band at low luminosities and moves toward the blue with increasing luminosity (the so-called blue band). The mean slope of this is determined by the complicated interplay among the above physical factors and cannot be established a priori. Nonetheless, since the slope of the instability strip is opposite to that of the blue band defined by stars of a given composition in the core He-burning phase, the crossing of the two bands is likely to occur, unless unfavourable circumstances conspire to keep the whole core He-burning phase of stars of any mass in the red giant region, which is not the case. It goes without saying that even though a second and a third crossing are possible when the blue band develops, the most favourable case is given by those stars whose mass, chemical composition, etc. are such that the blue region coincides with or largely overlaps the instability strip. In such a case the instability strip is crossed on a nuclear time scale. The predicted locations in the HR diagram of stars of typical Pop I composition (Y = 0.28 and Z = 0.02) during the major nuclear-burning phases are shown in Figure 1. These models are calculated under standard assumptions for the treatment of the central convection, i.e. they ignore the effect of convective overshoot. The hatched regions visualize schematically where stars of all initial masses and ages spend most of their lifetime. As already recalled, stars more massive than M_{HeF} have a core He-burning lifetime of about 15 to 25 per cent of the main sequence lifetime and share it among the two bands (red and blue). The location of the core He-burning phase of stars more massive than say 15 $M_\odot$ but lighter than about 50 $M_\odot$ is uncertain as the effect of mass loss by stellar wind during the whole stellar life is dominant. Nowadays calculations tend to favour models that spend the whole He-burning phase in the red supergiant region. Stars less massive than M_{HeF}, whose evolution is shortly summarized below, experience core He-flash which can be started only when the core mass has grown to a critical value (dependent on the composition) independent of the stellar mass. The growth of the core to the critical value occurs while matter there becomes highly electron degenerate and the star is ascending the RGB. During this phase external convection penetrates deeper and deeper, and the total mass of the star is decreased by about 20 per cent due to mass loss by stellar wind. The core He-flash occurs at the tip of the RGB at about the same luminosity for all stars ($LogL/L_\odot$ = 3.3). The onset of the He-flash lifts degeneracy in the core and helium is thereafter burnt quiescently at higher temperatures and lower densities than at ignition. The constancy of the core mass implies that the lifetime of the following core He-burning phase (either as a HB or a red clump star) is independent of the mass. The typical duration of the phase is about 10^8 yr. The location of the core He-burning stars, corresponds to the horizontal branch or red clump in the HR diagram depending on the mass and chemical composition. Under favourable circumstances (mostly the metal content) the HB can intersect the instability strip giving rise to the RR Lyrae pulsators.

Figure 1. Theoretical HR diagram for classical models (no convective overshoot) evolved with mass loss by stellar wind in the domain of massive stars, and standard radiative opacities. The initial masses are given along the zero age main sequence. The hatched areas indicate the slow phases of nuclear burning. The initial chemical composition is $Y = 0.28$ and $Z = 0.02$. The location of the instability strip for the same chemical composition is indicated. The blue edge is for the fundamental mode, whereas the red edge is for the width $\Delta \log T_{eff} = 0.04$. L_{max}, L_0 and L_1 are three characteristic luminosities of the Cepheids (see text).

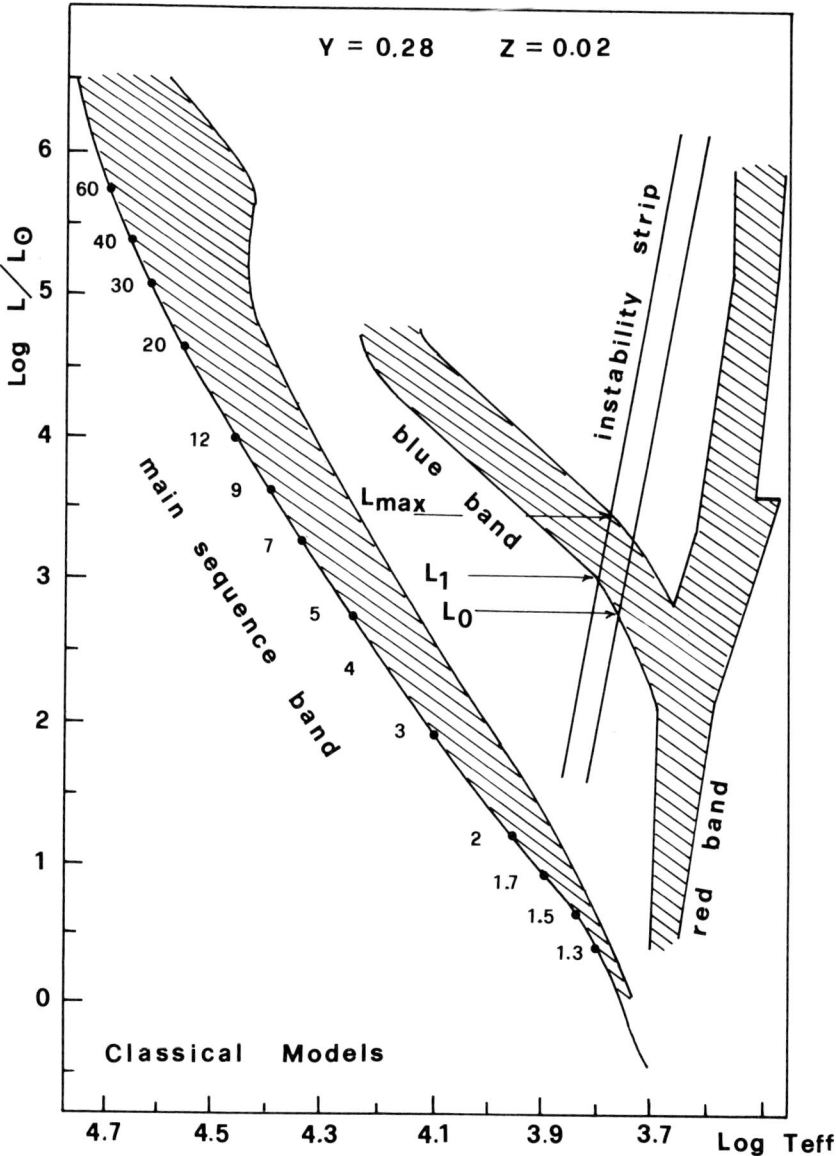

iv) <u>Later evolutionary phases</u>. All single stars less massive than M_{up} and heavier than say $0.7\ M_\odot$ become thermally pulsing AGB stars after exhausting helium in the core. AGB stars possess a C-O core, mostly oxygen with the new $^{12}C(\alpha,\gamma)^{16}O$ rate, within which electrons are higly degenerate, He- and H-burning shells separated by a thin layer of H-exhausted matter, and an extended convective envelope. In consequence of the high thermal conductivity of degenerate electrons and loss of energy by an outflow of neutrinos produced by plasma and related processes, temperatures in the core remain below the treshold value for carbon ignition. Temperatures in the core are maintained at the high value 10^8 K by the gravitational energy release from matter added to the core in virtue of the outward migration of the He- and H-burning shells. Sooner or later (for a more detailed discussion see Iben & Renzini 1983, 1984), nuclear burning in the He shell becomes thermally unstable. In brief, the nuclear burning does not occur at a steady rate, but the two shells alternate with one another as the major source of energy. For 90 per cent of the time the He-burning shell is inactive and the H-burning shell is the major source of energy. However, as the mass of the He-rich zone below the H-burning shell increases, the density and temperature at the base of this zone increase until the rate of energy production by the 3α-^{12}C reaction gets higher than the rate at which energy can be carried out by radiative diffusion. A thermonuclear runaway occurs. A thin convective layer is generated on the top of the He-burning shell. At first the energy goes into raising the temperature of and expanding the matter in and near the burning zone, and the material is pushed away in both directions. Matter at the base of the H-burning region is pushed out and cools to such low temperatures than the H-burning shell is temporarily extinguished. Eventually, matter at the He-burning region begins to cool as it overexpands and the rate of burning there drops dramatically. The convective layer disappears and a steady state is reached in which He-burning occurs quiescently at a slowly decreasing rate as the He-burning shell actually runs out of fuel. This quiescent phase lasts for about 10 per cent of the time elapsing betweeen successive outbursts. The material propelled out falls back and the H-burning shell eventually re-ignites. During this phase, by virtue of a complicated interplay among the outer convective envelope, the oscillatory nature of the shell burning, and the thin inner convective layer, material processed into the intershell region is brought into the outer convective envelope and exposed to the surface. The so-called third dredge up takes place. The pulsing regime of the AGB phase, which manifests itself in the HR diagram with continuous brightening of the star at about constant effective temperature, lasts until either the core has reached the critical value to ignite carbon with consequent supernova explosion or because of mass loss by stellar wind the whole envelope is removed and the star becomes a white dwarf. Therefore, from the standpoint of the evolution of AGB stars, mass loss plays an important role. This may occur in two forms: a stationary wind, whose rate has been approximated by various relationships involving basic stellar parameters like luminosity, mass, radius, etc, the most popular of which is the Reimers (1975) rate, and a so-called superwind which leads to the ejection of a planetary nebula shell (see Renzini & Voli 1981; Iben & Renzini 1984). From the standpoint of classical Cepheids, the AGB phase has no relevance, the only exception being the possibility that under favourable circumstances, past core He-burning and prior to the start of the thermally pulsing regime, during the so-called early AGB, a second blue loop develops. This is due to the complicated interplay between the contracting C-O core, the He-burning shell, and the structure of the outer envelope. The time scale of this crossing is about 10^3 yr.

1.3 Nuclear reactions

Although the rates of the major reactions involved in H- and He-burning are sufficiently known, the work of Fowler et al. (1967, 1975) and Harris et al. (1983) shows that many of them have changed over the years and that some uncertainty is possible. This is particularly true for the $^{12}C(\alpha,\gamma)^{16}O$ reaction whose cross section has been recently remeasured and increased by approximately a factor of five (Kettner et al. 1982) with respect to the Fowler et al. (1975) estimate. Although this rate has been somewhat reduced by the study of Fowler (1984), only a factor of three higher, settlement of this question has not yet been achieved. The presently accepted increase with respect to the classical value is a factor of 2 to 3. Nonetheless, it has been known for a long time (Iben, 1972) that any uncertainty in the rate of this reaction will immediately reflect on the evolutionary behavior of stellar models that become Cepheids. In fact, the larger the cross section, the greater is the extent to which carbon is converted into oxygen, and the longer the core He-burning lifetime. The greater the cross section, the further the H-exhausted core increases under the action of the H-burning shell, and the further the loop extends to the blue before rapid core contraction and envelope expansion set in and the evolution proceeds back to the red. As a result, for a given chemical composition, the blue band of core He-burning models turns out to be more inclined toward the blue than with the old value, since the shift to the blue tends in fact to be greater at higher masses. How this would affect the intersection with the instability strip is obvious. However, the effect of increasing the rate of $^{12}C(\alpha,\gamma)^{16}O$ is marginal toward the low end of intermediate mass stars and in the range of massive stars, where in these latter it is wiped out by the overwhelming effects of mass loss.

1.4 Radiative opacity

Among the various aspects of the input physics of model calculations, radiative (and conductive) opacities still suffer from a great deal of uncertainty, and to even small changes of which the properties of stellar models are extremely sensitive. To clarify the role played by each particular source of opacity to the building up of the total radiative opacity, Iben & Renzini (1984) have distinguished several temperature ranges (which are also somewhat dependent on the actual densities encountered in the stellar interiors), which are briefly summarized here. (i) At very low temperatures ($< 2.5 \times 10^3$ K), the opacity is dominated by transitions in molecular levels in species such as CO, TiO, H_2O, CN, CH etc. (ii) In the range 2.5×10^3 to 5×10^3 K, the major contribution to opacity comes from H^-, the number of required electrons being controlled by the abundances of elements of low ionization potential such as Mg, Si; (iii) In the range 5×10^3 to 5×10^5 the opacity is mostly given by bound-bound, bound-free and free-free transitions of H and He. (iv) At temperatures in the range 5×10^5 to 5×10^6 K, the bound-bound and bound-free transitions of elements from carbon to iron (shortly indicated as the CNO group of elements) dominate the opacity. (v) Finally, at temperatures higher than 5×10^6 K, the opacity is simply given by electron scattering. While the opacities of regions (iii) and (v) are probably not significantly in errors, those of regions (i) and (ii) are far more uncertain. However, these latter regions are always associated with convection, and their uncertainty is simply overwhelmed by the uncertainty in the treatment of convection. Region (iv) is by far the one presenting the major difficulty. In this case, the high number of electrons in each elemental species, the large number of ionization and excitation stages, the non-hydrogenic structure of the electronic configuration, the distortions of this induced by nearby electrons and ions that are difficult to model, all conspire to make the opacity of this temperature range very difficult to calculate, and of course, with the highest degree of uncertainty. This is the main reason why the characteristics of stellar models that are very sensitive to the so-called middle temperature opacity are still highly uncertain and a matter of debate. We will touch upon this topic in the course of this review.

1.5 Mass loss by stellar winds

To our knowledge, the effect of mass loss by stellar wind along the RGB phase on stars of intermediate mass and on Cepheid stars, in particular, has never been systematically investigated. The early studies of Lauterborn et al. (1971), and Lauterborn & Siquig (1974) have shown that mass loss may be able to suppress the formation of blue loops if more than 10% for a 5 $M_\odot$, 13% for a 7 $M_\odot$ and 20% for a 9 $M_\odot$ star of the initial mass is lost. Although the results are expected to depend somewhat on the initial chemical composition and details of the input physics, the implication is that a moderate degree of mass loss in intermediate mass stars can reduce the number of crossings from a few (say three) to only one (the first). With the current rates of mass loss (see Reimers 1975; Lamers 1981; de Jager et al. 1988) for red giant stars in the luminosity range of interest here, and the lifetime of the RGB typical of an intermediate mass star, mass loss is likely not to be of much importance. This is particularly true for classical models. The above conclusion no longer holds for Cepheids whose initial mass is greater than say 15 $M_\odot$, where mass loss may be important and it is already included in most model calculations (see Chiosi & Maeder 1986). Nonetheless, on one hand the observational determinations of mass loss rates are still very uncertain and tending to grow, and their dependence on basic stellar parameters (luminosity, mass, radius, etc.) poorly known, on the other hand stellar models potentially undergoing mass loss (along the RGB) have changed with respect to the classical ones for which the estimate of the effect of mass removal has been made. This means that the real amount of mass that can be lost by intermediate mass stars during their RGB phase is still highly uncertain. The calculations of Bertelli et al. (1985) for Pop I intermediate mass stars (Z = 0.02) with initial mass 5 $M_\odot$ and 9 $M_\odot$, in which different algorithms for the mass loss rates have been adopted, clarify the role played both by the algorithms themselves and the underlying evolutionary models. The following relations for the mass loss rate have been tested

$$\dot{M} = \alpha\, 1.27 \times 10^{-5}\, M^{-1}\, L^{1.5}\, T_{eff}^{-2} \tag{8}$$

$$\dot{M} = \eta\, 0.11\, L^{1.96}\, T_{eff}^{-3.54} \tag{9}$$

where M, L are in solar units, $\dot{M}$ in $M_\odot\, yr^{-1}$, and relation (8) is the Reimers (1975) law, whereas relation (9) is adapted from Waldron (1984). The parameter η has a well known meaning, whereas the parameter α has been introduced by Bertelli et al. (1985) to reconcile the rates from the two laws in the domain of massive stars ($\alpha = 1$ yields rates equal to those for $\eta = 1$). In classical models, little mass is lost during the RGB phase independently from the adopted algorithm, thus confirming the older studies quoted above (see the data in Tables 2a and 2b of Bertelli et al. 1985). On the contrary, in models with convective overshoot (to be discussed below), owing to their much higher luminosity, the amount of mass lost during the whole core He-burning phase critically depends on the adopted algorithm and the initial mass of the star. Stars of about 5 $M_\odot$ and lower, independently of the algorithm for the mass loss rate, lose only a few per cent and evolve in the HR diagram in a way similar to that of classical models. Stars of higher mass depending on the adopted algorithm, may lose an enormous amount of mass and accordingly drastically change their path in the HR diagram (see the data of Table 2a and 2b, and Figures 11a through 11c of Bertelli et al. 1985). This simply means that the problem of mass loss from intermediate mass stars is far from being settled and more work is needed. Another interesting possibility has been advanced by Willson (1987) and Willson & Bowen (1984), who suggest that in order to reduce the evolutionary masses to better agree with the pulsational masses (see below), and to avoid the difficulties encountered when mass loss occurs during the RGB phase, mass loss, somehow triggered by the pulsational instability itself, should occur while the star is within the instability strip. Evolutionary calculations including mass loss during the Cepheid stage confirm that Cepheids are trapped in the instability strip, mass loss can continue over a relatively long time scale, the total mass is significantly reduced, whereas the luminosity is about constant (Brunish & Willson, 1987). The rates of mass loss required by these model calculations are of the order of $7 \times 10^{-9}\, M_\odot\, yr^{-1}$ for a 5 $M_\odot$ star and $2 \times 10^{-7}\, M_\odot\, yr^{-1}$ for a 7 $M_\odot$ star. Unfortunately, efforts to observe Cepheid winds directly have so far given inconclusive results. Finally, it is worth recalling that mass loss not only affects the mass and the path in the HR diagram of a star, but also its appearance as a Cepheid. In fact, on one hand the mass is decreased (even though by a few per cent according to the classical algorithm of Reimers (1975), on the other hand the period-luminosity relation is changed. At given luminosity, the period is expected to increase at decreasing mass as a consequence of the period-mean density relationship (see the study by Stift (1984) on the frequency-period distribution).

1.6 Rotation and duplicity

Rotation. Main sequence progenitor stars of Pop I Cepheids rotate at speeds of about 250 km·sec^{-1} (Slettebak 1970), whereas red giant stars have low rotational velocities (Kraft 1970), implying that angular momentum is either lost or more likely transferred to inner layers of the star. Owing to the complexity of including rotation in stellar model calculations, since first 2D or 3D codes and second modelling of angular momentum transfer and coupling of it with convection are required, rotation is usually neglected in most of the stellar models currently available in literature. Despite the complexity of the problem, a few pioneering calculations exist (Kippenhahn et al. 1970; Meyer-Hofmeister 1972; Endal & Sofia 1976, 1978, 1979) showing from the standpoint of one dimensional calculations the effect of rotation on the evolution of intermediate mass stars. Somewhat contradictory results are obtained. While, they all agree that the core He-burning lifetime is lengthened by rotation, they disagree in the properties of the blue loop. Endal & Sofia (1976, 1978) find that the luminosity and tip effective temperature of the blue loop is increased and decreased, respectively, whereas the opposite is found by the other authors. The reasons for the disagreement cannot easily be singled out.

Duplicity. A large fraction (up to about 30%) of Pop I Cepheids are members of binary systems, with companions of similar mass. Depending on the separation, these stars can either evolve as if they were single objects (larger than a few 100 $R_\odot$) or mutually influence their structure and evolution by mass exchange. Unfortunately, very few calculations of binary systems of intermediate or massive stars are available in the literature, so little can be said on this topic. Likely, if a star fills its Roche lobe as a red giant, it will probably lose enough mass to prevent it from evolving further into the instability strip. Therefore this star can appear as a Cepheid only during its first crossing. The accreting mass companion will certainly evolve in a manner different from that of a single star having the companion's star initial or current mass. If this star crosses the instability strip, how the different evolutionary history would affect its appearance as a Cepheid cannot be foreseen (see Becker 1985).

2. EVOLUTION WITH CONVECTIVE OVERSHOOT

It has already been recalled that starting from the core H-burning stages the extension of convective cores in massive stars are affected by the H-semiconvection caused by the high radiation pressure and electron scattering opacity, whereas those of intermediate and low mass stars are affected by the He-semiconvection driven by an opacity difference between carbon-rich and carbon poor mixtures during the core He-burning phase. However, in addition to the problem of semiconvection, there is a more fundamental question concerning the mass extension of convective cores, i.e. the amount of convective overshoot from their borders. This in fact is always expected to occur at the boundary of a formally convective core, where the kinetic energy of convective elements carries them a finite distance into layers which are formally stable against convection. In other words, the classical criterion used to set the boundary of a core (the layer where the radiative and adiabatic temperature gradients are equal) simply means that the acceleration imparted to convective elements by the buoyancy forces becomes zero, whereas their kinetic energy does not. It goes without saying that the correct condition would be the one requiring the vanishing of the kinetic energy. In such a case further mixing may arise between the convective core and a part of its neighbouring radiative region causing a net increase in the mass extension of the fully mixed region. Though simply motivated from the standpoint of physical arguments, great controversy exists on the exact amount of convective overshoot. According to the theoretical formulations of this phenomenon proposed by different authors, the overshoot distance at the edge of the formally convective core goes from zero to about 2 pressure scale heights (see Chiosi 1986 for a review of the subject and exhaustive references). In any case, convective overshoot gives rise in stellar models to differences in structure and evolution that are much more substantial than those given by semiconvection. In addition to the above arguments of a merely theoretical nature, there are many observational facts requiring a deep revision of the basic physics, such as convective cores significantly bigger than usually assumed. Among others we recall:

a) The width of the main sequence band, which appears to be larger than predicted by standard models, and the number ratio of evolved to unevolved stars in galactic clusters (Maeder & Mermilliod 1981) and in the populous young globular clusters, like NGC 1866 in LMC (Becker & Mathews 1983; Chiosi et al. 1988), which is larger than expected.

b) The anomalous behaviour of several old open clusters, which in spite of a turnoff mass of about 1.5 $M_\odot$ do not exhibit an extended RGB indicating that the core He-flash does not occur but, on the contrary, He-burning takes place quietly (Barbaro & Pigattto 1984).

c) The lack of very luminous AGB stars in several clusters of the Magellanic Clouds, where they are expected on the basis of the turnoff mass. This fact suggests that M_{up} is much lower than predicted by classical models.

d) The long lasting discrepancy between evolutionary and pulsational mass together with the frequency-period distribution of Cepheid stars (see below).

2.1 Stellar models with convective overshoot

Among the possible formulations of convective overshoot, we have followed the method developed by Bressan et al. (1981), which is based on a non-local view of convection, uses the mixing length theory, and takes the mean free path of convective elements as a parameter to be fixed by comparing model results with the observational properties of star clusters. The mean free path is given by $l = \lambda H_p$, where H_p is the local pressure scale height and λ is a free parameter. The evolutionary models we are going to describe are for $\lambda = 1$, which has been found to give the most interesting results (Bertelli et al. 1985; 1986a,b; Bressan et al. 1986). In these models the rate of the $^{12}C(\alpha,\gamma)^{16}O$ reaction is assumed to be three times greater than the classical value of Fowler et al. (1975) as indicated by the recent measurements of the cross section (see Kettner et al. 1982).

i) **The core H-burning phase.** Stars whose initial mass is high enough to develop a convective core on the main sequence, ($M > 1.1\ M_\odot$), are strongly affected by convective overshoot. In fact, owing to their more massive convective cores, they run at higher luminosities and live longer than classical models. They also extend the main sequence band over a wider range of effective temperatures, this trend increasing with the stellar mass. Massive stars ($M > 40\ M_\odot$) would spread all across the colour-magnitude (CM) diagram, were it not for the contrasting effect of mass loss (see Bressan et al. 1981).

ii) **The core He-burning phase in massive and intermediate mass stars.** The overluminosity caused by overshoot during the core H-burning phase still remains during the shell H- and core He-burning phases. The mass of the H-exhausted core, M_{He}, and the mass of the C-O rich, He-burning convective core, are increased by overshoot. However, as a consequence of the higher luminosity, the lifetime of the He-burning phase (t_{He}) gets shorter in spite of the increase in the core mass. This, combined with the longer H-burning lifetime, t_H, makes the ratio t_{He}/t_H fairly low (from 0.12 to 0.06 when the stellar mass varies from 1.6 to 9 $M_\odot$). The location of the core He-burning models in the CM diagram can be schematically summarized as follows. In massive stars, where mass loss dominates throughout the stellar life, the He-burning phase takes place entirely in the blue for $M > 60\ M_\odot$, partly in the blue and partly in the red for stars in the range 60 to 20 $M_\odot$ or thereabouts, the lifetime of the two sub-phases being controlled by the mass loss rate. The blue loops of intermediate mass stars ($M < M_{HeF}$) are much less extended, whereas their dependence on chemical composition and nuclear reaction rates is the same as in classical models. This implies that the blue band of core He-burning models is less inclined in the HR diagram and, therefore, for a given chemical composition the intersection with the instability strip occurs at different luminosities, hence masses. Figure 2 shows the predicted locations in the HR diagram of stars of Pop I chemical composition ($Y = 0.28$ and $Z = 0.02$) calculated with convective overshoot ($\lambda = 1$).

iii) **The core He-burning phase of low mass stars.** HB stars while burning helium in the centre, possess a convective core whose mass extension may be affected by convective overshoot. Models of HB stars have been calculated according to the Bressan et al. (1981) algorithm with $\lambda = 1$, core mass equal to 0.475 $M_\odot$ and total mass in the range 0.6 to 1.4 $M_\odot$ (Bressan et al. 1986). The most important result of these calculations is that in presence of convective overshoot, semiconvection and/or breathing pulses of convection (see Castellani et al. 1985 for details) never develop. In particular, these models predict lifetime ratios in RGB, HB and AGB that are in closer agreement with the observational ones derived from star counts in globular clusters (see Buonanno et al. 1985, for a discussion of this topic). A detailed description of these models and a comparison with the observational data for globular clusters are given in Bressan et al. (1986) to whom we

Figure 2. The HR diagram of models with convective overshoot ($\lambda = 1$), standard radiative opacities, and chemical composition $Y = 0.28$ and $Z = 0.02$. The hatched areas show the phases of slow nuclear burning. No mass loss along the RGB phase of intermediate mass stars is taken into account. The value of the initial mass is indicated on the zero age main sequence. Note the much wider main sequence band and the less inclined blue band of He-burning models produced by the effect of convective overshoot. The instability strip is also drawn, and three characteristic luminosities at the intersection with the blue band of core He-burning models are shown (see text).

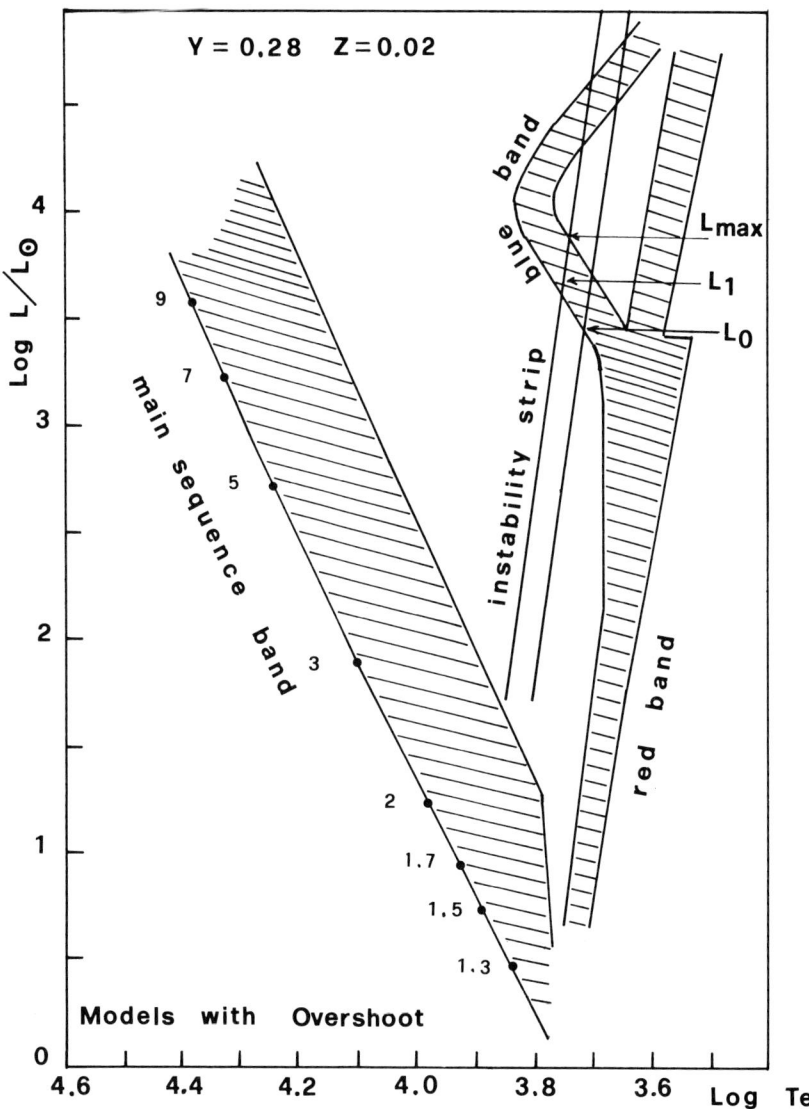

refer.

iv) The critical masses M_{mas}, M_{up} and M_{HeF}. By virtue of the larger masses of the He and C-O core left over at the end of core H and He-burning phases respectively, the relationships between the initial mass and the mass of the He and C-O core, which define the critical masses in question are different in models with convective overshoot. The most important result is that now both M_{up} and M_{HeF} are significantly lower than in classical models. The new masses are given in Table 1 for two initial chemical compositions. This means that no AGB and no prolonged RGB phases are expected for stars of initial mass above the new M_{up} and M_{HeF}, respectively. The impact of this result on the observational front is straightforward and of paramount importance.

v) The AGB phase. Over the past few years, the evolution of AGB stars has been the subject of many theoretical studies, recently reviewed by Lattanzio (1988a), aimed at understanding why carbon stars can be formed at lower luminosities than predicted by the classical models of Iben (1975a,b, 1976). The problem came when a comparison between theoretical AGB star distributions (Iben 1981; Renzini & Voli 1981) and observations of Magellanic Cloud stars (Blanco et al. 1978, 1980) showed that no carbon stars were expected less luminous than $M_B = -5$ or brighter than $M_B = -7.5$, in contrast with the observations indicating that carbon stars occur in the range $-3.5 < M_B < -6$. Since the evolution along the thermally pulsing AGB phase is essentially governed by the linear relationship between the H-exhausted core mass, or equivalently the C-O core mass, and the luminosity (see Iben & Renzini 1983 for all details), this is equivalent to say that the third dredge up must operate at lower core masses than suggested by the current theory. The series of recent models by Lattanzio (1986, 1987a, 1987b, 1988), Boothroyd & Sackmann (1988a-d) and Hollowell (1987, 1988) in the mass range 1-3 $M_\odot$, in which revised algorithms for semiconvection and breathing convection, the use of better opacities, the inclusion of mass loss by stellar wind, and the estimate of the dependence of the C-O core mass at the first thermal pulse on the initial chemical composition and initial mass of the star, concur to succeed in producing models of carbon stars of quite low luminosity. Models incorporating convective overshoot (see Chiosi et al. 1987) have not yet been evolved into the thermally pulsing regime, however we may foresee that they should give similar results, since they share many features in common with models calculated with semiconvection and breathing convection. Among other things, since they predict that M_{up} is much lower than the classical value, this would help removing all

Table 1

X	Z	M_{mas}	M_{up}	M_{HeF}
0.700	0.020	9	6	1.6
0.700	0.001	8	5	1.6

very bright AGB stars as indicated by the observational data (very few AGB stars brighter than $M_B = -6$).

2.2 Clusters of the LMC: NGC 1866 a test for convective overshoot

Although stellar models calculated with semiconvection and breathing convection look similar to those calculated with convective overshoot, the two types of model differ profoundly in many quantitative details. The characterizing features are the core H- and He-burning lifetimes and evolutionary brightening from the main sequence to the most advanced stages. These should immediately reflect on the morphology of CM diagrams of star clusters, namely the number of evolved to main sequence stars and associated luminosity functions. It goes without saying that the ideal laboratory where theories of stellar structure, in particular those concerning the domain of massive and intermediate mass stars, can be tested are the rich young clusters of the Magellanic Clouds. These in fact by virtue of their large number of stars allow statistically meaningful comparisons between theory and observations in samples of about coeval and chemically homogeneous objects even for the shortest lived evolutionary stages. To this purpose, Chiosi et al. (1988) have taken accurate Johnson BV CCD photometry of the cluster NGC 1866 of the Large Magellanic Cloud (LMC) and compared its CM diagram and main sequence star luminosity function with the theoretical predictions based on both classical models and models with convective overshoot. The results of the Chiosi et al. (1988) study are briefly summarized below.

NGC 1866 is a rich young cluster situated at about 2° North of the LMC centre, whose total mass is estimated in the range 3.6 to 8.5×10^4 $M_\odot$ (Heckman 1976). It is classified as a type III in the Searle et al. (1980) system, with integrated apparent visual magnitude V = 9.73, and colours (B-V) = 0.25 and (U-B) = -0.04 (van den Bergh 1981) and colour excess E(B-V) = 0.07 (Persson et al. 1983). Johnson B and V CCD photometry of 1517 stars in the central region of NGC 1866, and of 640 stars in a nearby field have been obtained with the RCA #5 CCD on the ESO 2.2m telescope and reduced using the ESO version of DAOPHOT. To define the region of the cluster frame that suffers from the least degree of uncertainty due to photometric incompleteness and contamination by field stars the following procedure was adopted. First, a composite CCD frame (cluster and field) was obtained superposing a few bright stars in common, and an iterative procedure was used to fix the cluster centre. Second, the surface brightness profile of the cluster was generated in a way similar to Meylan & Djorgovski (1987) and compared to the theoretical density profiles calculated from multi-mass, King-Michie models (Meylan 1987a,b, 1988) in the case of the Salpeter law with slope x = 2.35 and isotropical velocity dispersion. The cluster turned out to possess a concentration $c = \log r_t/r_c = 1.5$, a core radius $r_c = 6\rlap{.}''69$ and a tidal radius $r_t = 211\rlap{.}''6$. Furthermore, since the last point of the observed profile is at about $r = 2\rlap{.}'40$ to which a density contrast of about a factor of 10^3 corresponds, the radius $r_W = 2\rlap{.}'45$ is taken as the limiting radius beyond which the majority of stars are field objects. Third, the surface brightness profile was reconstructed from the photometric data (V and/or B magnitudes) of individual stars falling within circles of increasing radius centred on the cluster centre. The simultaneous comparison of the surface brightness profile, the density profile, and the surface brightness profile reconstructed from individual stars, shows that the ring confined between $0\rlap{.}'52$ and $0\rlap{.}'98$ is the one suffering from the least degree of incompleteness by crowding and other causes, and it can be taken to represent the cluster population.

The CM of the stars falling in the ring is assumed as the reference CM to which theoretical models ought to be compared. This is shown in Figure 3. In this CM diagram, three regions of particular interest can be defined, namely: the region of the main sequence stars (MS), the region of red giant stars brighter than 17th magnitude in V (BRGS), the region of red giant stars fainter than 17th magnitude in V (FRGS). The total number of stars is 656, of which MS = 594, BRGS = 39 and FRGS = 23. This CM diagram is photometrically complete to V = 19.2. The field is given by the region of the composite CCD frame outside r_w, whose area happens to be nearly identical to that of the cluster ring. After correction for photometric completeness (crowding) and contamination by field stars by means of the zapping technique of Mateo & Hodge (1986), the total number of stars is 1020 made of MS = 954, BRGS = 41, FRGS = 25. The group of BRGS likely represents the population of evolved stars belonging to the cluster, i.e. genetically linked to the populations of

Figure 3. The CM Diagram of the selected ring of NGC 1866. Capital letters A, B and C indicate three groups of stars discussed in the text.

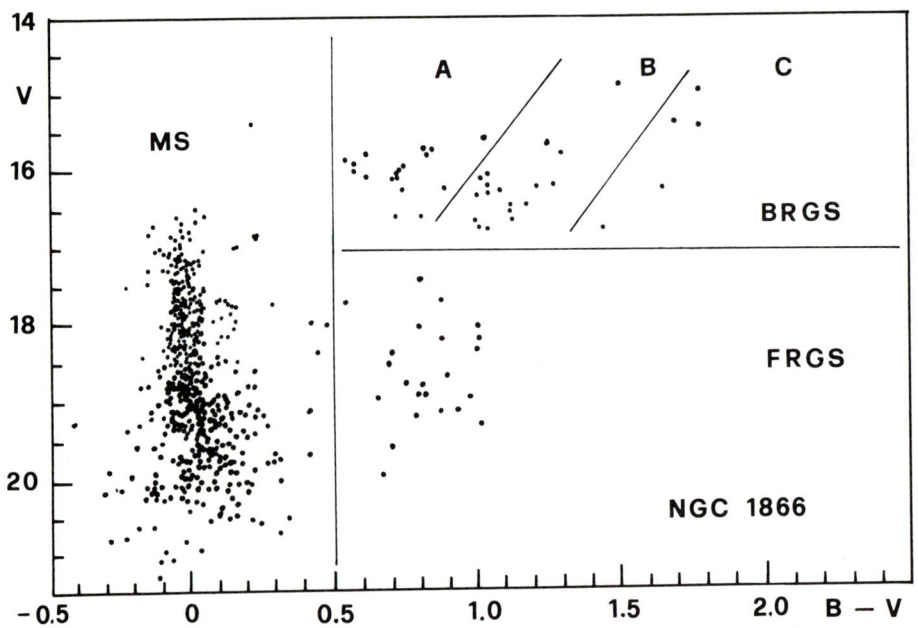

MS, whereas FGRS are likely field stars. NGC 1866 contains also several Cepheids: 11 well established objects (Walker 1987, and references therein) and about 10 more recent candidates (Storm et al. 1988), which are located in the middle of BRGS. The sample of BRGS can be split into three subgroups (labeled A, B and C). Group A and B contain almost equal numbers of stars and are located at the blue and red side of the instability strip, whereas group C is populated by stars (very few indeed) of the same luminosity but of much redder colour. Likely, these do not belong to the same population as those of groups A and B, but they are probably genetically related to FRGS.

The luminosity function for the MS stars is presented in Figure 4. This is the integrated number of stars counted from the tip of the main sequence band down to the current magnitude bin and normalized to the number of BRGS. The advantage of this type of luminosity function is that it can be simply related to the ratio of core H- to core He-burning lifetime.

On the theoretical side, a computer code has been developed for constructing synthetic CM diagrams, luminosity functions, integrated magnitudes and colours in the UBVRI bands as a function of the age, chemical composition, initial mass function, total number of stars, given number of evolved stars corresponding to the observed BRGS, and finally stellar evolution input. This algorithm also allows for

Figure 4. The integrated luminosity function for the main sequence stars normalized to the number of bright red giant stars (V < 17) likely representing the evolved counterpart of the main sequence population (see text for more details).

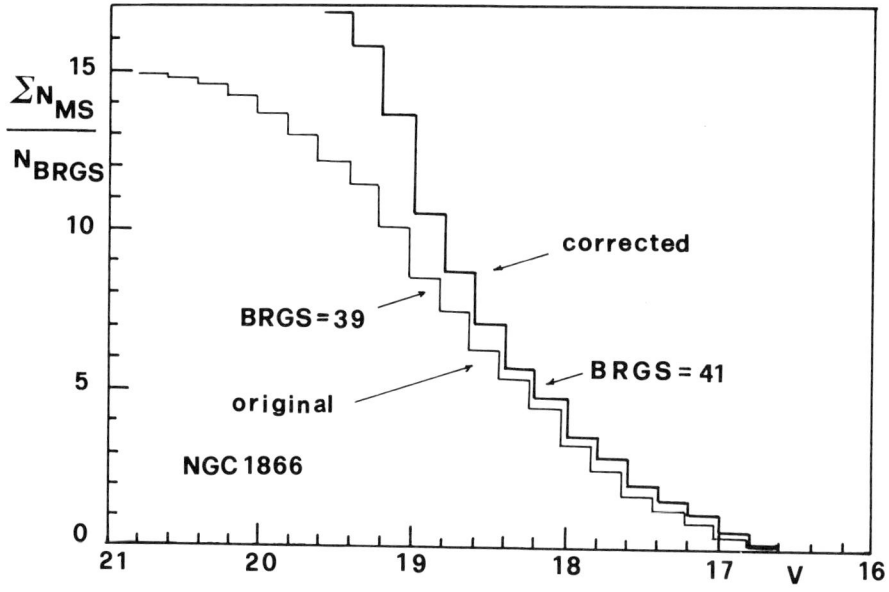

dispersion in age and fluctuations of a stochastic nature in the initial mass function.

The comparison between theory and observations based on several fiducial characteristics of the the CM diagram, such as the mean location of the main sequence band, its terminal luminosity, the mean luminosity of the blue and red giant stars, the tip of the blue loop (if present), and the location of the reddest stars (Hayashi limit) allows us to derive the age and chemical composition and to estimate the distance modulus in the two alternatives for the stellar evolution, namely classical models and models with overshoot. The simultaneous fit of the above constraints indicates that the appropriate chemical composition is $Y = 0.28$ and $Z = 0.02$ (or slightly less say $Z = 0.010-0.015$), the true distance modulus is $(m-M)_o = 18.6$ in good agreement with the value determined from the near infrared photometry of Cepheid stars (Welch et al. 1987), and the age is 200×10^6 yr for models with overshoot, or 70×10^6 yr for classical models. However, with these latter models the fit of the red giant stars is rather poor, as they turn out to be fainter than the mean luminosity of the loop and bluer than the Hayashi limit. Other combinations of age and chemical composition can be neglected as they would lead either to a very poor fit of the global properties of the CM diagram or to unacceptably low distance moduli. While the CM diagram does not clearly indicate the underlying evolutionary scheme, can be seen by looking at the integrated main sequence luminosity function normalized to the number of post main sequence stars. The comparison with the observational luminosity function strongly favours models with convective overshoot. In other words, the star counts in the clusters suggest that the ratio of core H- to He-burning lifetimes and the range of luminosity spanned by main sequence stars, whose masses are compatible with those of the evolved stars, must be smaller and wider, respectively, than given by the classical models. This is possible only if more efficient mixing occurs all over a star's evolutionary history and not only in stages beyond the main sequence phase. It is worth emphasizing that other mixing processes like semiconvection and/or breathing convection, would not be able to satisfy the observational demand, since they tend to increase the number of evolved stars with respect to those of the main sequence. Agreement with the classical models is marginally possible only by increasing the slope of the initial mass function above 4.

3. THE MASS-DISCREPANCY OF CEPHEID STARS

Since the review on Cepheid masses by Cox (1980) is still fully relevant, we will not go into any detail here. It has long been debated whether the masses determined from stellar evolution theory (L-M relation) agree with those derived from pulsation theory (P-M-L-T_{eff} relation) (see Iben, 1974; Iben & Tuggle, 1972, 1975). In general, pulsational masses (M_{pul}) are estimated to be 30 to 40% lower than evolutionary masses (M_{evol}) of the same luminosity. The apparent disagreement between M_{evol} and M_{pul} has been interpreted as an indication of 1) significant mass loss at some point between the main sequence and the Cepheid stage, 2) systematic errors in the determination of the distance scale with respect to the classical values of Sandage & Tammann (1969), possibly an underestimate of the distance to the Hyades, and in the derivation of the T_{eff} assigned to a Cepheid, 3) errors in the pulsation theory and 4) uncertainties in model calculations (particularly as these are influenced by uncertainties in opacity).

Cox (1980), in his review, has distinguished six methods of determining the mass of Cepheids: 1) the evolutionary mass (M_{evol}), 2) the theoretical mass (M_{the}), introduced by Cox et al. (1977) and based on the simultaneous solution of four equations for the unknowns M, R, L and Q_i, where $Q_i = P_i (M/R)^{1/2}$, 3) the pulsational mass (M_{pul}), 4) the bump Cepheid mass (M_{bump}), 5) the double and triple mode mass (M_{beat}) and 6) the Baade-Wesselink radius mass (M_{BW}). The bump masses are calculated by modeling the Hertzprung progression (dependence on period of a secondary maximum in the light curves of Cepheids with period P_0 in the range 6 to 18 days). They require good light curves and phase coverage. The beat masses are derived from the periods P_0 and P_1 (fundamental and first overtone) by means of mass luminosity relations as functions of P_1/P_0, which are very sensitive to the physical properties of models producing them. Finally the Baade-Wesselink radius masses are obtained from the Q_i (P_i, M, R) relations, where Q_i are from theory, P_i from observations, R is the radius measured by the Baade-Wesselink method. It goes without saying that each method is affected by many uncertainties inherent both to the observational and theoretical quantities required to derive the mass.

Cox (1980) discusssing the available data came to the conclusion that the "primary" mass problem ($M_{evol} > M_{pul}$) could be solved by varying the intrinsic luminosity of Cepheids (longer distance modulus) and by lowering the Cepheid temperatures with improved reddening and temperature calibrations. According to Cox, M_{evol} is normally in agreement with all the others but the bump Cepheid masses and the double and triple mode Cepheid masses. However, recent developments on the subject somewhat weaken Cox's conclusion. First, even though the Haydes distance modulus has been increased to 3.4 mag (Hanson, 1979; Vandenbergh & Bridges 1984), there is the well known complication related to the Hyades zeropoint in the distance scale, in that it remains uncertain how the distance to other clusters should be corrected for the effect of the high metallicity of the Hyades. This would immediately affect the distance hence absolute visual magnitude assigned to individual Cepheids. Second, the thorough discussion by Pel (1985) on the fundamental parameters of Cepheids points out that they are still far from being fully assessed, and that the problem of the mass discrepancy, albeit not very large, is nevertheless significant (see the cases of U Sgr and S Nor for which the use of revised data has again moved M_{evol} and M_{pul} apart). Finally, Simon (1986) and Böhm-Vitense (1986) have rediscussed the problem of the beat and bump Cepheid masses. Using the Schmidt (1984) distance scale, along with Baade-Wesselink masses, Böhm-Vitense (1986) finds the following relationships among the various masses: for P > 6 days, $M_{evol} > M_{BW} = M_{pul} > M_{bump}$, whereas for P < 6 days, $M_{evol} > M_{BW} > M_{pul} > M_{beat}$. Simon (1982, 1987), examining the period ratios (P_1/P_0 for beat Cepheids and P_2/P_0 for bump Cepheids), points out that not only are the masses discrepant but also the observed period ratios are lower than those predicted by the theory. While the discrepancy in the period ratios can be corrected by lowering the mass of a model at given luminosity, the period ratios are also very sensitive to the interior structure. The M_{beat} and M_{bump} could be raised to M_{evol} only by accepting that the current model structure is varied for instance by changes in the opacity law such as an increase of a factor of 2 to 3 in the middle temperature region (see below). It transpires from the above discussion that M_{evol} is usually taken as the standard of comparison for the various mass determinations despite the fact that it is perhaps the most difficult to obtain, since it requires the observational determination of P, L or T_{eff} and the assumption of a mass-luminosity relationship from stellar models. Among the various causes for the mass discrepancy, substantial mass loss along the RGB phase and/or uncertainties or differences in the evolutionary schemes have been advocated. We have already pointed out that under the current mass loss rates and their parameterizations, little mass can be lost, certainly insufficient to remove the discrepancy in question. The adoption of unusually high rates of mass loss, on one hand may solve the problem, but on the other hand can give rise to devastating effects on the evolutionary behaviour of intermediate mass stars thus making hard to interpret even the overall properties of Cepheid stars. If mass loss is supposed to be negligible, one way out of the problem of mass discrepancy is that for a given luminosity (for instance in stages during the second crossing) models intrinsically possess a lower mass. This is indeed characteristic of models with overshoot. In fact, at any given initial mass, the tracks cross the instability strip at higher luminosity than classical models, or conversely, at any given luminosity the correspondent Cepheid mass is significantly lower (Matraka et al. 1982; Bertelli et al. 1985). To illustrate the point, in Figure 5 we plot the luminosity versus M_{pul} using the data compiled by Cox (1980) together with the theoretical mass-luminosity relation for classical models (Becker et al. 1977) and models with overshoot [see the relations (13a) and (13b) below]. Although the scatter is large, it can be seen that classical models hardly match the bulk of data, because most of the stars appear to be either too luminous for their masses or too light for their luminosity (pulsational masses grossly in error?) or the classical relationship has to be shifted toward lower masses. It goes without saying that both coordinates are highly uncertain as indicated by the shifts that two prototype Cepheids, U Sgr and S Nor, would suffer when Pel's (1985) data are adopted. Amazingly enough, after correction these two stars fall along the mass-luminosity relationship for models with overshoot. Shifts of the same size are likely to be expected for all stars in the sample. The conclusion from all this can only be that the solution of the mass discrepancy requires improvements on both the observational and theoretical side, even though the net advantage offered by models with overshoot ought to be borne in mind.

4. THE FREQUENCY-PERIOD DISTRIBUTION

It is clear that in a statistically large sample of stars the intersection of the instability strip with the blue band of core He-burning models will give rise to a population of Cepheids whose periods distribute according to a specific law that is easy to understand. The period distributions, number of stars per period bin, $N(\log P)$, observed in the solar vicinity, Milky Way, M31, SMC, and LMC and shown in Figure 6 (Becker et al. 1977), are characterized by three parameters, the short period cutoff P_0, the period P_1 at the maximum of the distribution, and the rate at which Cepheids with $P > P_1$ decay in number with respect to the period. It is soon evident that the $N(\log P)$ distribution varies from one system to another both in overall shape and the characteristic periods P_0 and P_1. The underlying reasons for these variations can be understood in a simple way.

Following Becker et al. (1977), Iben & Tuggle (1975) and Serrano (1983), we start supposing that all Cepheids possess the same chemical composition (indeed this is not the case), which implies that only one blue band of core He-burning models and one instability strip (both are known to depend on the chemical composition) can be used. The

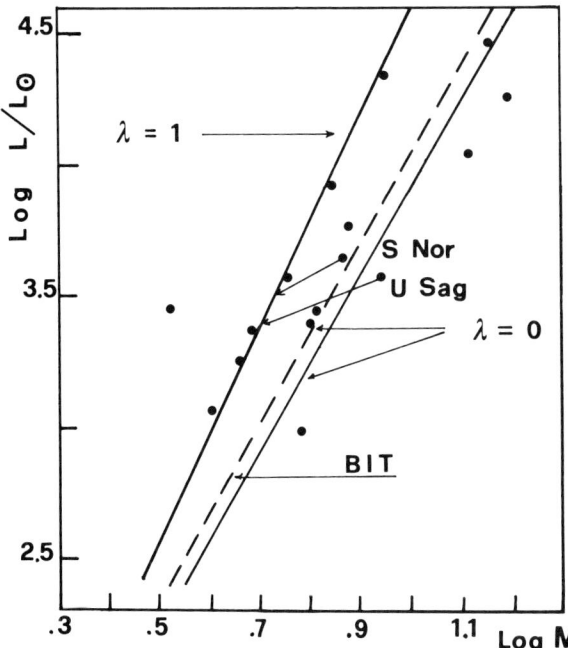

Fig. 5. Luminosity versus mass relationship for Cepheid stars. The full dots show the luminosities and pulsational masses of Cox (1980). The arrows indicate the revision made by Pel (1985) for U Sgr and S Nor. Standard models of Becker et al (1977), and those with convective overshoot are shown.

intersection of the two bands define two characteristic luminosities L_0 and L_{max}. L_0 corresponds to the luminosity of the model (mass) whose blue loop is extended enough to reach the red edge of the instability strip, whereas L_{max} is the last model (mass), above which the blue loop

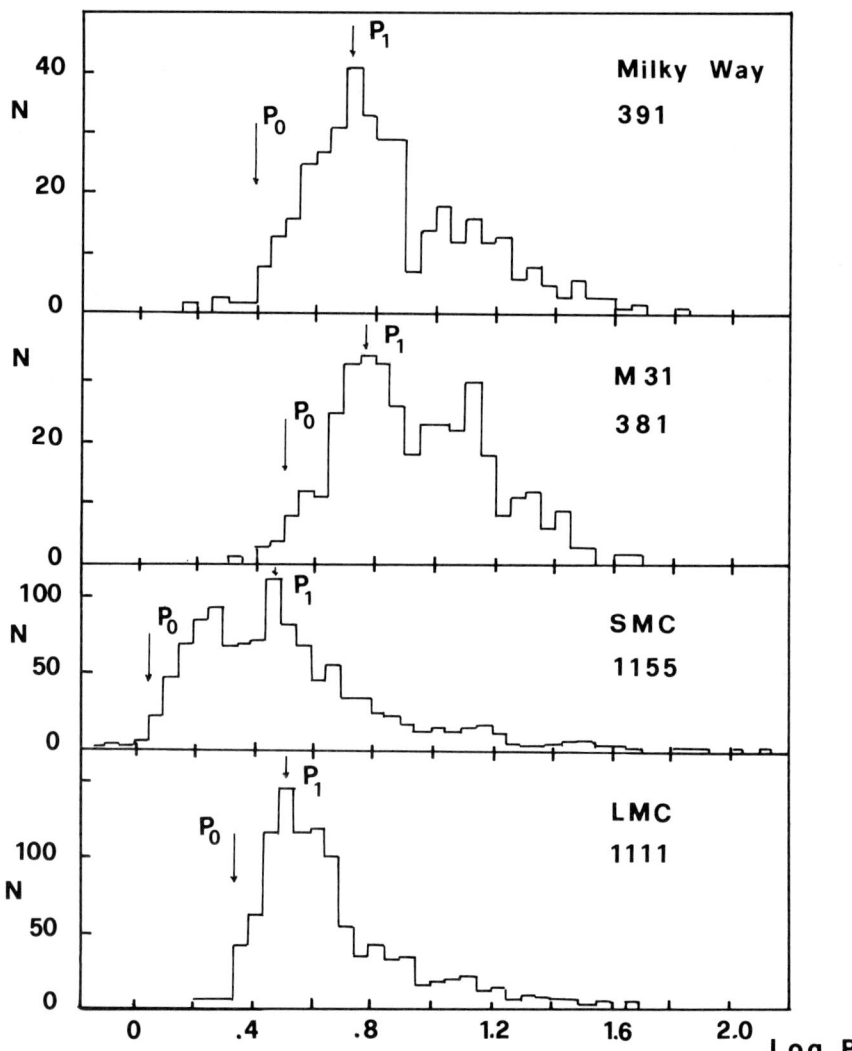

Figure 6. The frequency-period distributions, $N(\log P)$, defined for Cepheid stars in the Milky Way, M31, SMC and LMC are shown. In each panel the total number of stars in the sample is given. The low period cutoff P_0 and the period P_1 at the maximum of the distribution are indicated by the vertical arrows. This figure is adapted from Becker et al. (1977).

extends so far into the blue that the instability strip cannot be crosssed on nuclear time scales. L_{max} is determined by the intersection of the blue edge of the instability with the blue band. It goes without saying that a star (mass) would exist whose blue loop has the appropriate extension to spend all the blue He-burning phase within the instability strip. The luminosity of this particular models is indicated by L_1. Let $\Psi(t)$ be the total stellar birthrate (in some units dictated by the observational sample of Cepheids) and $\Phi(M)$ the initial mass function assumed to be independent of time and chemical composition. Then $\Psi(t')\Phi(M)dM\,dt'$ is the number of stars born between t' and $t'+dt'$ and mass between M and M+dM. Hence the number of Cepheids with mass between M and M+dM is given by

$$dN = \Psi(t-\tau_M)\Phi(M)\Delta t(M)\,dM \tag{10}$$

where τ_M is the lifetime of a star of mass M from ZAMS to the edge of the instability strip and $\Delta t(M)$ is the crossing time of the instability strip for this stellar mass. Since for all relevant masses τ_M is short compared to the typical time scales of star formation, we may write

$$dN/d\log P = \Psi(t)M\Phi(M)\Delta t(M)\,\ln 10\,d(\log M)/d(\log P). \tag{11}$$

The incorporation of the pulsation theory starts from the general relationship between P, L, T_{eff} and M for a Cepheid pulsating in the fundamental mode anywhere in the instability region given by Iben & Tuggle (1975), and the period-luminosity relationship of Becker et al. (1977) for models in the second transit, which to a first approximation is nearly independent of composition,

$$\log L = 2.51 + 1.25 \log P, \tag{12}$$

where L is in solar units and P in days.

Finally, with the aid of a mass-luminosity relationship we may determine $d(\log M)/d(\log P)$. The mass-luminosity relation for models in the blue band of the core He-burning depends on the underlying evolutionary scenario and it is very sensitive to chemical composition, opacity, nuclear reaction rates, mass loss, mixing mechanisms, and finally mass range. In the following we will use the following relations which based on models with convective overshoot, the new rate for the $^{12}C(\alpha,\gamma)^{16}O$, and no mass loss by stellar wind:

$$\log L = 0.924 + 3.61 \log M, \tag{13a}$$

for Z = 0.020 and $4 < M < 9\,M_\odot$ and

$$\log L = 1.511 + 3.22 \log M, \tag{13b}$$

for Z = 0.001, $2 < M < 9\,M_\odot$, where L and M are in solar units. Similar relations for classical models can be found in Becker et al. (1977) or constructed from the models of Becker (1981). The instability strip is

assumed to have a constant width in effective temperature
$W = \log(T_{eff})_{BE} - \log(T_{eff})_{RE}$, where BE and RE stand for blue and red
edge, respectively. The width W is however very uncertain, ranging from
the theoretical value of 0.04 (Becker et al. 1977) to the more recent
determination of 0.15 (Pel & Lub, 1978; Dupree, 1980).

The $\Delta t(M)$ relation must be derived from theoretical models. Like the
mass-luminosity relationship, it depends on the underlying evolutionary
scenario and details of the input physics of model calculations. In the
following we adopted the values derived from models with convective
overshoot. Finally, the Salpeter law with slope x= 2.35 is adopted in
the N(logP) distribution presented below.

Although the study of Becker et al. (1977) clearly shows the potential
capability of the theory to reproduce the observed N(logP) distribution
in seven galaxies of the Local Group, when a detailed comparison is made
the agreement is rather poor. In fact, they find that the short period
cutoff, P_0, and the maximum peak, P_1, in the distribution occur at
progressively larger values of logP when galaxies are ordered according
to their heavy element abundance. However, on one hand their P_0's and
P_1's are too low compared to the observational values, on the other hand
to explain the excess of long period Cepheids, they had to adopt a two
component birthrate. The secondary component (that of long period, hence
massive stars) would have an amplitude 5 to 50 times greater than the
primary component. Serrano (1983) re-examining this problem concludes
that the excess of long period Cepheids in the galactic N(logP)
distribution may be explained by a gradient of star formation combined
with a gradient in metallicity. The observational data for the solar
vicinity and the two Magellanic Clouds together with the predictions
from classical models are summarized in Table 2. The metallicity
assigned to each galaxy is a mean value of current estimates in
literature. Regardless of the excess of long period Cepheids, the
disagreement between observational P_0's and P_1's and theoretical
predictions is hard to reconcile on the basis of classical models.
Models with overshoot can alleviate the discrepancy since they predict
longer P_0's. This can be easily understood looking at the partial
derivatives of P with respect to L, T_{eff}, M in the P-L-T_{eff}-M
relationship of Iben & Tuggle (1975), where the period increases at
increasing luminosity, whereas it decreases at decreasing effective
temperature and mass. For any given mass, models with overshoot run at
higher luminosity and lower effective temperature. However, even these
latter models predict high P_0's. To date, two ways out have been
suggested to increase $\log P_1/P_0$, namely: i) a much wider instability
strip; ii) an increase in the slope of the blue band, or in other words
a blue band not extending too much to high effective temperatures. The
first possibility has been already considered by Bertelli et al. (1985),
but it turned out to be insufficient unless W gets very large. A blue
band of core He-burning models more vertical in the HR diagram and
closer to the Hayashi limit can be obtained under favourable
circumstances (higher opacity and metallicity, more efficient mass loss,
etc.). Nevertheless, it can be easily understood that both W and the
slope of the blue band affect P_1 more than P_0. Since models with

overshoot and metallicity appropriate for the solar vicinity almost matched P_1, what is really needed is to decrease P_0, hence include stars of lower mass and at the same time to obey the constraint of P_0 and P_1 being increasing toward the galactic centre, where the metallicity is higher than in the solar neighborhood. This is easily achieved by allowing for a dispersion in chemical composition as already suggested by Becker et al. (1977). We find that adopting a gaussian dispersion in metallicity with $\sigma_Z = 0.0015$ both the periods P_0 and P_1 for the solar vicinity can be matched and the N(logP) distribution observed in the Milky Way, LMC and SMC can be accounted for. The results for three galaxies are given in Table 3. In Figure 7 we show the N(logP) distribution expected for the LMC together with its observational counterpart.

Finally, we would like to point out that both the new models and the (small) dispersion in metallicity concur to offer an explanation of the N(logP) distribution without invoking a two component birthrate. In fact, in models with overshoot the blue band of core He-burning tends to move back toward the Hayashi line at relatively higher masses ($M > 9\,M_\odot$), the tendency being enhanced by increasing the metallicity and/or the opacity (see below). This would allow for a wider range of masses spending a large fraction, if not all, of the blue band lifetime within the instability strip. Another interesting consequence of the

Table 2 (*)

Y	Z	λ	log P_1	log P_0	M_{min}	Notes
	0.020		0.90	0.30		Solar Vicinity
	<0.020		0.75	0.30		Milky Way
	0.009		0.52	0.20		LMC
	0.002		0.48	0.10		SMC
		Classical Models				
0.280	0.001		-0.28	<-0.33		(a)
0.280	0.010		-0.38	-0.61		
0.280	0.030		-0.73	-0.81		
0.280	0.020		-0.08	-0.26		
0.360	0.020		0.68	0.60		
		Models with overshoot				
0.280	0.020	0	0.45	0.37		(b)
0.280	0.020	0.5	0.85	0.77		
0.280	0.020	1.0	1.02	0.94	5.46	
0.299	0.001	1.0	0.06	-0.02	2.05	

(*) Width of the instability strip log T_{eff} = 0.04
(a) Becker et al. (1977)
(b) Bertelli et al. (1985)

fact that the blue core He-burning band of models with overshoot, after a maximum diplacement toward the blue, moves back to the red giant region is that it may naturally lead either to a broad N(logP) distribution or even to a bimodal distribution of periods, which is not due to inhomogeneities in the initial chemical composition but it is intrinsic to stellar models. As a matter of fact, the distributions observed in SMC, M31 and perhaps the Milky Way seem to suggest a bimodal N(logP) distribution. However, whether or not those observational distributions are really bimodal cannot be easily assessed at the present time, and therefore this point is left to future investigation.

5. EVOLUTION WITH VARIED RADIATIVE OPACITY

Since the early studies of Henyey et al. (1965), Fricke et al. (1971), Robertson (1972), Johnson & Whittaker (1975), Carson & Stothers (1976), to name a few, it has been known that changing the opacity used either in stellar interiors or in stellar atmospheres can

Table 3

Site	Z	Z	logP	logP
Milky Way	0.020	0.006	0.90	0.30
LMC	0.009	0.007	0.52	0.37
SMC	0.002	0.001	0.48	0.28

Figure 7. The theoretical N(logP) distribution from models with overshoot and dispersion in metallicity (thin dashed line). This is compared with the N(log P) defined for the LMC (thick line).

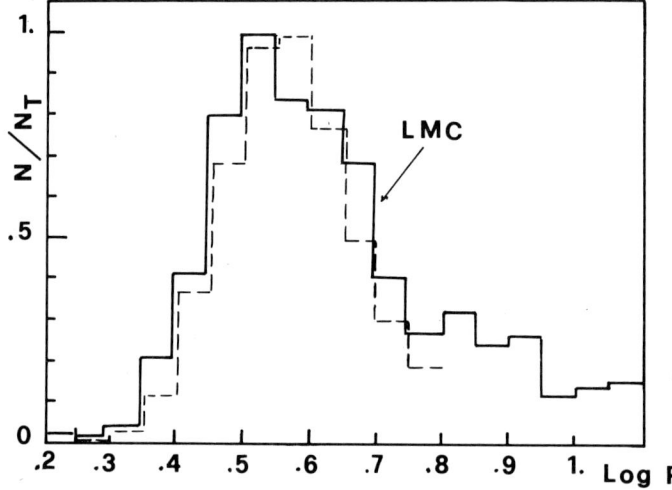

have a great impact on a star's evolutionary history and path in the HR diagram. Almost any property of the star's structure depends on opacity including luminosities, lifetimes, occurrence of semiconvection, extension of the blue loops, etc. With regard to Cepheids, the most important question addressed in the past was whether the Carson (1976) or the Los Alamos opacities (Cox & Stewart 1965, 1970) are to be used (see Carson & Stothers, 1976; Cox & Tabor, 1976, respectively). The primary difference in behaviour between the two opacities is that Carson's opacity has a larger contribution from He and (mostly) CNO elements in the temperature range $6.5 < \log T < 4.3$ and moderate densities (10^{-4} to 10^{-5} gr·cm^{-3}). Since to date the controversy has not been settled from the standpoint of the fundamental physics going into the calculation of radiative opacities, the question has been addressed from an empirical point of view, and at least two arguments have been advocated to suggest an increase in the heavy element opacity, in particular that due to CNO elements. These are the relative numbers of supergiant stars per spectral type (Bertelli et al. 1984) and the mass anomaly and period ratios of bump Cepheids (Simon, 1982). They are briefly summarized below.

Supergiant stars. The effects of the Carson opacity on models of massive stars have been investigated by Stothers (1976), and Stothers & Chin (1977, 1978). The most interesting results of those calculations were that the locus of the zero age main sequence progressively shifted toward cooler effective temperature, and the main sequence band systematically broadened by increasing the metallicity, decreasing the mixing length parameter used to describe the outermost convective zone, and increasing the mass of the star. For a suitable combination of the three parameters the main sequence band may even extend across the whole HR diagram. Though interesting, Carson's opacities and, in turn, the models for massive stars based on their usage never received a general consensus. Among others, the main objection resided in the predicted location of the zero age main sequence which does not agree with the location of luminous early type stars obtained from the so-called hot temperature scale, transferring spectral types into effective temperature and bolometric corrections, currently in use. The question was subsequently reopened by Bertelli et al. (1984), who starting from the suggestion by Simon (1982), to be discussed below, and the criticism of Iben & Renzini (1984) on current radiative opacities in general, proposed a modification to standard radiative opacities, mimicking the bump in the Carson opacity, to reconcile the observational frequency of luminous OB type and supergiant stars with theoretical expectation. Star counts in fact indicate that a large fraction (up to 40%) of stars fall outside the region of core H-burning (even allowing for mass loss by stellar winds and convective overshoot), whereas the theoretical expectation is 10 to 20%. It appears that either star counts are still severely biased by incompleteness and/or selection effects or the theoretical main sequence band ought to extend at least up to spectral type A0 (see Chiosi & Maeder 1986 for a thorough discussion of this point). The following modification to the opacity was proposed by Bertelli et al. (1984):

$$\kappa = \kappa_{CS} \{ 1 + X \exp [-10 (5.8 - \text{Log} T)^4] \} \tag{14}$$

where κ_{CS} stands for the Cox and Stewart opacity, and X is an adjustable parameter. As Carson's opacities show the bump in the range of temperature $5.4 < \text{Log } T < 6.2$, and are higher at lower densities, the bell-shaped bump given by relation (14) was centered at $\text{Log } T = 5.8$, whereas no dependence on density was considered. Bertelli et al. (1984) conclude that models characterized by mass loss at a moderate rate, overshoot from the convective core, and opacity of the CNO ionization layers slightly enhanced with respect to the standard one ($X = 2$) can easily account for a much wider main sequence band and for the star counts.

Cepheid stars. The mass anomalies and period ratios encountered by double-mode (admixture of both fundamental and first overtone pulsations) and "bump" Cepheids (the bump often appears in the light curves of Cepheids with periods in the range 7-11 days), which cannot be accounted for by classical models, led Simon (1982) to suggest that an increase by a factor of 2 to 3 in the heavy element opacity could help to remove the mass anomaly.

However, Simon's (1982) and Bertelli's et al. (1984) suggestions have been questioned by Magee et al. (1984), who claim that such a large increase in opacity is incompatible with atomic physics. An opposite conclusion has been reached by Iglesias et al. (1987), who argue that an improved treatment of the atomic physics can significantly increase the Rosseland mean opacities of metals in astrophysical mixtures. This possibility has been recently reinforced by Andreasen (1988), who claims that enhanced metal opacity is needed to reduce the Cepheid period ratio discrepancies. The above contrasting conclusions simply indicate that the topic is still a matter of vivid debate, and therefore it is not implausible that the true radiative opacities will eventually turn out to be larger than usually accepted.

In order to clarify the role played by variations in the contribution of heavy elements to the radiative opacity, we may use the following relation proposed by Renzini (1984b) and adopted by Bertelli & Bressan (1988) in the calculations of stellar models we are going to present,

$$\kappa = \kappa_{XY} + A \Delta Z \sigma [1 + X f(\rho, T) \exp \{-4 \alpha \log (T/T_0)\}], \tag{15}$$

where κ_{XY} is the opacity of a metal-free mixture of hydrogen and helium. The term $A \Delta Z$ gives the contribution of heavy elements to the classical radiative opacity. This term is evaluated, at each value of ρ and T, subtracting the opacity of a metal-free mixture with given X and Y from the opacity of a mixture having the same X and Y but a metal abundance Z. σ is a parameter by which the contribution of heavy elements can be varied all over the domain of temperatures and densities. Furthermore, $f(\rho, T)$ is a suitable function defining in the ρ-T plane a band along which the opacity enhancement, given by the exponential term, is allowed to occur, mimicking the opacity bump proposed by Bertelli et al. (1984). T_0 is the central value of the temperature interval relative to the ultimate ionization of the CNO group of elements ($T_o = 10^6$ K). The requirement of matching the opacity increase proposed by Simon (1982) and Bertelli et al. (1984) allows us to determine α, $f(\rho,T)$ and X.

Effects of a bump in the opacity. Assuming $\sigma = 1$ and $X = 2$ to 3, relation (15) simulates a bump in the opacity centered on the temperature T_o and suitably modulated in density by means of $f(\rho,T)$. The effects of a bump on the properties of massive stars have been already described above. They do not depend on whether or not convective overshoot is allowed to occur. The behaviour of intermediate mass stars is more complicated. The core H-burning phase of classical models is not affected to any appreciable extent by a bump in the opacity, whereas the core He-burning phase is more sensitive to it. In fact the calculations show that either very extended loops may occur (Robertson 1972) or that very minor variations in the luminosity and effective temperature can be seen (Becker 1985). On the contrary, core He-burning models with overshoot respond very much to this type of enhancement in the opacity (even of modest size). Almost independently of the efficiency of convective overshoot, hence increase in the mass of the convective core, the loops are strongly inhibited for all masses greater than about 5 $M_\odot$ when the metallicity is about solar (Z = 0.02). At decreasing metallicity, the loops develop again though much narrower at increasing mass. The luminosity of these models is about the same as in models with overshoot and standard opacity. As a consequence of this, the C-shaped blue band of core He-burning models, that was already seen to occur in models with overshoot and standard opacity, is now found to be extremely sensitive to the opacity enhancement and, for given values of X and T_o, to the metallicity. The apex of this band, which was very blue and luminous in classical models without convective overshoot (see Figure 1), and which became less luminous and much redder in models with overshoot and standard opacity (see Figure 2), may now occur at even fainter luminosities and cooler effective temperatures, depending on the opacity enhancement and metallicity of the star. If these numerical experiments are confirmed by detailed opacity calculations, the impact on the properties of Cepheid stars and, in particular, on the N(logP) distribution is of paramount importance. It goes without saying that for appropriate metallicities (within the range of plausible values) the instability strip is either easily crossed twice on nuclear time scales or intersected by the blue band over a range of luminosities (masses) much wider than ever before. Although such a possibility is already offered by models with overshoot but standard opacities, provided that the right combination of convective overshoot and metallicity is adopted (in general, substantial overshoot and/or very low metallicities were needed), now due to the varied opacity it may occur for "normal" overshoot and metallicities. Therefore the changes advocated by Simon (1982) are potentially permissible. We like to speculate that the bimodality in the N(logP) distribution perhaps seen in the Milky Way, SMC and M31 could be the result of this variation in the opacity.

Effects of an overall increase in the heavy element opacity. This can be simulated by assuming in relation (15) that $\chi = 0$ and $\sigma \neq 0$ (say 2 or 3). This type of variation would have very severe consequences. In brief, the luminosity and effective temperature of the models both in core H- and He-burning phases are expected to decrease, the lifetimes to increase, and loops to be almost always suppressed unless unacceptably low metallicities are assumed. Models of this type have been calculated by Nasi & Forieri (1988) in the range of massive stars (incorporating mass loss and convective overshoot) and used to simulate the observational distribution of luminous stars in the colour magnitude diagram. In this mass range, the effects of this type of opacity are marginal and indeed the models have been found to match reasonably well the observational data. On the contrary, no systematic analysis of the effects given by this type of opacity variation on the properties of intermediate and low mass stars has been made.

6. SURFACE COMPOSITION OF CEPHEIDS

With few rare exceptions, the vast majority of Cepheid stars have undergone the first dredge-up while ascending the RGB, or in other words external convection penetrates the inner layers that in the past experienced nuclear processing with consequent changes of the surface abundances. Classical calculations of stellar models (see Becker & Iben, 1979) predict that, as a consequence of the dredge-up, the surface abundance of ^{14}N should increase by a factor of two to three and the abundance of ^{16}C should decrease by 30% to 40% (see also Iben & Renzini, 1984). The surface abundances of Cepheid stars are therefore particularly interesting as they can be used as a probe of the interior structure and efficiency of convective mixing. In relation to this we recall that: i) Cepheids have main sequence progenitors typically rotating fast enough that enhancement in mixing and abundance variations at the surface are likely to be expected; ii) classical models, for which the above abundance variations have been calculated, can now be replaced by models with convective overshoot, which by virtue of their larger internal mixed and nuclear processed regions can more easily show abundance variations at the surface. Luck & Lambert (1981) were the first to measure the surface CNO abundances of Cepheids with the conclusion that the distribution of CNO elements indicates that mixing of matter in which considerable amounts of O (as well C) have been processed into N (so-called ON processing) far in excess of the classical prediction. Becker & Cox (1982) discussed the implications of how stellar models would be altered to account for these observations. However, Luck and Lambert's result was questioned by Iben & Renzini (1984), who rediscussed the data and came to the conclusion that there is no evidence that ON-cycled material has been exposed to the surface, while there is some evidence that CN materials are present, however in agreement with the canonical predictions for the first dredge-up. Models incorporating convective overshoot have not yet been analysed from the standpoint of surface abundances.

Finally, we would like to conclude this section, touching upon another crucial problem related to the chemical composition of Cepheid stars, namely the dependence of the period-luminosity (PL) and period-luminosity-colour (PLC) relations on the helium and metal abundances. These relations in fact play an important role in establishing the extragalactic distance scale. This problem has been addressed by Stothers (1988), who has derived PL and PLC relations using relevant data from stellar evolution, atmosphere, and pulsation theory incorporating the composition dependences for all quantities. As expected, the PLC relation in B and V magnitudes is much more sensitive to chemical composition than is the PL relation. The new calibrations are used to derive the distance moduli of the Magellanic Clouds. These corrected for the abundance effects are $(m-M)_o = 18.51 + 0.06$ for LMC and $(m-M)_o = 18.8 + 0.06$ for SMC.

ACKNOWLEDGEMENTS
This work has been financed by the National Group of Astronomy (GNA) and the Italian Space Research Program (PSN) of the National Council of Research of Italy (CNR).

REFERENCES
Andreasen, G.K. (1988). Astron. & Astrophys.,201, 72.
Barbaro, G. & Pigatto, L. (1984). Astron. & Astrophys.,136, 355.
Becker, S.A. (1981). Ap.J.Suppl.,45, 478.
Becker, S.A. (1985). In Cepheids: Theory and Observations, ed. B.F.
    Madore, p. 104 Cambridge: Cambridge University Press.
Becker, S.A. & Cox, A.N. (1982). Ap.J.,260, 707.
Becker, S.A. & Iben, I. (1979). Ap.J.,232, 831.
Becker, S.A., Iben, I. & Tuggle, R.S. (1977). Ap.J.,218, 633.
Becker, S.A. & Mathews, G.J. (1983). Ap.J.,270, 155.
Bergh, S. van den, (1981). Astron. & Astrophys.Suppl.,46, 79.
Bertelli, G., Bressan, A. & Chiosi, C. (1985). Astron. & Astrophys.,130,279.
Bertelli, G., Bressan, A. & Chiosi, C. (1985). Astron. & Astrophys.,150, 33.
Bertelli, G., Bressan, A., Chiosi, C. & Angerer, K. (1986a). Astron. & Astrophys.Suppl.,66, 191.
Bertelli, G., Bressan, A., Chiosi, C. & Angerer, K. (1986b). In The Age of Star Clusters, ed. F. Caputo, Mem.Soc.Astron.It.,57, 427.
Blanco, B.M., Blanco, V.M. & McCarthy, M.F. (1978). Nature,271, 638.
Blanco, B.M., McCarthy, M.F. & Blanco, V.M. (1980). Ap.J.,242, 948.
Böhm-Vitense, E. (1986). Ap.J.,303, 262.
Boothroyd, A.I. & Sackmann, I.J. (1988a). Ap.J.,328, 632.
Boothroyd, A.I. & Sackmann, I.J. (1988b). Ap.J.,328, 641.
Boothroyd, A.I. & Sackmann, I.J. (1988c). Ap.J.,328, 653.
Boothroyd, A.I. & Sackmann, I.J. (1988d). Ap.J.,328, 671.
Bressan, A. & Bertelli, G. (1988). in preparation.
Bressan, A., Bertelli, G. & Chiosi, C. (1981). Astron. & Astrophys.,102, 25.

Bressan, A., Bertelli, G. & Chiosi, C. (1986). In The Age of Star
    Clusters, ed. F. Caputo, Mem.Soc.Astron.It.,57, 411.
Brunish, W.M. & Willson, L.A. (1987). In Stellar Pulsation, eds. A.N.
    Cox, W.M. Sparks & S. G. Starrfield, p. 27. Berlin:
    Springer-Verlag.
Buonanno, R., Corsi, C.E. & Fusi-Pecci, F. (1985). Astron. &
    Astrophys.,145, 97.
Carson, T.R. (1976). Ann.Rev.Astron.Astrophys.,14, 95.
Carson, T.R. & Stothers, R. (1976). Ap.J.,204, 461.
Castellani, V., Chieffi, A.. Pulone, L. & Tornambé, A. (1985).
    Ap.J.,296, 204.
Chiosi, C. (1986). In Nucleosynthesis and Stellar Evolution, 16th
    Saas-Fee Course, eds. B. Hauck et al., p. 199. Geneva:
    Geneva Observatory.
Chiosi, C., Bertelli, G. & Bressan, A. (1987). In Late Stages of Stellar
    Evolution, eds. S. Kwok & S.R. Pottasch, p. 239. Dordrecht:
    Reidel.
Chiosi, C., Bertelli, G., Meylan, G. & Ortolani, S. (1988). Astron. &
    Astrophys., submitted.
Chiosi, C. & Maeder, A. (1986). Ann.Rev.Astron.Astrophys.,24, 329.
Cox, A.N. (1980). Ann.Rev.Astron.Astrophys.,18, 15.
Cox, A.N. (1985). In Cepheids: Theory and Observations, ed. B.F. Madore,
    p. 126. Cambridge: Cambridge University Press.
Cox, A.N., Deupree, R.G., King, D.S. & Hodson, S.W. (1977). Ap.J.,214,
    L127.
Cox, A.N. & Stewart, J.N. (1965). Ap.J.Suppl.,11, 22.
Cox, A.N. & Stewart, J.N. (1970). Ap.J.Suppl.,19, 243.
Cox, A.N. & Tabor, J.E. (1976). Ap.J.Suppl.,31, 271.
Cox, J.P. (1980). Theory of Stellar Pulsation. Princeton, New Jersey:
    Princeton University Press.
Deupree, R.G. (1977). Ap.J.,211, 509.
Deupree, R.G. (1980). Ap.J.,236, 225.
Endal, A.S. & Sofia, S. (1976). Ap.J.,210, 184.
Endal, A.S. & Sofia, S. (1978). Ap.J.,220, 290.
Endal, A.S. & Sofia, S. (1979). Ap.J.,232, 531.
Fowler, W.A. (1984). Rev.Mod.Phys.,56, 149.
Fowler, W.A., Caughlan, G.R. & Zimmerman, B.A. (1967).
    Ann.Rev.Astron.Astrophys.,5, 525.
Fowler, W.A., Caughlan, G.R. & Zimmerman, B.A. (1975).
    Ann.Rev.Astron.Astrophys.,13, 69.
Fricke, K., Stobie, R.S. & Strittmatter, P.A. (1971). M.N.R.A.S.,154,
    23.
Hanson, R.B. (1979). In The HR Diagram, eds. A.G.D. Philip & D.S. Hayes,
    p. 154. Dordrecht: Reidel.
Harris, M.J., Fowler, W.A., Caughlan, G.R. & Zimmerman, B.A. (1983).
    Ann.Rev.Astron.Astrophys.,21, 165.
Heckman, T.M. (1976). Ap.J.Suppl.,36, 451.
Henyey, L.G., Vardja, M.S. & Bodenheimer, P. (1965). Ap.J.,142, 841.
Hollowell, D.E. (1987). In Late Stages of Stellar Evolution, eds. S.
    Kwok & S. R. Pottasch, p. 239. Dordrecht: Reidel.

Hollowell, D.E. (1988). Ph. D. Thesis, University of Illinois.
Iben, I. (1972). Ap.J.,178, 433.
Iben, I. (1974). Ann.Rev.Astron.Astrophys.,12, 215.
Iben, I. (1975a). Ap.J.,196, 525.
Iben, I. (1975b). Ap.J.,196, 549.
Iben, I. (1976). Ap.J.,208, 165.
Iben, I. (1981). Ap.J.,246, 278.
Iben, I. & Renzini, A. (1983). Ann.Rev.Astron.Astrophys.,21, 271.
Iben, I. & Renzini, A. (1984). Physics Reports 105, no. 6, 329.
Iben, I. & Tuggle, R.S. (1972a). Ap.J.,173, 135.
Iben, I. & Tuggle, R.S. (1972b). Ap.J.,178, 441.
Iben, I. & Tuggle, R.S. (1975). Ap.J.,197, 39.
Iglesias, C.A., Rogers, F.J. & Wilson, B.G. (1984). Ap.J.,322, L45.
de Jager, C., Nieuwenhuijzen, H. & van der Hucht, K.A. (1988). Astron. & Astrophys. Suppl.,72, 259.
Johnson, H.R. & Whittaker, R.W. (1975). M.N.R.A.S.,173, 523.
Kettner, K.U., Becker, H.W., Buchmann, L., Görres, J., Kräwinkel, H., Rolfs, C., Schmalbrock, P., Trautvelter, H.P. & Vlieks, A. (1982). Z.Phys.A-Atoms and Nuclei,308, 73.
Kippenhahn R., Meyer-Hofmeister, E. & Weigert, A. (1970). Astron. & Astrophys.,5, 155.
Kraft, R.P. (1970). In Spectroscopic Astrophysics, ed. G.H. Herbig, p. 385. Berkeley: University of California Press.
Lamers, H.J.G.L.M. (1981). Ap.J.,245, 593.
Lattanzio, J.C. (1986). Ap.J.,311, 708.
Lattanzio, J.C. (1987a). In Late Stages of Stellar Evolution, eds. S. Kwok & S.R. Pottasch, p. 235. Dordrecht: Reidel.
Lattanzio, J. C. (1987b). Ap.J.,313, L15.
Lattanzio, J. C. (1988a). In Evolution of Peculiar Red Giant Stars, in press.
Lattanzio, J. C. (1988b). In Origin and Distribution of the Elements, ed. G.J. Mathews, p. 398. Singapore: World Scientific.
Lauterborn, D., Refsdal, S. & Weigert, A. (1971). Astron. & Astrophys.,10,97.
Lauterborn, D. & Siquig, R.A. (1974). Ap.J.,191, 589.
Luck, R.E. & Lambert, D.L. (1981). Ap.J.,245, 1018.
Maeder, A. & Mermilliod, J.C. (1981). Astron. & Astrophys.,93, 136.
Magee, N.H., Mertz, A.T. & Huebner, W.T. (1984). Ap.J.,283, 264.
Mateo, M. & Hodge, P. (1986). Ap.J.Suppl.,60, 893.
Matraka, B., Wassermann, C. & Weigert, A. (1982). Astron. & Astrophys.,107, 283.
Meyer-Hofmeister, E. (1972). Astron. & Astrophys.,16, 282.
Meylan, G. (1987). Astron. & Astrophys.,184, 144.
Meylan, G, (1988). Astron. & Astrophys.,191, 215.
Meylan, G. & Djorgovski, S. (1987). Ap.J.,322, L94.
Nasi, E. & Forieri, C. (1988). Astron. & Astrophys., submitted.
Pel, J. W. (1985). In Cepheids: Theory and Observations, ed. B.F. Madore, p. 1. Cambridge: Cambridge University Press.
Pel, J.W. & Lub, J. (1978). In The HR Diagram, eds. A.G.D. Philip & D.S. Hayes, p. 229. Dordrecht: Reidel.

Persson, S.E., Aaronson, M., Cohen, J.G., Frogel, J.A. & Matthews, K. (1983). Ap.J.,266, 105.
Reimers, D. (1975). Mem.Soc.Roy.Sci. Liege, 6 serie, tome 8,p. 369.
Renzini, A. (1984a). In Observational Tests of the Stellar Evolution Theory, eds. A. Maeder & A. Renzini, p. 21. Dordrecht: Reidel.
Renzini, A. (1984b). private communication.
Renzini, A. & Voli, M. (1981). Astron. & Astrophys.,94, 175.
Robertson, J.W. (1972). Ap.J.,177, 473.
Sandage, A. & Tammann, G.A. (1969). Ap.J.,157, 683.
Searle, L., Sargent, W.L.W. & Bagnuolo, W.G. (1980). Ap.J.,179, 427.
Serrano, A. (1983). Rev. Mexicana Astron. Astrophys.,8, 131.
Schmidt, E.G. (1984). Ap.J.,285, 501.
Simon, N.R. (1982). Ap.J.,260, L87.
Simon, N.R. (1986). preprint.
Simon, N.R. (1987). In Pulsation and Mass Loss in Stars, eds. R. Stalio & L.A. Willson, p. 27. Dordrecht: Reidel.
Slettebak, A. (1970). In Stellar Rotation, ed. A. Slettebak p. 3. Dordrecht: Reidel.
Stellingwerf, R.F. (1978). A.J.,83, 1184.
Stellingwerf, R.F. (1979). Ap.J.,227, 935.
Stift, M.J. (1984). Astron. & Astrophys.,140, 445.
Storm, J., Andersen, J., Blecha, A. & Walker, M.F. (1988). Astron. & Astrophys.,190, L18.
Stothers, R. (1976a). Ap.J.,209, 800.
Stothers, R. (1976b). Ap.J.,225, 939.
Stothers, R. (1981). Ap.J.,329, 712.
Stothers, R. & Chin, C.W. (1977). Ap.J.,211, 189.
Stothers, R. & Chin, C.W. (1978). Ap.J.,225, 939.
Vandenberg, D.A. & Bridges, T.J. (1984). Ap.J.,278, 679.
Waldron, W.L. (1984). In The Origin of Non-Radiative Heating/Momentum in Hot Stars, eds. A.B. Underhill & A.G. Michalitsianos, p. 2358. Washington, D.C.: NASA.
Walker, A.R. (1987). M.N.R.A.S.,225, 627.
Welch, D.L., McLaren, R.A., Madore, B.F. & McAlary, C.W. (1987). Ap.J.,321, 162.
Willson, L.A. (1987). In Pulsation and Mass Loss in Stars, eds. R. Stalio & L.A. Willson, p. 285. Dordrecht: Reidel.
Willson, L.A. & Bowen, G.H. (1984). Nature,312, 429.

PROGRESS TOWARD AN IMPROVED EQUATION OF STATE AND OPACITY
FOR STELLAR ENVELOPES

Dimitri Mihalas
Department of Astronomy, University of Illinois

Abstract. A brief description is given of a large
international project for computing new stellar envelope
opacities and equation of state data.

INTRODUCTION

For the past four to five years a large project to recompute opacities for stellar envelopes has been underway. Approximately 20 physicists and astrophysicists are contributing to this effort, the largest group being located in the UK (under the leadership of M.J. Seaton and P.G. Burke), a much smaller group in the US, and a few individuals in other countries. The non-US team members are responsible primarily for producing the requisite atomic data, while the US group has been working on the equation of state (EOS), and on combining cross-sections with the EOS to produce opacities.

For over 25 years the astrophysical community has relied almost exclusively on opacity calculations done at Los Alamos, and it is natural to ask "Why should this job be done over?". There are a number of partial answers to this question. First, we all have simply accepted the Los Alamos results without (because the codes and databases are not in the public domain) being able to evaluate them critically. Yet we know that many phases of stellar evolution depend sensitively on the opacity, and so long as we have only one set of data available to us, we cannot objectively assess the implications of alternative methods of calculating the opacity for computations of stellar structure, evolution, and pulsation. Second, there have long been some annoying discrepancies in Cepheid models and observations, and the existing evidence points towards the need for a significant increase in the opacity in regions of the (ρ, T) plane near where He, C, N, and O are ionizing. Third, it has been only very recently that numerical methods for calculating atomic structure and the optical properties of atoms have been able to produce these data quickly. As a result, only now is it possible to produce an essentially complete set of atomic data of very high quality. Fourth, the computers are now large enough and fast enough to permit us to resolve the spectrum, including lines, over the entire Rosseland-mean window. Finally, if we succeed with this project, astronomers will have, for the first time, a facility (hardware + software + documentation + data) that will allow them to carry out the calculations themselves for any mix they desire.

THE ATOMIC DATA

Wave functions and energy levels are being computed by the R-matrix method for all ions of all astrophysically abundant elements up through Fe. At present all elements up through Ne are complete, and work on third period elements (up through Ar) is underway. Initial studies on Fe have begun. These calculations give a complete set of oscillator strengths for all lines through upper principal quantum number = 10; beyond that we use hydrogenic values. A characteristic of this work is that continuum cross-sections are typically riddled with sharp resonances. As a result, the distributions of differential oscillator strength is not even roughly hydrogenic; this one fact could lead to large changes from presently accepted values of opacities.

When these computations are completed, we shall have an atomic dataset unprecedented in both quality and extent. However we must also point out a basic limitation of our approach: all of the atomic data apply, strictly speaking, only to isolated atoms, i.e. atoms (and ions) whose internal structure is not significantly altered by the presence of the other particles in the plasma. As a result, we cannot realistically treat high-density material, such as exists in stellar cores. In practice we have chosen an upper density bound of 0.01 grams per cubic centimeter; this limit is quite adequate for treating, e.g. the pulsational properties of Cepheid and RR Lyrae envelopes.

EQUATION OF STATE

Our equation of state is based on a free-energy minimization method. The free energy includes contributions from (1) the translational motions of classical nuclei; (2) the internal excitation states of molecules, atoms, and ions; (3) the translational motions of semi-degenerate electrons; and (4) Coulomb interactions among all charged particles.

Our implementation of this method differs from others like it in the literature in two important respects. First, we allow each atom or ion to have many hundreds of bound states (the uppermost of which are hydrogenic). We have used all available empirical energy levels, and supplemented them where necessary with values obtained by extrapolating quantum defects up series. Then in order to truncate the internal partition function we introduce occupation probabilities $w(i)$ which depends on the binding energy $E(i)$ of state i and a weighted sum of ionic charges times the numbers of corresponding ions. To calculate $w(i)$, D. G. Hummer made a thorough quantum mechanical investigation of the rate of Stark ionization of a bound electron by an imposed fluctuating field. The form of $w(i)$ is such that w approaches unity for strongly bound (i.e. low-lying) states, and w approaches zero for weakly bound states lying close to the ionization continuum. The effective quantum number at which w makes the transition from near unity to near zero is a monotonically decreasing function of the ion charge density.

Second, because w(i) is chosen to be continuous (indeed differentiable to all orders) atomic states do not vanish abruptly as they do in other formulations, hence discontinuities in the free energy (which would produce delta functions in the pressure and/or internal energy) are avoided. In fact, all four contributions to the free energy are analytical and differentiable. As a result it is possible to compute all secondary thermodynamic properties (specific heats, adiabatic gammas, etc.) analytically, without numerical differentiation. The evaluation of these formulae, though laborious, yields very smooth values, and provides a powerful numerical check on the whole calculation.

At present we have calculated EOS tables for 6 mixtures of 15 elements from H through Fe; in all 205 distinct species of particles (including hydrogen molecules, hydrogen molecule ions, negative hydrogen ions, and free electrons) are considered. We have chosen 6 representative mixtures ranging from super-metal-rich ([Fe/H] = +0.5), through solar, to extremely metal poor ([Fe/H] = -2). We have been able to check some of our results against Los Alamos computations. In general the agreement is good, though, as noted by Pesnell, the hydrogen ionization loci in the temperature-density plane computed from the two theories do not coincide. This discrepancy is a direct result of the two different methods of handling the internal partition function, and suggests that there could be significant differences in our opacity values, compared to Los Alamos, near ionization loci.

We are now prepared to release EOS data to "friendly users". They can be obtained by sending a tape to D. Mihalas for a UNIX "tar" format or to D. G. Hummer for a blocked ASCII format (e.g. for VMS VAX). The UNIX tapes can be written on either standard 1/2" tape or a 1/4" tape cartridge (preferred); ASCII tapes must be 1/2". We hope that users will keep us informed of problems and errors they encounter.

OPACITIES

A first draft of the opacity code is virtually completed. Still needed is a quick way to compute Stark profiles, and code to compute multigroup means and an archival set of opacity distribution functions. These tasks are being worked on now.

In our opacities we attempt to include effects of the occupation probabilities used in the EOS on the optical properties of the gas. These effects can be quite important, when, for example, a sequence of lines converging on an ionization limit is blurred into a pseudocontinuum as the upper states are destroyed, and what normally would be a bound-bound transition between states "l" and "u" becomes a bound-free transition because the upper state no longer exists as a bound state. A similar effect by which bound-free transitions are converted into free-free also occurs. We expect the occupation probabilities will affect our opacities in certain regions of the temperature-density plane, a point that we shall study at a later date.

At present we have carried out only a few experimental calculations for H, He, C, and some mixes of these elements. We hope to add N and O by the end of the year. If all goes well we may be able to produce a table for, say, the solar mix by the end of 1989, though I suspect 1990 may be more realistic.

ACKNOWLEDGEMENTS

This research has been supported in part by grant number AST 85-19209 from the National Science Foundation to the University of Illinois. I am pleased to be here representing my close coworkers Yu Yan, David Hummer, and Anil Pradhan, and all of our overseas colleagues (too numerous to list here). The cooperative spirit which has pervaded this project from its outset has made it much easier carry on in the face of the daunting realities of the work to be done.

STELLAR MASS LOSS AND PULSATION

L.A. Willson
Astronomy Program, Physics Department
Iowa State University, Ames IA 50011, USA

Abstract. Mass loss at rates sufficient to alter the evolution of stars is known to occur during the pre-main sequence evolution of most stars, on the main sequence for massive stars, and during advanced evolutionary phases when the luminosity is high and the effective temperature is low. While most investigations of the effects of mass loss on stellar evolution have assumed continuous (parametrized) mass loss laws apply, there is increasing evidence that mass loss rates are substantially higher for stars that are pulsating with large amplitude and/or in selected modes. Some new insights into the mass loss that terminates the AGB evolution of intermediate mass stars, and leads to the formation of planetary nebulae, come from recent detailed studies of the mass loss process from the Mira variables.

INTRODUCTORY COMMENTS

Any systematic study of mass loss from stars must include three parts: detailed calculations of the mechanisms that drive the mass loss; observations of the mass flux, temperature structure and velocity structure of the winds; and the inclusion of the mass loss in stellar evolution calculations. Current studies mostly fall short of these ideal goals; only for the case of the late asymptotic giant branch (AGB) Mira variables are at least some results available in all three categories.

Mass loss: general considerations

To lift material from the surface of a star and drive it away as a wind requires the input of both energy ($L_{\dot{M}} \approx 0.5 \dot{M} (v_e^2 + v_\infty^2)$) and momentum flux ($\approx \dot{M} v_\infty$). The condition $L_{\dot{M}} < L_*$ gives

$$\dot{M} \leq 4.9 \times 10^{-8} LR\, M^{-1} (1 + v_\infty^2 / v_e^2)^{-1} \qquad (1)$$

where L, R, and M are in solar units. For the case where the momentum is provided by radiative driving, the momentum condition

$$\dot{M} v_\infty = (4\pi/c) \int_{r_0}^{\infty} r^2 dr \int_0^{\infty} d\nu\, \mathbf{X}_\nu\, F_\nu \qquad (2)$$

also applies, with $\mathbf{X}_\nu$ = total extinction coefficient (Mihalas 1978, §2-3). For an optically thin envelope, $F_\nu = F_\nu$ (star) + F_ν (cs envelope) $\approx F_\nu$ (star), and Equation 2 reduces to $\dot{M} v_\infty = \langle \tau \rangle_F L_*/c$ (with

$<\tau>_F$ = flux-averaged optical depth). Knapp (1986; Knapp et al. 1982) defined a factor $\beta = L/(\dot{M}cv_\infty)$ that is < 1 for most radiatively driven stellar winds. It is possible to get $\beta > 1$ if, for example, there is an opaque circumstellar shell with a transparent cavity near the star, allowing photons to scatter several times across the cavity, so that a dust particle in the inner shell sees F_ν (cs envelope) $\gtrsim F_\nu$ (star). Such a configuration arises naturally for an OH-IR source with a moderately warm central star; then the dust forms at some distance above the stellar photosphere, and the opacity increases abruptly where the dust forms.

Virtually all studies of the effects of mass loss on stellar evolution have assumed that there is a continuous parametrizable mass loss relation that applies to nearly all stars, such as the relation (Reimers 1975):

$$\dot{M} = \eta_R \; 4 \times 10^{-13} \; LR \; M^{-1} M_\odot \cdot yr^{-1}. \tag{3}$$

This is the same as equation (1) if $\eta_R = 8 \times 10^4 \; \alpha (1 + v_\infty^2/v_e^2)^{-1}$, where α is a measure of the efficiency of the wind (if $v_\infty \ll v_e$, $\alpha = L_M/L_*$). Observed mass loss rates correspond to a wide range of α, from $\alpha \lesssim 10^{-6}$ for the sun to $> 10^{-2}$ for some pre-main sequence stars.

Reimers originally obtained equation (3) with $\eta_R = 1$ from a fit to observed mass loss rates for red giants. Subsequent testing of this relation in evolutionary models indicated that this overestimated the average mass loss rate, and currently favored representative values of η_R are $\sim 1/4$ to $1/3$ (see e.g. Reimers 1987). Note that these values are sufficiently low as to imply very little mass loss over most of the evolution of intermediate and low mass stars. For example, for $\eta_R = .25$, at most $\sim 0.01 \; M_\odot$ is expected to be lost on the first giant branch (RGB), while the existence of blue horizontal branch stars and RR Lyrae stars requires that at least $0.1 \; M_\odot$ be lost from stars on the RGB (Rood 1973). It has become quite clear that there is no single value of η_R that can satisfy observational constraints, but rather, that there are "episodes" of enhanced mass loss that are most important in the evolutionary picture.

Pulsation and mass loss

Hydrodynamical models for the atmospheres of pulsating stars have been calculated by Hill (Hill & Willson 1979, Willson & Hill 1979 collectively = HW), by Wood (1979), and by Bowen (1988) for the Mira variables; an analytic understanding of these calculations is described in some detail by HW and by Willson & Bowen (1985, 1986a, 1986b collectively = WB). A few explorations of the effects of pulsation on the atmospheres of other categories of variable stars have been made by Hill (1972, 1975) for RR Lyrae stars and δ Scuti stars, and by Bowen (unpublished) for Cepheids and RR Lyrae stars; but these so far lack essential details of the thermal and radiative physics, so only adumbrate the results of more detailed calculations to come.

Essential results from the model calculations, as understood by the analytic theory, reveal that pulsation quite generally leads to an increased scale height in the stellar atmosphere, lifting substantial amounts of material to an altitude where other mechanisms can act more effectively to drive it away. Pulsation is capable of greatly increasing the density in the outer atmosphere even for stars with small static scale heights. The increase in the density at a given altitude for a given set of stellar parameters depends on the pulsation mode and on the driving amplitude (or, more physically, on the fraction of the pulsation damping that is contributed by the atmosphere). The increase in the scale height in the region containing atmospheric shocks depends mainly on the pulsation mode. The density at which the shocks form depends also on the driving amplitude: stronger driving in a given mode $\longrightarrow$ deeper shocks $\longrightarrow$ higher densities in the outer atmosphere.

Pulsation alone can drive mass loss (for example by also heating the atmosphere to drive a thermal wind) but in most cases where substantial mass loss occurs there is also an important second mechanism involved. Examples of second mechanisms that have been at least tentatively identified include radiative forces ("radiation pressure") on dust in Miras, rapid rotation in Be stars and main sequence A stars, radiative forces due to the high opacity in UV resonance lines in O and B stars, and the absorption of Lyman line and continuum emission (originating from shocks or in an extended calorisphere; WB) in Cepheids (Willson & Bowen 1984 = WB*; Brunish & Willson 1987 and in preparation).

IMPORTANT MASS LOSS STAGES IN STELLAR EVOLUTION

There are only a few classes of stars for which there is clear, unambiguous evidence that important mass loss is occurring: pre-main sequence objects, luminous early-type stars, and highly evolved red giants or supergiants. Winds have also been observed coming from R CrB stars, planetary nebula nuclei, Be stars and Wolf-Rayet stars, but it is not yet clear whether these represent mass loss sufficient to affect the (already rapid) evolution of these objects. Winds have been proposed to occur (as the result of pulsation) for Cepheids, RR Lyrae stars, and main sequence A stars (WB*; Willson, Bowen & Struck-Marcell 1987 = WBS) but these winds have yet to be confirmed by direct measurement.

Pre-main sequence mass loss

Direct observations of stellar winds plus the almost certain necessity for protostars to shed angular momentum during their evolution to the main sequence provide persuasive evidence that stars lose substantial mass before reaching the main sequence. As yet there has been no detailed investigation of the possibility that pulsation plays a role in this process, although the Hayashi tracks for stars with $M < 1.5 M_\odot$ lie within the region of Mira instability and pulsation is known to drive substantial mass loss from the (post-main sequence) Miras. There are two complications facing investigations concerning the possible role of pulsation in pre-main sequence evolution and mass loss (Willson 1988): (1) the stars are not directly observable until they leave the Mira region (Stahler's (1983) "birth line" lies on or near the blue edge of the Mira strip as determined from post-main sequence Miras), complicating observations, and (2) rotation is likely to play an important role both by affecting the pulsation mode(s) and in enhancing or modulating the mass loss, greatly complicating the theory.

Main sequence mass loss

On the upper main sequence, O and early B stars ($M > \sim 5\ M_\odot$) have observed mass loss rates high enough to affect the evolution of at least the more massive ($> 10\ M_\odot$) stars (Chiosi & and Maeder 1986). The mechanism in this case is the transfer of momentum from the radiation field to the gas via resonance line absorption by abundant ions (Castor, Abbott & Klein 1975). The rapidly rotating Be stars also show substantial if episodic mass loss, and the possibility that this is closely linked to the pulsation properties has been considered (Smith 1988; Smith & Penrod 1985; Willson 1986).

Stars between 1+ and 2+ solar masses (early A to mid F spectral types) lie on the main sequence in the region of the "Cepheid" instability strip. Some of these are observed to pulsate as δ Scuti stars but most have small amplitudes or no apparent variability, even though theoretical calculations indicate that these should be quite pulsationally unstable (Stellingwerf 1979). Stellingwerf noted that one explanation for the lack of obvious pulsation is that there is atmospheric damping associated with the driving of substantial winds from these stars. WBS suggested that mass loss associated with pulsation combined with rapid rotation in this part of the main sequence could cause stars to evolve down the main sequence. Such evolution would invalidate the standard method of dating clusters that are older than about a billion years, possibly providing a solution to the problem that standard dating of globular clusters yields ages of 14-18 Gyr while other methods of dating the galaxy and the universe appear to favor ages < 12 Gyr (WBS). This hypothesis would also provide a novel explanation for the blue stragglers as being those stars that (due to slow rotation or lying to the blue of the instability strip) did not share in the general downward migration of the rapidly rotating, pulsating and hence mass-losing stars. It would predict (or account for) the observed deficiency of A and early F stars (that indicates a deficiency of stars with masses between about 1.2 and 2 to 3 $M_\odot$ if standard mass-spectral type calibrations are used). The level of mass loss needed to produce evolution down the main sequence is sufficiently low that direct detection may be difficult, depending on the temperature of the outflowing gas. Many A stars have been found to have infrared excesses in the IRAS survey, though these may be accounted for by the presence of a remnant proto-planetary disk of relatively large solid grains (Aumann et al. 1984; Gillett 1986).

Post-main sequence mass loss

For evolved low mass, Population II, stars there is evidence for mass loss near the tip of the red giant branch; without such mass

loss, there would be no RR Lyrae stars (Rood 1973). This mass loss may
be associated with a brief stage of Mira-like pulsations. Mass loss
associated with RR Lyrae pulsation may also play a significant role in
populating the blue horizontal branch (WB*), causing the blueward
evolution of RR Lyrae stars to occur at higher luminosities than
expected, and concentrating the RR Lyrae stars along the blue edge and
fundamental blue edge of the instability strip. RV Tauri stars show
Mira-like, somewhat irregular, pulsations; many RV Tauri stars also show
IR excesses indicating substantial current or recent past mass loss
(Gehrz & Woolf 1970; Raveendran & Rao 1988).

Do Cepheids lose mass?

Stars with initial masses between 3 to 5 $M_\odot$ and about 15 $M_\odot$
pass through the Cepheid instability strip during their core-He burning
blueward loops (Becker 1981); the possibility that significant mass loss
occurs as a result of Cepheid pulsation for at least the shorter period,
lower mass Cepheids was suggested by WB*. It is not yet clear whether
such mass loss affects the evolution of the Cepheids: Theoretical
models confirm that it is possible to "trap" the Cepheids in the
instability strip with mass loss, to reduce the masses substantially and
thus possibly solve the "Cepheid Mass Anomaly" (Cox 1980 — but see also
Andreason 1988), and that this reduces the rate of change of the periods
to better agreement with those observed (Brunish & Willson 1987 and in
preparation). However, the critical mass loss rate required to produce
these effects is relatively high (from 10^{-7} $M_\odot \cdot yr^{-1}$ for a 5 $M_\odot$, 3-5 day
model to 10^{-5} $M_\odot \cdot yr^{-1}$ for an 8 $M_\odot$, 10 day model). Observations in the
radio and infrared spectral regions seem to exclude such high rates of
flow either as a fully ionized plasma or as a cool, dusty wind (Butler
1987, MacAlary & Welch 1986, Welch & Duric 1988). There is some
indication of a blueshifted, steady line component at about 30-50
$km \cdot sec^{-1}$ in the UV spectra (Schmidt & Parsons 1984), consistent with an
outflow in which the absorption of line radiation (Lα, particularly)
from deep-lying shocks plays a role in driving the flow; this is
consistent with a wind that is neither ionized nor dusty. More detailed
modeling of the Cepheid atmospheres and winds is needed in order to (1)
establish the expected rate of mass loss and (2) allow for the
derivation of reliable mass loss rates or limits from the spectral line
features.

For stars with initial masses less than about 5$M_\odot$, stars that ultimately
become white dwarfs with M $\lesssim$ 0.7 $M_\odot$, the most important post-main
sequence mass loss occurs near the tip of the AGB. This mass loss stage
is closely associated with Mira pulsation, and will be discussed in
detail in the next section.

A SYSTEMATIC COMPARISON OF THEORY WITH OBSERVATIONS: MIRAS AND MASS LOSS ON THE ASYMPTOTIC GIANT BRANCH

It is appropriate to devote the largest portion of this
review to the Miras and the end of the AGB for two reasons: (1) this is
the best-studied and most obvious case of pulsation-related mass loss

and (2) the mass loss at this stage has important consequences for fundamental problems of astronomy. AGB stars are believed to be the primary source of the grains or the seed-nuclei for interstellar grains, that in turn play an essential role in star formation. Carbon stars are very common and very luminous, and thus may provide a very important "standard candle" for extragalactic distance determinations (Richer 1988). AGB processes determine the mass functions for white dwarfs and for planetary nebulae. An understanding of the processes that terminate the AGB evolution of low/intermediate mass stars is essential for models of nucleosynthesis and galactic chemical evolution. AGB stars are the prime source for s-processed material; stellar mass loss during and following the AGB phase returns this material to the interstellar medium. Also, supernovae provide the r-process elements, and the AGB mass loss process probably determines the minimum mass progenitor for supernovae; only when the mechanism is well understood can we expect to determine how this limiting mass varies with metallicity or initial helium abundance.

Properties of individual Miras

The properties of individual Miras should ultimately be determined by matching observed stellar characteristics and spectra to a model that includes both the dynamical effects and the essential radiative transfer to predict radiative fluxes, line profiles, angular diameters and light curves for Miras. Efforts to achieve this synthesis have been undertaken by several investigators, and preliminary results are promising. Beach, Willson & Bowen (1988 = BWB) used dynamical models by Bowen to investigate the effects of the extended atmosphere on angular diameter measurements. Bessell (1989) has calculated spectral energy distributions from simplified dynamical models by Wood, and finds good agreement with the observations.

Comparisons of the dynamical models with the observed velocity variations in the Miras have provided essential evidence that the Miras almost certainly are fundamental mode pulsators. For a given (observed) period distribution, fundamental mode models have smaller radii and hence higher gravities. Shock amplitudes as large as those observed occur only in fundamental mode models (Bowen 1988; see also WB and HW). The greatly extended atmospheres of the Miras, resulting from the combined effects of pulsation and radiative forces, lead to apparent angular diameters that may exceed the "theoretical radius" (where $\tau_{Rosseland} \sim 2/3$) by factors of 2 or more (BWB); this misled early investigators into believing that the stars were overtone pulsators based on "observed" $Q = P\sqrt{\rho}$ values.

A study of the MgII emission from Mira variables, carried out with the International Ultraviolet Explorer (IUE) satellite (Willson 1988; Brugel et al. 1986, 1987), provides further confirmation that the nearby Miras are fundamental mode pulsators. The decay of the MgII emission agrees well with the predictions for fundamental mode models (as interpreted, so far, with simplified emission and radiative transfer treatments). According to this interpretation, the MgII emission that is seen is generated by the shock as it passes between 1.5 or 2 stellar radii and 3 to 5 stellar radii; thus the MgII emission is a potentially valuable probe of the conditions in the dust-forming part of the stellar atmosphere that the Bowen models suggest plays a critical role in the mass loss process.

Systematic study of AGB evolution with mass loss

By comparing the (parametrized) results of evolutionary and atmospheric models with observed (mostly statistical) properties of the Miras we can determine which mass loss processes are most important as a function of metallicity and progenitor mass. Preliminary results of such a study are presented here; a more detailed report is under preparation by Willson, Kowalsky, Kirpes and Preston.

Required for such a systematic study are: evolutionary tracks ($L\{M, T, Z, l/H\}$); evolution rate ($d(\log L)/dt$ or L vs. M_{Core}); pulsation relations (P_0 and $P_1\{M, R, Z, l/H\}$); the locus of Mira onset ($L_{on}\{M, L, T, Z, l/H\}$); a mass loss function ($\dot{M}\{L, M, R, Z, \text{mode}, l/H\}$); and an initial AGB mass function (the I(AGB)MF). The most uncertain relations are the onset locus (L vs T and/or M, Z) of Mira pulsation, the mass loss function, and the I(AGB)MF. The only unobservable (and artificial) parameter is the ratio of convective mixing length to scale height, l/H.

Constraints on the evolution of AGB stars in L and M from observations include several that are observationally well determined: the period-luminosity relation, the period histogram, the relation between the limiting wind flow speed and the period or luminosity, the initial/final mass relation and the mass functions of white dwarfs and of the central stars of planetary nebulae. Also relatively well determined observationally (although tricky to interpret) are the relative fraction of M, S and C stars as a function of P, L and/or T, and the distribution with period of stars with surface Technetium (indicating recent dredge-up of s-processed material; Little-Marenin 1989). Less well determined observationally, but desirable from the theoretical standpoint, are the relations between progenitor mass and period, and the effective temperatures vs. period or luminosity. The shock velocity amplitudes are useful at the level of deciding the mode of pulsation; present data do not provide closer constraints because the geometrical correction factor v_r/v_{obs} (to correct for integration of the line profile over the stellar disk) is quite uncertain.

Mira parameters that are consistent with the period-luminosity relation, some choice of published effective temperatures, and the initial/final mass relation can be chosen either for fundamental or for overtone pulsation (Willson & Kowalsky 1986). However, as we have indicated, atmospheric models computed for the overtone case fail to reproduce the observed shock amplitudes. In the discussion that follows only the fundamental mode case will be considered.

Before the observed properties of Mira variables can be compared with the predictions, it is essential to sort the true Miras (fundamental mode AGB stars) from their close relatives (overtone pulsators and/or first giant branch variables). Again, the dynamical atmosphere models and the analytic theory give us guidance. Shock amplitudes in excess of 20-30 km·sec^{-1} occur only in fundamental mode pulsators; it is these large shocks that have allowed the identification of the mode of the Miras (WB, HW). Only for a few of the nearest stars have detailed high-resolution near-IR spectra allowed us resolve the velocities of the deep shock. However, for overtone pulsation, regardless of the driving amplitude, the shock amplitude will be too small to produce a lot of hydrogen Balmer emission. For fundamental mode pulsation, all but the lowest amplitude or lowest gravity cases are expected to show H Balmer emission. Thus the presence of H Balmer emission lines over a portion of the pulsation cycle is probably the most useful criterion for selecting F mode pulsators. Luminosity provides a criterion for selecting AGB stars, since most lie above the maximum luminosity of the first giant branch. The presence of Technetium lines in the stellar spectrum confirms that we are dealing with a thermally-pulsing AGB star; but some TP-AGB stars may not show Technetium.

Evolutionary models for AGB stars have been recently reviewed by Iben & Renzini (1983). AGB stars alternate between H shell burning and He shell flashes, with the pulse length and the duration of the interpulse phase a function of the stellar properties. A troubling source of uncertainty in the modeling is the parametrized mixing length theory of convection; the evolutionary tracks shift to cooler temperatures as the mixing length/scale height ratio is decreased. Unfortunately observations only weakly constrain T_{eff} as well for the stars near the tip of the AGB, the Miras, due to the complications introduced by their highly extended atmospheres (BWB).

A parametrized fit to evolutionary tracks for AGB stars is given by (Iben 1984):

$$R = 312 \ (L/10^4)^{0.68} \ (1.175/M)^{0.31S} \ (Z/0.001)^{0.088} \ (l/H)^{-0.52} \qquad (4)$$

where L, R, and M are in solar units; S = 1 for M > 1.175, 0 for M < 1.175; and (l/H) is the ratio of the mixing length to the pressure scale height.

Pulsation calculations for Miras have been computed by Ostlie (1982; Ostlie & Cox 1986) and by Wood (1979; Fox & Wood 1982, 1985). For fundamental mode pulsation Ostlie found

$$\log P_0 = -1.92 - 0.73 \log M + 1.86 \log R \qquad (5)$$

and for first overtone pulsation

$$\log P_1 = -1.60 - 0.51 \log M + 1.59 \log R. \qquad (6)$$

The location of the blue edge of the instability strip for Mira pulsation is not easily derived from the theoretical studies, as linear studies show both overtone and fundamental modes to be pulsationally unstable over a broad range of L and T_{eff}. Taking the mode of pulsation of the Miras to be the fundamental mode, making the risky assumption that the mode with the greatest linear growth rate will be preferred, and assuming that the onset of Mira pulsation corresponds to the transition from overtone to fundamental mode, yields a theoretical Mira onset line, as is indicated in Figure 1. However, it is not clear that the linear growth rates contain the essential physics of mode selection, and there are not enough models to allow even this crude method to be used for the lower metallicity stars, so we treat the onset line as a mostly free parameter.

Figure 1. Representative evolutionary tracks from Equation 4, based on Iben's (1984) parametrization, together with mode switching loci based on linear growth rates from Ostlie (1982). Stars to the right of the "mode switch" line would be more likely to pulsate in the fundamental mode, to produce hydrogen emission lines, and hence to be classified as Mira variables.

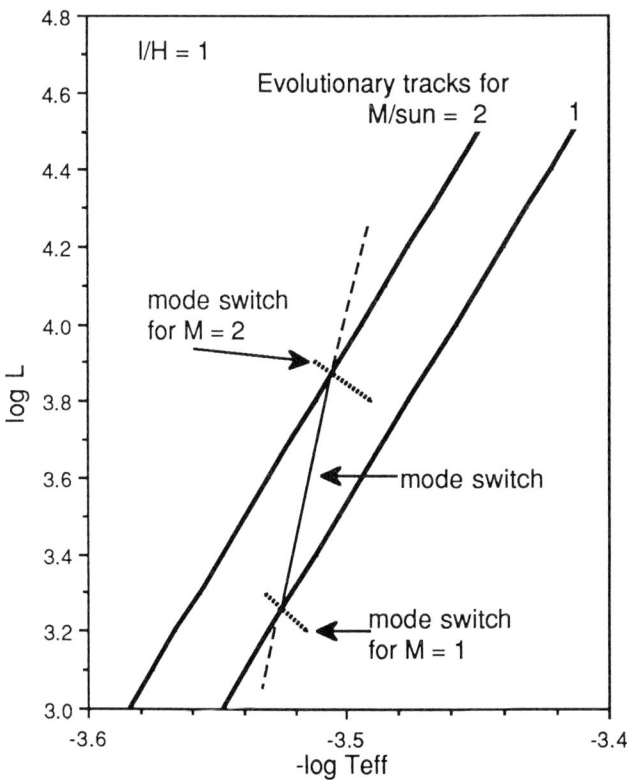

The "topography" of terminal AGB evolution. AGB stars follow a well-determined luminosity-core mass relation during their H-shell burning "interpulse" phases, given roughly by $L = 6 \times 10^4 (M_c - .52)$, and as a consequence evolve at a nearly constant rate in logL (e.g. Iben & Renzini 1983 = IR). This, coupled with the importance of mass loss during this phase, led Wood & Cahn (1987) to suggest the use of a plot of log Ṁ vs. log L to show essential aspects of late AGB/Mira evolution. Such a plot is shown in Figure 2, again for the case $l/H = 1$ and for solar metallicity. Included in this plot are lines of constant period (fundamental mode); the mode-switching line from Figure 1; and lines of constant T_{eff}. Miras are expected to occupy some portion, possibly all, of the region between the O→F line and the $M = M_{core}$ line, with most of the Miras to be found where the evolution is slowest (i.e. where the mass loss rate is lowest, because without mass loss they all evolve to the right with roughly the same $d(\log L)/dt$). From this diagram it is clear that we may expect to find Miras at a given period with a range of M and L. Also, the relationship between the average L or $M_{progenitor}$ and the period will depend on the distribution of Miras in this region — particularly, on the mass loss function that affects the rate of evolution vertically in the diagram, and on the I(AGB)MF.

Figure 2. The "mode switch" line from Figure 1, lines of constant fundamental mode period = 200, 300, 400 and 500 days, lines of constant effective temperature = 2950, 3100K, and the core mass corresponding to each mass and luminosity have been computed using Equation 4 and $l/H = 1$. Stars evolve to the right in this plot with $d(\log L)/dt \sim 10^{-6}$ yr^{-1} and downwards with $d(\log \dot{M})/dt = \dot{M}/M$.

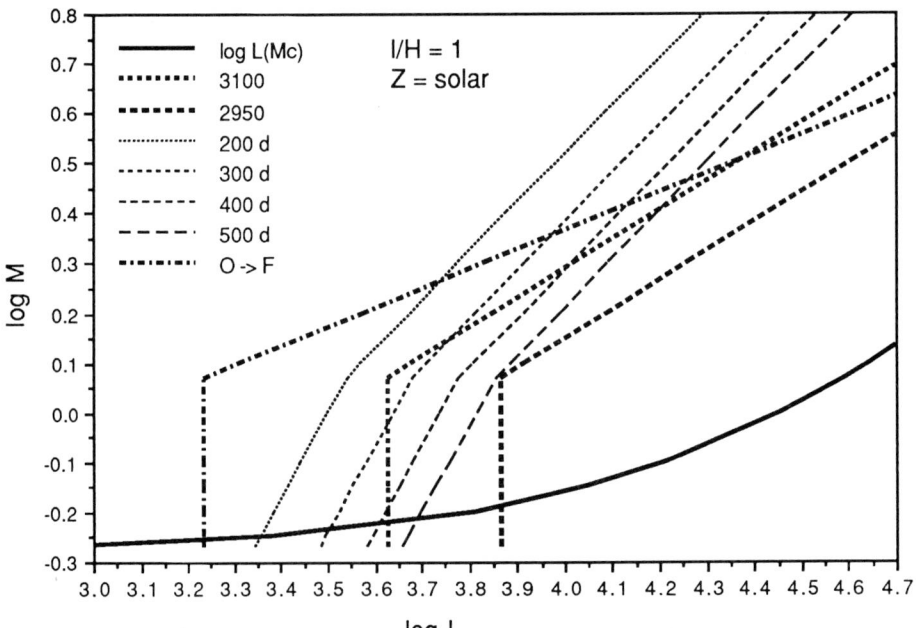

The log M vs. log L plot for a population with $Z = 0.1\ Z_\odot$ is shown in Figure 3; in this case there is no theoretical mode-switching line indicated, because there are not sufficient published models to use to derive one. It is clear that the effect of lowering Z is to raise the luminosity at a given M, P. It is also clear that the typical effective temperatures will be higher for lower Z stars compared with higher Z stars with the same P, L or M, L.

The effect of a decrease in l/H for either composition is to lower the effective temperature at a given M, L or M, P, and to lower L for a given M, P. Since the expected mass of a Mira at a given P is less than the progenitor mass (constrained by observations) and the expectation that these stars are on their way towards $M = M_{core}$, M is essentially fixed and either an increase in l/H or a decrease in Z will raise the expected average $<L>(P)$ and $<T_{eff}>(P)$.

Since the dependence of the mass loss rate on stellar parameters for Miras is unknown, we approximate the effect on the evolution by assuming that the mass loss increases fairly abruptly when some critical combination of stellar parameters is reached, so that before this occurs, $d(\log M)/dt \ll d(\log L)/dt$ and after, $d(\log M)/dt \gg d(\log L)/dt$. This describes a "deathline" for the Miras; as they will spend relatively little time where $d(\log M)/dt$ is large.

Figure 3. As for Figure 2, but with $Z = 0.1\ Z_\odot$. Note that a given (M,P) has higher L for lower Z, and that Teff is also higher at a given (L,P) or (M,P).

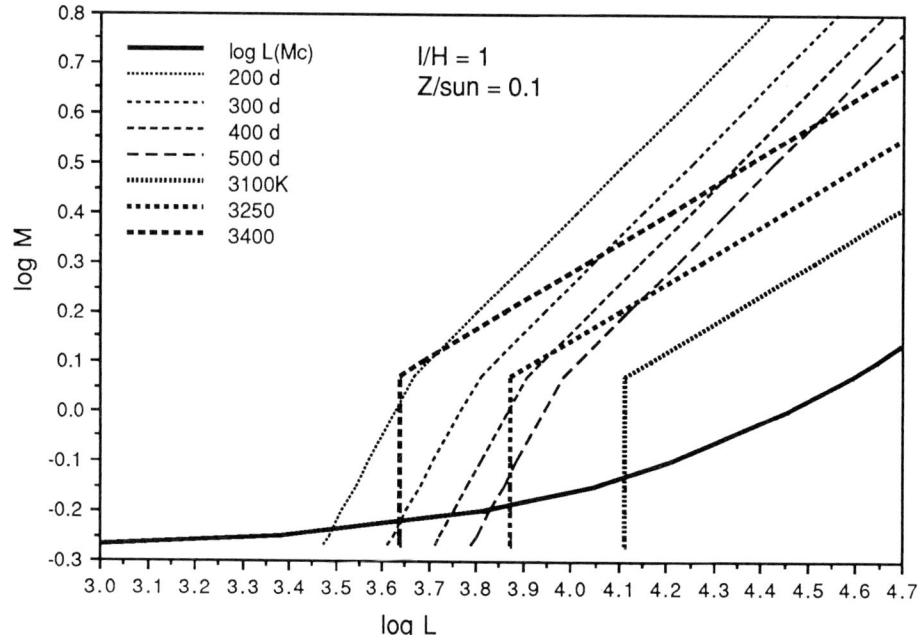

Thus most Miras will be found to the left/above the deathline, and the mass of the resultant white dwarf for a given star will be only slightly larger than M at the deathline. This approach is motivated by several arguments favoring the existence of a relatively abrupt increase in the mass loss rate, termed the "superwind", at the end of the AGB.

The superwind phase. Studies of the relative number of white dwarfs in the solar neighborhood, of the occurrence of white dwarfs in relatively young clusters, and of the frequency of supernovae all indicate that stars with initial masses up to 5 to 8 $M_\odot$ eventually become white dwarfs with masses mostly less than 1 $M_\odot$ and that the terminal white dwarf mass for a 3-5 $M_\odot$ star is ~ 0.7 $M_\odot$ (Weidemann 1984; see also IR). Comparisons of evolutionary models with observations indicate that most of this mass is lost while the stars are near the end of their AGB evolution. Models that incorporate steady, parametrized mass loss according to Reimers' relation cannot fit all the observational constraints with a single, constant, value for η_R: if η_R is too small, the calculations predict too many high luminosity AGB stars and too many high mass white dwarfs, and if η_R is too large, they predict too many low mass white dwarfs and not enough AGB stars. Iben (1984) pointed out that an average $\eta_R = 1$ would prevent essentially all stars from reaching the AGB, and thus contradicts the fact that there are S stars, C stars and stars with atmospheric Technetium. In other words: any constant η_R leads to too steep a dependence of the maximum AGB luminosity on the initial mass of the star. The best agreement with observational constraints is obtained by assuming that the mass loss rate increases more precipitously than LR/M near the end of the AGB. This enhanced mass loss "episode" has been termed the "superwind" by Renzini (1981a,b; see also IR).

The rapid increase in the mass loss rate that the observations suggest occurs near the end of the AGB evolution may in fact be necessary in order for a planetary nebula to be formed. Schonberner's (1983, 1987) evolutionary models indicate that a star will leave the AGB when its envelope mass is sufficiently small, with $M_{ef} \sim 0.01$ $M_\odot$ for a final core mass ~ 0.55 $M_\odot$. The effective temperature of the post-AGB star depends mainly on its envelope mass. If only nuclear burning altered the envelope mass, then the star would take so long to evolve from 3000K ($M_{env} \sim 0.01$ $M_\odot$) to 5000K ($M_{env} \sim 0.001$ $M_\odot$) that the mass ejected during the AGB stage would be long gone, and no planetary nebula would form. Schonberner (1987, 1989) invoked a continued "superwind" during the evolution from 3000 to 5000K, to reduce the envelope mass more rapidly. From our present understanding of the mass loss processes on the AGB, however, the mass loss rate should decrease abruptly as T_{eff} decreases, so this assumption is unrealistic. However, if the mass loss at the end of the AGB is sufficiently rapid, such that $t_M = M_{ef}/\dot{M} < t_{KH}$ (where M_{ef} is the final mass of the envelope), then the star will evolve blueward only at the rate given by t_{KH}; while the star is adjusting to accommodate the mass loss, the envelope mass may be reduced to an amount that corresponds to an effective temperature > 5000K. This may be easier for the more massive core remnants to achieve: Schonberner's models indicate that the difference between the envelope mass that induces blueward motion and that corresponding to T = 5000K decreases with increasing remnant masses.

The mechanism(s) of the superwind. Two factors determine $\dot{M}$ in the dynamical models by Bowen: the effective temperature and the "driving amplitude" (= the model parameter that governs the rate at which mechanical energy resulting from pulsation is being dissipated in the envelope). In the atmospheric models the effects of the internal driving are included through the imposition of an arbitrary inner "piston" boundary condition: at a position well below the photosphere but above the region of internal driving, the material is assumed to be moving with an imposed period, amplitude, and wave-form (usually sinusoidal). While the velocity structure of the lower atmosphere depends mostly on the period (or more precisely, on $Q = P\sqrt{\bar{\rho}}$) and on the escape velocity, the density throughout the atmosphere also depends sensitively on the driving amplitude (Bowen 1988; WB). The mass loss rate is in turn very sensitive to the density distribution. For example, a Bowen model with $P = 400$ days, $M = 1.2\ M_\odot$, $T = 2925K$ driven at 3 km·sec^{-1} had $\dot{M} = 9 \times 10^{-8}\ M_\odot \cdot yr^{-1}$; the same model, driven at 4 km·sec^{-1}, had $\dot{M} = 2.5 \times 10^{-6}\ M_\odot \cdot yr^{-1}$. The first mass loss rate is low enough to have little effect on the evolution; the second would dominate the evolution.

Generally, for a fixed piston amplitude, the mass loss rate increases exponentially near some temperature T_c which is between 2900 and 3000K for model and dust parameters that produce reasonably well-behaved atmospheric motions (for example, for piston amplitudes of 3-4 km·sec^{-1} in models similar to those described in Bowen 1988). The Mira models with $T > T_c$ have small mass loss rates and thus evolve mainly in L, while the ones with $T < T_c$ evolve mainly in M. This means that $T = T_c$ is a kind of "deathline" for Miras, where $\dot{M}$ passes M_*/t_{ev}. Physically, the existence of such a "critical temperature" is probably the consequence of the assumption that dust forms where the local radiative equilibrium temperature reaches some critical value; the ratio R_{dust}/R_{phot} is determined almost entirely by T_{eff}.

The critical temperature at which the mass loss rate becomes $\sim M_{env}/t_{evol}$ depends sensitively on the driving amplitude in the atmospheric models. Ostlie (1982) found that the linear growth rates became very large when t_{KH}/P became small (< 10 to 100 depending on M_*). While linear growth rates are not necessarily reliable indicators of the limiting non-linear behavior, Ostlie did also find that non-linear models with small t_{KH}/P were very unstable and tended to grow to very large amplitudes. Also, when the linear growth rates are large compared to the period, the star can probably oscillate in a more irregular way: the fundamental mode can increase and decrease in amplitude over only a few cycles, or there may be a more chaotic behavior. There are a number of stars whose light curves indicate that this is occurring (Cadmus et al. 1989). Such changes in the pulsation behavior are likely to enhance mass loss; at least, in the dynamical atmospheres, quick changes in the driving piston amplitude or waveform produce transients that tend to eject more material. Thus it is reasonable to consider the locus where the growth rates become large as a potential "deathline". The linear growth rates of the fundamental mode in the Ostlie models exceed 100%/cycle when the ratio $t_{KH}/P_0 < 1.815 - 1.400 \log M$. The line labeled "$L_{crit}$" in Figures 4 and 5 shows where this value of t_{KH}/P_0 is reached.

Baud & Habing (1983) proposed that the mass loss process be modeled with a mass loss rate that is inversely proportional to the envelope mass, so that the rate increases rapidly near the end of the AGB evolution. This was motivated by IRAS observations of AGB and post-AGB objects, that seemed to show a shortlived phase of high mass loss rates immediately preceding the end of the AGB phase. They proposed that $\dot{M}_{BH} = \dot{M}_{Reimers} M_*/M_{env}$. The location where $\dot{M}_{BH} = M_{env}/t_{ev}$ for the case with $\eta_R = \frac{1}{4}$ is also marked in Figures 4 and 5. While their relation is motivated by observations rather than by a physical mechanism, it clearly gives a similar "deathline" to the ones derived from the dust formation and the growth rate vs. t_{KH}/P considerations.

Figure 4. Possible "deathlines" for solar composition AGB stars, computed using Equation 4 with $l/H = 1$. Teff = 2950K is a rough indicator of where the dust-driven mass loss becomes destructive in Bowen models with moderate driving. The line labeled "L_{crit}" is the locus where the linear growth rate of fundamental mode pulsation is expected to become large, based on the Ostlie (1982) models. The line labeled "BH" shows where $\dot{M}_{BH} = \eta_R 4 \times 10^{-13} LR/M \ M_\odot \cdot yr^{-1} = M_{env}/t_{ev}$ for the case $\eta_R = \frac{1}{4}$. Each deathline maps (roughly) progenitor masses into final masses.

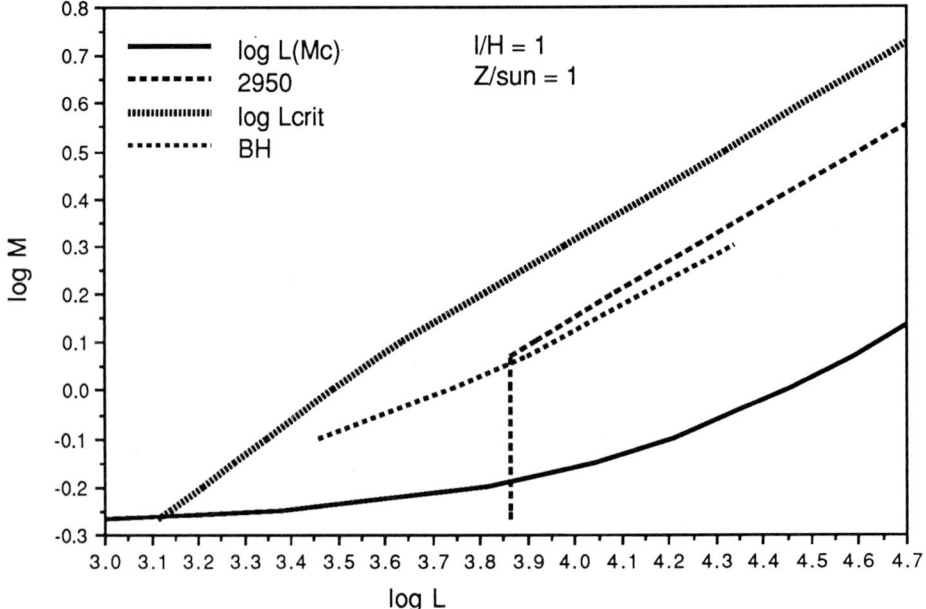

The three fundamental mode "deathlines" for Pop. I Miras are shown in Figure 4 for solar composition, and in Figure 5 for $Z = 0.1\ Z_\odot$, for $l/H = 1$ and assuming that the same values of t_{KH}/P_0 will be critical for both values of Z. For Pop. I stars (and if $l/H > 1$), the growth rate instability is probably the most important factor, because T_{eff} is still >3000K when stars reach this line. However, a slightly smaller l/H results in a diagram where $T_{eff} = 2950$ occurs before the critical t_{KH}/P_0 is reached, so that the "dusty death" of a regular pulsator is still a viable possibility for Pop. I. For the lower Z Miras the "dusty death" line T = 2950K has moved to higher L and longer P, and thus the growth rate instability is probably even more important. I am not saying that the dust is unimportant in these cases; just that it is the growth in amplitude or irregularity of the oscillations that determines when the combined pulsation/dust mechanism becomes destructively efficient.

Note that Figure 5 indicates that there may be a class of supernovae in low metallicity populations that is less common or altogether absent in Pop. I: supernovae from stars of approximately 5 $M_\odot$ that reach the tip of the AGB (log L = 4.7 or M_{core} = 1.4) before they have a chance to

Figure 5. As for Figure 4 but for $Z = 0.1\ Z_\odot$. Also shown is the locus determined by the observed period-luminosity relation for the LMC, based on m_{bol} vs. P from Glass et al. (1987) with an assumed LMC distance modulus of 18.5.

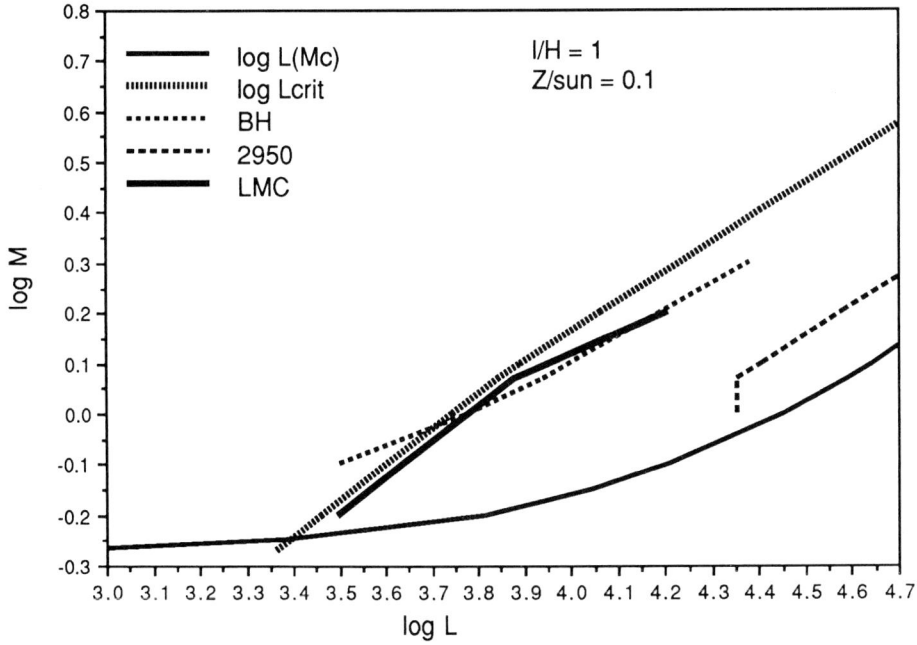

experience the superwind phase. This occurs for two reasons: (1) as can be seen in Figure 4, all the potential deathlines move to higher L at a given M as the metallicity decreases; and (2) the ability of the atmospheric material to form sufficient dust to support the dusty wind will be less for lower metallicity stars (see also Jura 1986).

Are planetary nebulae "ejected"? It has been widely suggested that the AGB is terminated by the sudden "ejection" of a planetary nebula, either as the result of some sort of dynamical instability in the pulsating envelope, or as a consequence of a He shell flash. If by "ejection" is meant the removal of the envelope on a dynamical timescale, then the following facts argue that PNe are not "ejected":

First, if we assume that the excess mass of AGB stars is lost abruptly, after a gentle mass loss stage characterized for example by a Reimers-type relation with constant η_R, then choices of pre-Mira mass loss relations that account for the population and limiting luminosities of AGB stars lead to predictions of planetary nebula masses that are, on the average and in the extreme, higher than are observed. PN masses average $\sim 0.1\ M_\odot$ and extend to a maximum of $\sim 1\ M_\odot$, while the envelope mass needing to be removed exceeds 1-2 $M_\odot$ in the stars with the more massive progenitors (e.g. Peimbert 1981). Of course one can get around this by carefully adjusting the pre-ejection mass loss function; but to get the right mass distribution in this case requires rather a lot of adjusting of the parametrized relations, with η_R increasing rapidly with initial stellar mass, to leave just the right envelope mass to be "ejected" at the end.

Second, if this "ejection" is to occur simply as the result of instabilities in the envelope, there is only enough energy to eject the lower mass planetaries; to eject 1 $M_\odot$ of material in a dynamical time of $\leqslant 10$ years requires a mass loss rate of 0.1 $M_\odot \cdot yr^{-1}$; to accelerate this to a terminal velocity of 20 km·sec^{-1} would require 100% of the luminosity of the star (from Equation 1). While there is enough energy in a He shell flash to eject planetaries up to 1 $M_\odot$, careful studies of the properties of PNN and PNe indicates that at most a few percent of these are formed while the star is burning He (Iben 1984; Schonberner 1989).

Third, observational studies of IRAS sources, OH-IR sources and proto-PNe, are providing constraints on the maximum mass loss rates of AGB stars, and these cluster around $10^{-4}\ M_\odot \cdot yr^{-1}$ (Baud & Habing 1983; Bedijn 1986, 1987). This is close to the maximum mass loss rate that can be driven by the combination of pulsation and radiative driving that accounts well for the mass loss from the Miras; it corresponds to the case with $\beta = \dot{M}/Lcv_\infty \sim 10^{1\pm 1}$. Note that these maximum observed mass loss rates give characteristic mass loss time scales $(M_{env}/\dot{M}) \sim t_{KH} \gg t_{dyn}$.

I rather prefer the idea that the mass loss rate increases continuously though perhaps precipitously during the stage of Mira/OH-IR pulsation, and that the observable planetary nebula is the result of the sweeping up of this outflow by the later, faster wind from the central stellar remnant, as originally explored by Kwok et al. (1978). This high rate of terminal AGB mass loss fits the requirements of Renzini's "superwind". It also fits well with observations of some objects that have been tentatively identified as the immediate progenitors of planetary nebulae: the non-variable OH-IR sources as identified by Olnon et al. (1984). According to an analysis of IRAS observations by Bedijn (1986, 1987), these non-variable OH-IR sources typically consist of a relatively warm central star (5-10,000 K or more) surrounded by a detached, outflowing remnant AGB wind. Characteristic ages for these, deduced from the flow speeds and the size of the cavity, are of the order of the Kelvin-Helmholtz time scale for the late AGB precursor, and deduced terminal mass loss rates of $\sim 10^{-4} M_\odot \cdot yr^{-1}$ are consistent with $t_M \sim t_{KH} \sim 10$ to 100 t_{dyn} at the end of the AGB.

With steady mass loss, the mass of the planetary nebula is equal to the mass of the envelope at a time Δt before the end of the AGB (equal to $\int \dot{M} dt$ over Δt). If the interval between the end of the AGB and the formation of the planetary nebula is sufficiently small, $\Delta t \approx R_{neb}/v_{wind}$. With 10^{-4} $M_\odot \cdot yr^{-1}$ terminal AGB mass loss rate and $v_{wind} = 10$ km·sec^{-1}, the mass of the nebula when it reaches R = 0.1 pc is ≤ 1 $M_\odot$ — less if M did not stay at 10^{-4} $M_\odot \cdot yr^{-1}$ for the entire $\Delta t = 10^4$ years or if there was a significant delay between the end of the wind and the beginning of the PN stage. A delay of 5000-7000 years is predicted from Schonberner's models if the mass loss process stops when $T_{eff} \sim 5000K$; this would result in a PN mass of just a few tenths of a solar mass. Thus the observed range of PN masses is quite consistent with the Kwok et al. colliding wind mechanism and the observed properties of terminal AGB winds, combined with Schonberner's post-AGB evolutionary models, as long as the star is able to stay cool, continue to pulsate and continue losing mass until $M_{env} \sim 0.001$ $M_\odot$.

CONCLUSIONS

The connection between stellar pulsation and mass loss for most classes of pulsating variable stars is in an early stage of investigation, and much work remains to be done to test the extent to which pulsation is implicated in important mass loss episodes for stars. Tantalizing results have been obtained from theoretical calculations for mass loss from Cepheid variables and from main sequence A and F stars; however, such mass loss has not yet been confirmed by observations.

For the Mira variables and the closely-related OH-IR stars, current modelling and recent observations (particularly from IRAS) have led to the identification of key mechanisms involved in stripping the envelopes from AGB stars to produce planetary nebulae and (mostly low mass) white dwarfs. The mass loss that produces planetary nebulae is seen to be a natural consequence of the pulsation combined with the production of dust in the stellar atmosphere. A prediction of this picture is that the mass loss process should allow a star of low metallicity to reach a higher final AGB luminosity than a higher metallicity star with the same initial mass, and that as a result there may be a class of relatively low mass "AGB supernovae" that occur only in low metallicity populations.

Dynamical "ejection" is neither needed nor supported by observations; however the increasingly unstable pulsation behavior of models with decreasing t_{KH}/P_0 probably plays a key role in the production of the terminal "superwind". Planetary nebulae are proposed to be possible when the final AGB mass loss occurs sufficiently rapidly that $t_{\dot{M}} < t_{KH}$, so that the star's envelope mass is brought below the minimum AGB envelope mass before the star has time to adjust its internal structure to have higher a T_{eff}.

BIBLIOGRAPHY

Andreasen, G.K. (1988). Astron. & Astrophys.,201, 72.
Aumann, H.H., Gillett, F.C., Beichman, C.A., de Jong, T., Houck, J.R., Low, F.J., Neugebauer, G., Walker, R.G. & Wesselius, P.R. (1984). Ap.J.,278, L23.
Baud, B. & H.J. Habing (1983). Astron. & Astrophys.,127, 73.
Beach, T.E., Willson, L.A. & Bowen G.H. (1988). Ap.J.,329, 241.
Becker, S.A. (1981). Ap.J.Suppl.,45, 33.
Bedijn, P.J. (1986). In Light on Dark Matter, ed. F. P. Israel, p. 119. Dordrecht: Reidel.
Bedijn, P.J. (1987). Astron. & Astrophys.,186, 136.
Bessell, M.S. (1989). In Peculiar Red Giants, IAU Colloquium 106, ed. H.R. Johnson, in press.
Bowen, G.H., (1988). Ap.J.,329, 299.
Brugel, E.W., Beach, T.E., Willson, L.A. & Bowen, G.H. (1987). In The Symbiotic Phenomenon, IAU, Colloquium 103, p. 67.
Brugel, E.W., Willson, L.A. & Cadmus, R.R. (1986). In New Insights in Astrophysics, Proc. Joint NASA/ESA/SERC Conference, London, 14-16 July 1986, SP 263, p. 213.
Brunish, W.M. & Willson, L.A. (1987). In Stellar Pulsation, Lecture Notes in Physics 174, eds. A.N. Cox, W.M. Sparks & S.G. Starrfield, p27. Berlin: Springer-Verlag.
Butler, C.J. (1987). In Circumstellar Matter, eds. I. Appenzeller & C. Jordan, 227.
Cadmus, R.R. Jr., Willson, L.A. & Sneden, C. (1989). In preparation.
Castor, J.I., Abbott, D.C. & Klein, R.I. (1975). Ap.J.,195, 157.
Chiosi, C. & Maeder, A. (1986). Ann. Rev. Astron. Astrophys.,24, 329.
Cox, A.N. (1980). Ann. Rev. Astron. Astrophys.,18, 15.
Fox, M.W. & Wood, P.R. (1982). Ap.J.,259, 198.
Fox, M.W. & Wood, P.R. (1985). Ap.J.,297, 455.
Gehrz, R.D. & Woolf, N.J. (1970). Ap.J.,161, L213.
Gillett, F.C. (1986). In Light on Dark Matter, ed. F. P. Israel, p. 61. Dordrecht: Reidel
Glass, I.S., Catchpole, R.M., Feast, M.W., Whitelock, P.A., Reid, I.N. (1987). In Late Stages of Stellar Evolution, eds. S. Kwok & S.R. Pottasch, p 51. Dordrecht: Reidel.
Hill, S.J. (1972). Ap.J.,178, 793.

Hill, S.J. (1975). A.J.,80, 1044.
Hill, S.J. & Willson, L.A. (1979). Ap.J.,229, 1029.
Iben, I. Jr. (1984). Ap.J.,277, 333.
Iben, I. Jr. & Renzini, A. (1983). Ann. Rev. Astron. Astrophys.,21, 271.
Jura, M. (1986). Ap.J.,301, 624.
Knapp, G.R. (1986). Ap.J.,311, 731.
Knapp, G.R., Phillips, J.G., Leighton, R.B., Lo, K.Y., Wannier, P.G., Wooten, H.A. & Huggins, P.J. (1982). Ap.J.,252, 616.
Kwok, S., Purton, G.R. & Fitzgerald, P.M. (1978). Ap.J.,219, L125.
Little-Marenin, I.R. (1988). In Peculiar Red Giants, IAU Colloquium 106, ed. H.R. Johnson, in press.
McAlary, C.W. & Welch, D.L. (1986). A.J.,91, 1209.
Olnon, F.A., Baud, B., Habing, H.J., de Jong, T., Harris, S. & Pottasch, S.R. (1984). Ap.J.,278, L41.
Ostlie, D.A. & Cox, A.N. (1986). Ap.J.,311, 864.
Ostlie, D.A. (1982). PhD Thesis, Iowa State University.
Peimbert, M. (1981). In Physical Processes in Red Giants, eds. I. Iben & A. Renzini, p. 409. Reidel: Dordrecht.
Raveendran, A.V. & Rao, K. (1988). Astron. & Astrophys.,193, 259.
Reimers, D. (1975). In Problems in Stellar Atmospheres and Envelopes, eds. Baschek, B., Kegel, W.H., Traving, G., p. 229. Berlin: Springer Verlag.
Reimers, D. (1987). In Circumstellar Matter, IAU Symposium 122, eds. I. Appenzeller & C. Jordan, p. 307. Dordrecht: Reidel.
Renzini, A. (1981a). In Physical Processes in Red Giants, eds. I. Iben & A. Renzini, p. 165. Dordrecht: Reidel.
Renzini, A. (1981b). In Effects of Mass Loss on Stellar Evolution, eds. C. Chiosi & R. Stalio, p. 319. Dordrecht: Reidel.
Richer, H.B. (1989). In Peculiar Red Giants, IAU Colloquium 106, ed. H.R. Johnson, in press.
Rood, R.T. (1973). Ap.J.,184, 815.
Schmidt, E.G. & Parsons, S.B. (1984). Ap.J.,279, 202.
Schonberner, D. (1983). Ap.J.,272, 708.
Schonberner, D. (1987). In Late Stages of Stellar Evolution, eds. S. Kwok & S. R. Pottasch, p. 337.
Schonberner, D. (1989). In Peculiar Red Giants, IAU Colloquium 106, ed. H.R. Johnson. In press.
Smith, M.A. & Penrod, G.D., (1985). In Relations Between Chromospheric/Coronal Heating and Mass Loss in Stars, eds. Stalio & Zirker, p. 394. Italy: Tabographis.
Smith, M.A. (1988). In Pulsation and Mass Loss, eds. L.A. Willson & R. Stalio, p. 251. Dordrecht: Reidel.
Stahler, S.W. (1983). Ap.J.,274, 822.
Stellingwerf, R.F. (1979). In Nonradial & Nonlinear Stellar Pulsations, Lecture Notes on Physics 125, eds. H.A. Hill & W.A. Dziembowski, p. 50. Berlin: Springer Verlag.
Weidemann, V. (1984). Astron. & Astrophys.,134, L1.
Welch, D.A. & Duric, N. (1988). A.J.,95, 1794.
Willson, L.A. (1986). P.A.S.P.,98, 37.
Willson, L.A. (1988). In Pulsation and Mass Loss, eds. L. A. Willson & R. Stalio, p. 285. Dordrecht: Reidel.

Willson, L.A. & Bowen, G.H. (1984). Nature,312, 429.
Willson, L.A. & Bowen, G.H. (1985). In Relations Between Chromospheric/Coronal Heating and Mass Loss in Stars, eds. R. Stalio & J.B. Zirker, p. 127. Italy: Tabographis.
Willson, L.A. & Bowen, G.H. (1986a). Irish Astronomical Journal,17, 249.
Willson, L.A. & Bowen, G.H. (1986b). In Cool Stars, Stellar Systems, and the Sun, Lecture Notes in Physics Vol. 254, eds. M. Zeilik & D. M. Gibson, p.385. Berlin: Springer-Verlag.
Willson, L.A. & Hill, S.J. (1979). Ap.J.,228, 854.
Willson, L.A. & Kowalsky, P. (1987). In Late Stages of Stellar Evolution, eds. S. Kwok & S. R. Pottasch, p. 277. Dordrecht: Reidel.
Willson, L.A., Bowen, G.H. & Struck-Marcell, C. (1987). Comments on Astrophysics,XII, pp 17.
Wood, P.R (1979). Ap.J.,227, 220.
Wood, P.R. & Cahn, J.H. (1977). Ap.J.,211, 499.
Wood, P.R., Bessell, M.S. & Fox, M.W. (1983). Ap.J.,272, 99.

STRUCTURE AND EVOLUTION OF THE MILKY WAY GALAXY

Gerard Gilmore
Institute of Astronomy, Cambridge

Rosemary F.G. Wyse
Department of Physics and Astronomy
The Johns Hopkins University

Abstract. The combination of chemical abundance, kinematic, and age data for stars near the sun provides important information about the early evolution of the Galaxy. We review available data, with some new analysis, to show that the sum of all available information strongly suggests that the extreme population II subdwarf system formed during a period of rapid collapse of the proto-Galaxy. This subdwarf system now forms a flattened, pressure-supported distribution, with axial ratio ~2:1. The thick disk formed subsequent to the subdwarf system. At least the metal-poor tail of the thick disk is comparable in age to the globular cluster system. The thick disk is probably kinematically discrete from the Galactic old disk, though the data remain inadequate for robust conclusions.

1 INTRODUCTION

One of the most important aspects of current non-stellar astrophysical research to which studies of variable stars contribute invaluable information is the study of the structure and evolution of the Milky Way Galaxy. In this paper we provide a brief overview of some active areas of Galactic research to which knowledge of the kinematics, chemical abundances, ages, and spatial distribution of pulsating variable stars make a fundamental contribution.

In principle, an understanding of the formation and early evolution of the Galaxy is a well-defined theoretical problem. All one requires is a detailed knowledge of the spectrum of perturbations in the early universe and their subsequent evolution; an understanding of the physics of star-formation in a variety of environments, with particular emphasis on a prediction of the distribution of orbital elements of those intermediate mass massive-star binaries which will evolve to supernovae; a description of the hydrodynamics of a proto-galaxy, particularly including the effects of a high supernova rate, the efficiency of mixing of the chemically enriched ejecta, and the incidence of thermal and gravitational instabilities; the growth and transport of angular momentum and their effect on the growth of a disk; and the effects of a time-dependant gravitational potential on the dynamics of any stars formed up to that time. In practise, there remain some limitations in our understanding of at least some of these physical processes. Hence, it is still useful on occasion to try to deduce the important physics involved in galaxy formation from observations of those old stars which were formed at the time of the formation of the Milky Way, and whose present properties contain some fossil record of the Galaxy's history.

We discuss some of this information here, with emphasis on results relevant to the evolution of all galaxies. In §2 we present evidence from stellar chemical abundance and kinematic data that the oldest stars in the Galaxy formed during a period of rapid collapse of the proto-Galaxy, while §3 summarises some recent results regarding the shape of the stellar distribution in the Galactic spheroid, and §4 discusses current data which suggest that the thick disk is an old, discrete component of the Galaxy.

2 THE TIMESCALES OF GALACTIC FORMATION

The kinematic properties of stars in the Galaxy are related, through the gravitational potential Φ, to their spatial distribution. The scale length of the spatial distribution is determined by the total energy of the stellar orbits, as well as by the gradient of the potential. The shape of the spatial distribution depends on the relative amounts of angular momentum (rotational), and pressure (stellar velocity anisotropy) balance to the potential gradients. The total orbital energy and angular momentum of the gas which will become a star depend on the maximum distance from the centre of the Galaxy which it ever reached, the angular momentum of its orbit at that time, the depth of the potential well (generated by both dark and luminous mass) through which it fell, the fraction of the total orbital energy which was dissipated before star formation, and the subsequent dynamical evolution of the stellar orbit. That is, the present kinematic properties of old stars in the solar neighbourhood are determined in part by the initial conditions in the proto-galaxy, and in part by the physics of galaxy formation. Hence, local kinematic studies can help to determine both the detailed physics of galaxy formation and the distribution of gravitating mass in the Galaxy.

The chemical abundance of the ISM at any time depends on the local history of formation and evolution of stars sufficiently massive to have created new chemical elements, and the mixing of local gas with more distant material. This more distant gas may or may not itself be enriched, so that the time-dependance of the chemical abundance of newly forming stars depends on both the local and the global star formation rates, the rate of infall of primordial gas, and the efficacy of mixing in the ISM. Thus, while the chemical abundance of newly formed stars is a timepiece, this chronometer need not be a smooth or even a single-valued function of chronological time.

Clearly, however, the distribution function of stellar kinematics, chemistry and age contains a wealth of information on the distribution of protogalactic gas, the dissipational and star-formation history of that gas, the subsequent dynamical history of the resulting stars, and the Galactic gravitational potential.

2.1 Kinematics and chemistry of old stars

The dynamics of any large stellar system are governed by the collisionless Boltzmann equation (CBE):

$$\frac{Df}{Dt} \equiv \frac{\partial f}{\partial t} + \frac{\partial \overline{x}}{\partial t} \cdot \frac{\partial f}{\partial \overline{x}} + \frac{\partial \overline{v}}{\partial t} \cdot \frac{\partial f}{\partial \overline{v}} = \frac{\partial f}{\partial t} + \overline{v} \frac{\partial f}{\partial \overline{x}} - \nabla\Phi \frac{\partial f}{\partial \overline{v}} = 0, \quad (1)$$

where f is the phase space density at the point $(\overline{x},\overline{v})$ in phase space (i.e. there are $f(\overline{x},\overline{v}) d^3\overline{x} d^3\overline{v}$ stars in a volume of size $d^3\overline{x}$ centered on $\overline{x}$ with velocity in the volume of size $d^3\overline{v}$ about $\overline{v}$). The collisionless Boltzmann equation is satisfied by **any** stellar population. If there exist several identifiable populations in the system, the CBE is satisfied by each of them separately. This arises because stars do not interact except through long-range gravity forces, which are being described through a smooth background potential. Consequently, f does not have to describe the entire Galaxy; one can concentrate on any subsample of stars, and apply the collisionless Boltzmann equation to it.

If we have a steady-state tracer population, and a time-independent potential, as the large-scale field in the Milky Way presumably is, we can set

$$\partial f/\partial t = 0. \quad (2)$$

For present purposes, the Galaxy is adequately described as being rotationally symmetric, so that it is convenient to write out the collisionless Boltzmann equation in cylindrical polar coordinates (r,Φ,z) in which $z = 0$ is the disk plane of symmetry, with corresponding velocity components (v_r, v_Φ, v_z):

$$v_r \frac{\partial f}{\partial r} + v_z \frac{\partial f}{\partial z} + \left(K_r + \frac{v_\Phi^2}{r}\right)\frac{\partial f}{\partial v_r} - \frac{v_r v_\Phi}{r}\frac{\partial f}{\partial v_\Phi} + K_z \frac{\partial f}{\partial v_z} = 0 \quad (3)$$

where the accelerations $\dot{v}_r, \dot{v}_\Phi, \dot{v}_z$ have been equated to the forces that cause them, Φ-gradients in f and in the potential have been set to zero, and K_r and K_z are the components of the gravity force.

In view of the intractability of the general problem of solving the CBE, one proceeds in general by taking velocity moments. Multiplying through by v_z and by v_r and integrating over all velocity space produces the Jeans' equations:

$$\nu K_z = \frac{\partial}{\partial z}[\nu \sigma_{zz}] + \frac{1}{r}\frac{\partial}{\partial r}[r \nu \sigma_{rz}] \quad (4)$$

$$\nu K_r = \frac{1}{r}\frac{\partial}{\partial r}[r \nu \sigma_{rr}] + \frac{\partial}{\partial z}[\nu \sigma_{rz}] - \frac{\nu \sigma_{\Phi\Phi}}{r} - \frac{\nu}{r}\langle v_\Phi \rangle^2, \quad (5)$$

where $\nu(r,z)$ is the space density of the stars, and $\overline{\overline{\sigma}}(r,z)$ their velocity dispersion tensor (i.e. $\sigma_{ij} = \langle v_i v_j \rangle - \langle v_i \rangle\langle v_j \rangle$). Note that the velocity dispersions σ_{ij} are thus **squared** velocities, not r.m.s.

values.

For present purposes, we rewrite the radial moment equation (5) in terms of observables in the Galactic plane (z = 0) to get:

$$v_c^2 - \langle v_\phi \rangle^2 = \sigma_{\phi\phi} - \sigma_{rr} - \frac{r}{V}\frac{\partial(V\sigma_{rr})}{\partial r} - r\frac{\partial \sigma_{rz}}{\partial z}$$

$$= \sigma_{rr}\left\{\frac{\sigma_{\phi\phi}}{\sigma_{rr}} - 1 - \frac{\partial \ln(V\sigma_{rr})}{\partial \ln r} - \frac{r}{\sigma_{rr}}\frac{\partial \sigma_{rz}}{\partial z}\right\} \quad (6)$$

In this relation v_c is the circular velocity (i.e. $v_c^2 = r(\partial\Phi/\partial r) \equiv -rK_r$ where we adopt a locally flat rotation curve with v_c = 220 km·sec^{-1} here), $\langle v_\phi \rangle$ is the mean rotation velocity of the relevant sample of tracer stars, which has velocity dispersions $\sqrt{\sigma_{rr}}$, $\sqrt{\sigma_{\phi\phi}}$, and $\sqrt{\sigma_{rz}}$ and radial spatial density distribution $V(r)$, remembering that r is the **planar** radial coordinate. The quantity $v_c - \langle v_\phi \rangle \equiv v_a$ is usually called the asymmetric drift.

2.2 The asymmetric drift

Equation (6) relates measurable local moments of the stellar distribution function to global properties of the Galaxy. In order to understand its application, we discuss each term briefly.

$\sigma_{\phi\phi}/\sigma_{rr}$: The velocity dispersions at z = 0 of old disk stars are probably best estimated from the nearby, spectroscopically-selected K+M dwarfs with good parallax distances. These give $\sigma_{rr}:\sigma_{\phi\phi}:\sigma_{zz}$ = $39^2:23^2:20^2$ (Wielen 1974). For spectroscopically-selected low metallicity field stars the relevant values (Carney & Latham 1986) are $\sigma_{rr}:\sigma_{\phi\phi}:\sigma_{zz} = 128^2:96^2:93^2$. The first term in equation (6) then becomes $\sigma_{\phi\phi}/\sigma_{rr} = 0.35$ for the old disk, and $\sigma_{\phi\phi}/\sigma_{rr} = 0.56$ for the low-abundance field stars.

$\partial \ln(V\sigma_{rr})/\partial \ln r$: If one assumes that galactic disks have constant thickness independent of distance from the galactic centre, as suggested by photometric observations, and combines that assumption with the assumption that the shape of the velocity ellipsoid is independent of position, or more specifically that $\sigma_{rr} \propto \sigma_{zz}$, then

$$\partial \ln(V\sigma_{rr})/\partial \ln r = 2 (\partial \ln V/\partial \ln r) = -2rh_r^{-1} \quad (7)$$

The derivation of equation (7) assumes a form for the velocity ellipsoid. For simplicity, we restrict discussion of spheroidal distributions to an isothermal spheroid, σ_{rr} = constant, so that

$$\partial \ln(V\sigma_{rr})/\partial \ln r = \partial \ln V/\partial \ln r \quad (8)$$

$(r/\sigma_{rr})(\partial \sigma_{rz}/\partial z)$: The term involving σ_{rz} describes the orientation of the velocity ellipsoid, and has no general analytic solution. It is discussed in detail in Kuijken & Gilmore (1989). Here we consider two limiting cases only. If the potential is that of an infinite constant surface density sheet the velocity ellipsoid will be diagonal in cylindrical-polar coordinates, and point always at the Galactic minor axis, so that $\sigma_{rz} \equiv 0$. This is the assumption most commonly adopted. An alternative idealisation is to assume that the potential is dominated by a spherical mass distribution, so that the velocity ellipsoid points always at the Galactic centre. In this case it is straightforward to show that, for the velocity ellipsoid parameters from Wielen noted above,

$$\sigma_{rz} = (3rz/\{4z^2+r^2\})\sigma_{zz}$$

so that

$$(r/\sigma_{rr})(\partial \sigma_{rz}/\partial z) \approx 3 \, (\sigma_{zz}/\sigma_{rr}). \tag{9}$$

For the velocity dispersions quoted above, equation (9) provides, for the old disk,

$$(r/\sigma_{rr})(\partial \sigma_{rz}/\partial) \cong 0.75,$$

while for the spheroidal field stars the numerical value is 0.47. The true value for the σ_{rz} term is quite uncertain, though of large amplitude. Neglect of this term, as often done, is unjustified.

For an exponential disk of radial scale length h_r, and a sample of stars observed in the solar neighborhood, equation (6) therefore becomes

$$v^2 - \langle v_\phi \rangle^2 = \sigma_{rr} \, (2\{d_*/h_r\} - 1.4) \tag{10}$$

where d_* is the distance of the sun from the Galactic centre (~ 7.8 kpc, Feast 1987). Alternatively, for a spheroid with a power-law density distribution with exponent γ, $\nu(r) \propto r^{-\gamma}$, we have

$$v_c^2 - \langle v_\phi \rangle^2 = \sigma_{rr} \, (\gamma - 0.9) \tag{11}$$

Thus a stellar tracer population which belongs to an exponential distribution with scale length ~ 3.5 kpc (a plausible value for the Galactic old disk) will follow a similar asymmetric drift relation to that of a tracer population which is part of an isothermal distribution describing an r^{-4} spheroidal density distribution. It is also interesting to note that the largest allowed radial velocity dispersion, corresponding to zero net rotation, for such a stellar system is $\sim v_c/\sqrt{3}$, or ~ 130 km·sec^{-1}. For a tracer population with any smaller radial velocity dispersion, equation (6) describes the interplay between the pressure (velocity anisotropy) support and the angular momentum (rotation velocity) support to the spatial distribution. A stellar system with velocity dispersion larger than ~ 130 km·sec^{-1} has larger total energy and will form a more extended system. One might also assume it would have formed from less dissipated material, which is of course the clue to the **physical** significance of equation (6).

The relevant observational data are shown in Figure 1, where the data points shown have been either collated from or calculated from data available in the identified references. It is apparent that all data for tracer samples with a Galactic rotation velocity greater than about 50 km·sec^{-1} are consistent with a single density distribution, with the marginally significant exception of the metal-rich globular cluster system, whose radial velocity dispersion is rather low. This datum is however somewhat more uncertain than most of the other data shown, due to distance and reddening uncertainties. Observational selection effects can have a very substantial effect on the appearance of the diagram, and have not been considered at all adequately. Obvious examples include explaining the apparent systematic difference in the

Figure 1. The relation between rotation velocity relative to the local standard of rest and the radial velocity dispersion for recently studied tracer samples. The model lines are for different solutions of equation (6). The data and models are described more fully in Gilmore et.al. (1989). The tendency for the data to cross the model lines at low V_{rot} indicates that the oldest stars in the Galaxy formed during a period of dissipational collapse.

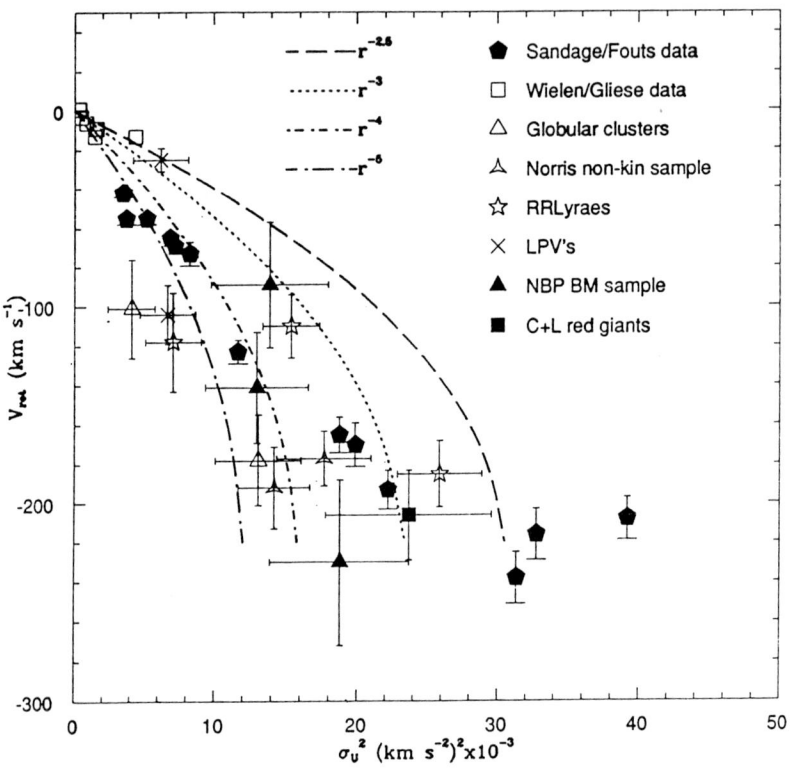

deduced radial velocity dispersion between spectroscopically- and
kinematically-selected samples for the highest velocity stars, which are
far from the regions of phase space expected to be affected strongly by
the selection criteria, and allowing for the fact that stars on high
angular momentum high energy orbits will always lie beyond the solar
circle, and so will not be sampled in local surveys.

The mean density law consistent with the majority of the data with
significant angular momentum corresponds to an isothermal spheroidal
distribution with a power law with index ~ -4.5, or an exponential disk
with scale length 3.1 kpc. The tendency for the tracer populations with
the lowest mean rotational velocities to have larger radial velocity
dispersions than consistent with this density profile is of considerable
significance, if real. (We note that systematic distance uncertainties
move the data roughly parallel to the body of the data with smaller σ_{rr},
so are unlikely to be relevant. With the exception of the data for the
metal-poor RR Lyrae stars, however, there are very few stars in the bins
with the highest values of σ_{rr}.) The stars with the highest radial
velocity dispersions are also the most metal-poor (see below) and hence
those which presumably formed first in the Galaxy. If they really do
form a more extended spatial distribution than more metal-rich stars,
then one may conclude that these stars formed earlier in the collapse of
the proto-galaxy than more metal-rich stars, from less dissipated gas,
and hence preserve a fossil record of the star-formation and dissipation
history of the protogalaxy during the first condensation of the Galaxy
from the expanding background. The validity of this conclusion rests
almost entirely on the data for low metallicity RR Lyrae stars at
present. Confidence in the distance scale and kinematics for these
stars is clearly of considerable importance.

Although the asymmetric drift arguments above provide strong evidence
that star formation continued **during** a period of dissipational collapse,
there is no information in this relation regarding the **rate** of this
collapse. For this one requires another clock.

2.3 Correlations between kinematics and chemistry

Stellar chemical abundance is a clock which measures age in
units defined by the lifetimes of massive stars. Stellar orbital energy
in the Galaxy is a measure of the amount by which the proto-Galactic gas
had collapsed out of the background Hubble expansion and cooled, prior
to star formation. Thus the existence of a correlation between stellar
kinematics and [Fe/H] would allow one to relate the chemical
evolutionary timescale to the dynamical timescale.

The existence or otherwise of an abundance gradient in the spheroid is
often cited as an important diagnostic of the timescale of galaxy
formation. In general this is **not** correct. The presence of an
abundance gradient means that dissipation was an important process
during formation of the stellar component of the spheroid, but does not
necessarily define timescales. Imagine a cloud of gas orbiting in the
proto-Galaxy. Star formation in this cloud is presumed to increase the

chemical abundance with time, so that stars formed at later times are increasingly more metal rich than stars formed earlier. If the total orbital energy of the cloud is unchanging, all stars formed will have the same orbit as the cloud, there will be no correlation of abundance with orbital parameters. Only if the enriched gas is continually transferred onto lower energy orbits will an abundance gradient exist. In modern terminology, dissipationless collapse does not create abundance gradients; dissipation is essential.

The important parameter which must be added to this argument to determine timescales for the collapse is the timescale for loss by cooling of the dissipated energy. Idealised models of protogalaxies suggest this cooling timescale is less than a dynamical collapse time (see Gilmore et al. 1989 for a review). Thus dissipation will not necessarily slow a collapse significantly beyond a dynamical timescale. Thus, the existence of an abundance gradient determines whether or not dissipation was significant during star formation and collapse, but does not constrain the rate of the collapse.

The existence of a correlation between chemical abundance and kinematics in stars at present however requires not only that such a correlation was set up during the relevant epoch of star formation, but also that the stellar orbital energy has not been totally rearranged since star formation. That is, one expects a tight correlation between rotation velocity and stellar chemical abundance only if the star formation occurred during a dissipational collapse, and also if there has not been an efficient violent relaxation of the Galaxy since the metal-poor stellar tracer population formed. The fact that the velocity ellipsoid of metal-poor stars near the sun is anisotropic clearly shows that such violent relaxation has not been completely efficient (if it happened at all), so that at least some memory of conditions in the proto-Galaxy remains. Nevertheless, the existence or otherwise of a tight correlation between asymmetric drift and chemical abundance is not the clean test of the relative timescales of star formation and dissipation which it is often assumed to be.

2.4 Correlations between abundances and ages

Calibration of the abundance enrichment rate onto a time scale which is calibrated independently of the collapse rate, i.e. in years, is necessary to provide direct evidence for the timescale of Galactic formation. In practise only stars near the main-sequence turnoff have surface gravities which change sufficiently rapidly and monotonically that reliable comparison with evolutionary tracks is possible, although some useful information on a combination of age and chemical abundance can be derived from the colour of field giant stars (e.g. Sandage 1987). For **single** stars near the turnoff the comparison of uvbyβ photometry with theoretical isochrones is by far the most reliable and precise age-dating technique available.

If independent abundance estimates are available, then any photometric measure of the temperature of the hottest turn-off stars will measure the age of the **youngest** star in a tracer population. It is this method which is utilised to determine ages for globular clusters, where it also seems that all the member stars are coeval. A similar technique can be applied to field stars (cf. e.g. Gilmore & Wyse 1987), and is illustrated in Figure 2. The important conclusion from Figure 2 is that essentially all stars with [Fe/H] ≲ -0.8 are, insofar as is measurable, the same age as the globular cluster system. More metal rich stars have a bluer turnoff, implying that **at least some** of these stars are younger. The **distribution** of ages is however unmeasurable from a turnoff colour. Some information on the age distribution for stars with [Fe/H] ≳ -0.8 is provided by studies of open clusters. These form a system with a very large scatter in the age-metallicity plane; clusters exist near the sun with solar abundance and an age of 12 Gyr (NGC 6791, Janes in

Figure 2. B-V vs [Fe/H] for stars observed by Laird et al. (1988; points) and turnoff colours for globular clusters with good CCD data. The solid line is a 15 Gyr oxygen enhanced isochrone scaled from those of Vandenberg & Bell (1985) to match 47 Tuc.

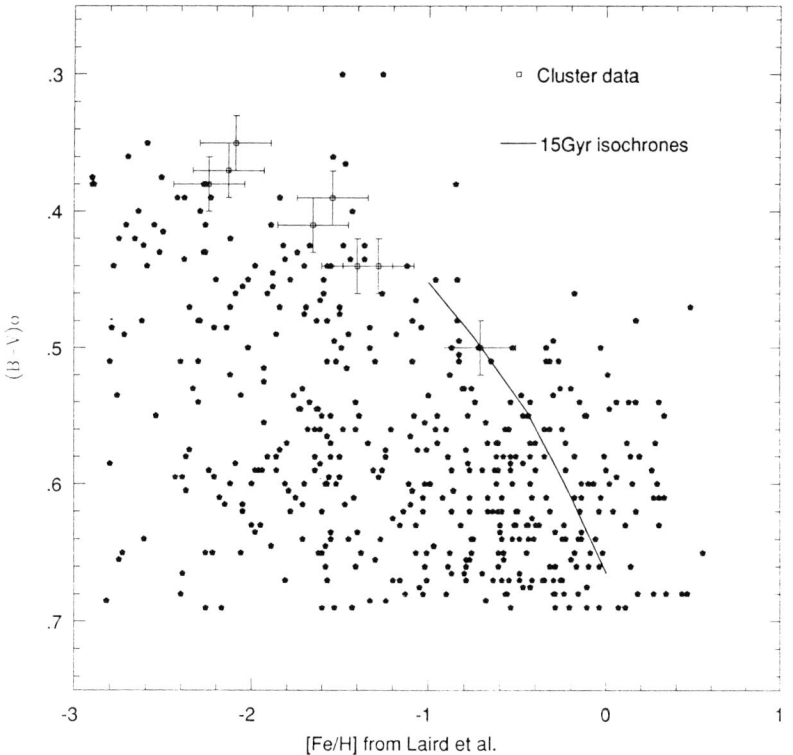

preparation) and with [Fe/H] ~ -0.5 but an age of only a few Gyr
(e.g. Melotte 66). Thus any attempt to deduce an age for a stellar
population from the turnoff colour of the **bluest** field stars with
metallicity ≳ -0.8 dex (Norris & Green 1989) is fundamentally
unreliable.

2.5 The timescales of galactic chemical evolution

In attempting to deduce the rate of star formation and
dynamical evolution in a proto-galaxy, it is desirable to have available
a clock whose rate can be calibrated, and which runs sufficiently fast
to resolve the dynamical evolutionary timescales. Such a clock is
provided by stellar evolution of high-mass stars, while the fossil
record of the clock is observable in the chemical abundance enrichment
patterns in long-lived low-mass stars. Fortunately, there exists a
subset of common elements (most importantly oxygen) whose creation sites
are restricted to very massive stars, and another subset (most
importantly iron) which is also created in lower mass stars. Since the
evolutionary timescales for high- and low-mass stars span the timescale
range of interest in galaxy formation, the differential enrichment of
oxygen and iron provides an ideal clock to calibrate the rate of star
formation in the proto-Galaxy.

Oxygen to iron element ratios have been now been measured for a
sufficient number of stars to define the systematic trends in the data.
The important result for present purposes is that a significant change
of slope occurs in the relationship between the element ratio [O/Fe] and
[Fe/H] close to a metallicity where there also occurs a change in the
stellar kinematics, that is at [Fe/H] ~ -1. The [O/Fe] ratio is
observed to be approximately constant, independent of [Fe/H] for the
most metal-poor stars, -2.5 ≲ [Fe/H] ≲ -1, while [O/Fe] declines for the
more metal-rich stars, [O/Fe] ~ -½ [Fe/H]. Present data are summarised
in Figure 3, in which all scatter is considered by the relevant
observers to be consistent with observational error (Sneden et
al. 1989).

Assuming that [Fe/H] is a monotonically increasing function of time,
this behavior can be explained if the oxygen and iron in the more metal
poor stars have been produced in stars of the same lifetime, while for
the more metal-rich stars, although the oxygen and iron continue to be
produced together, an additional, longer timescale source now dominates
the iron production. Such behaviour is in good agreement with supernova
nucleosynthesis calculations, which show that oxygen is produced only in
Type II supernovae by massive stars ($M \gtrsim 20\ M_\odot$), while iron has a
contribution from both massive and low mass stars ($M \gtrsim 3\ M_\odot$, Type I
supernovae) thereby having an enhanced production once the much more
numerous, lower mass stars contribute to its nucleosynthetic yield
(Tinsley 1979; Matteucci & Greggio 1986; Wyse & Gilmore 1988). This
results in the ratio [O/Fe] decreasing systematically with increasing
metallicity [Fe/H].

If the arguments above contained the whole story over the history of star formation in the Galaxy, then it would have to be mere coincidence that the change in the predominant production mechanism of iron occurred close to a metallicity, or epoch, at which the stellar kinematics change from those of a pressure-supported system, which formed stars rapidly, to those of an angular momentum-supported system. Rather, the coincidence of the value of [Fe/H] at which the Galaxy changed from a pressure-supported system to an angular momentum-supported system, with the value of [Fe/H] at which the interstellar medium became diluted by the products of long-lived stars, provides a diagnostic of the relative star-formation and dissipation rates in the proto-Galaxy.

When sufficient data and reliable massive-star evolutionary models all the way through the supernova explosion, with corresponding elemental yields, become available it will be possible to quantify these arguments (subject to the assumption of a constant stellar IMF) and provide a real timescale (in years) for the periods of proto-galactic evolution which were dominated by collapse (possibly non-dissipational) on a dynamical timescale, and that period when angular momentum transport (in dissipational collapse) became an important physical process, and angular momentum support became the dominant dynamical process.

Figure 3. A compilation of oxygen to iron element ratio measurements from the literature. This figure is adapted from Wyse and Gilmore (1988). The smooth curve through the data for [Fe/H] $\gtrsim$ -1 shows the prediction of a simple model with constant supernova rates in the ratio 1.5:1.0 for Type I:Type II, resulting in twice as much oxygen as iron being produced per unit time.

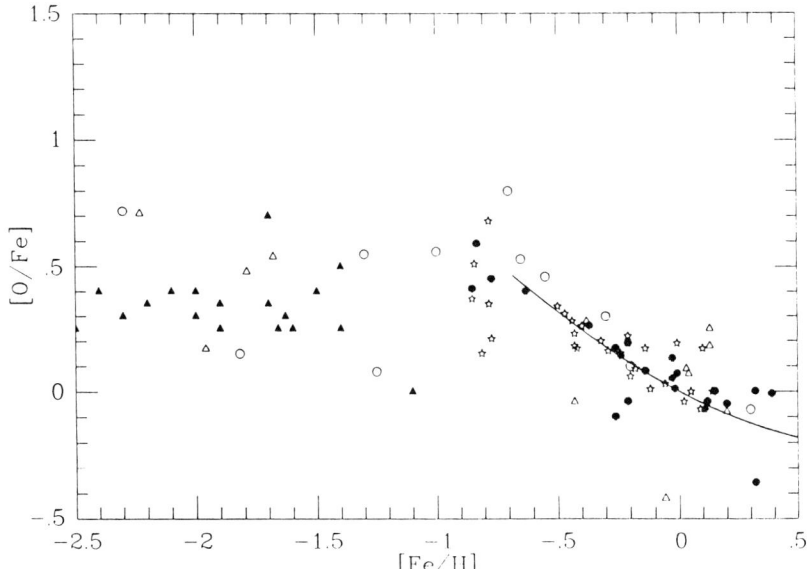

One conclusion which is relatively independent of the details of
elemental synthesis follows from the fact that features in both the
stellar abundance and kinematics occur more or less together, at
[Fe/H] $\lesssim$ -1. As the elemental yields are a fairly slow function of
progenitor mass, and hence lifetime, one does not expect discontinuities
in element ratios to occur in a situation where the star formation
proceeds at a reasonably constant rate. A discontinuity in kinematic
properties implies that the ratio of the dissipation rate to the star
formation rate also changes rapidly. A possible explanation is that at
metallicities [Fe/H] $\gtrsim$ -1.5 the efficiency with which a gas cloud cools
from $\sim 10^6$ K (a typical galactic virial temperature) increases markedly,
due to a transition of the dominant cooling mechanism from free-free
radiation, independent of metallicity, to line radiation, proportional
to the number density of metals. Thus a rapid increase in the
dissipation rate and collapse to a disk-like angular-momentum supported
structure is not implausible at a metallicity of $\sim$ -1 dex. It is not
crucial for these arguments that the breaks in kinematics and element
ratios occur at **exactly** the same metallicity.

In terms of models of the early chemical evolution of the Galaxy, one
must explain the fact that [O/Fe] is approximately constant, at three
times the solar value, for [Fe/H] $\lesssim$ -1, while decreasing smoothly for
[Fe/H] greater than this value, together with the fact that the mass of
stars with [Fe/H] $\lesssim$ -1 is only a few percent of the total stellar mass
of the Galaxy, as discussed in detail in §4 below. Clearly, the
approximate constancy of the [O/Fe] ratio independently of the [Fe/H]
ratio at low metallicities requires that essentially **all** the stars with
[Fe/H] $\lesssim$ -1 formed on a timescale less than that on which a significant
number of low mass (Type I) supernovae exploded. This timescale is
rather difficult to estimate precisely, due to uncertainties in the
mechanism of Type I supernovae and the fraction of all stars formed
which are in binaries of the type that may be expected to be precursors
(cf. Iben 1986); the lowest mass, and hence most numerous, progenitors
of CO white dwarfs have main-sequence masses and lifetimes of $\sim 5 M_\odot$ and
$\sim 2.5 \times 10^8$ yr respectively. Thus a reasonable estimate for the
characteristic time after which one expects dominance of iron from Type
I supernovae is $\lesssim 10^9$ yr (but bearing in mind that some Type I systems
may take a Hubble time to evolve). This general argument appears to be
the strongest direct evidence for a rapid formation timescale for the
Extreme Population II stars in the Galaxy.

3 THE SPATIAL STRUCTURE OF THE MILKY WAY GALAXY

Pulsating variable stars have traditionally been the most
commonly applied and most reliable tracer of the spatial structure of
the Milky Way Galaxy. One example of this type of analysis which is of
considerable current interest involves comparison of the shape of the
extreme Population II stellar system with the shape expected from
dynamical analysis of the kinematics of local metal-poor stars. The
shape of the non-thin disk stars is important for its implications for
the early stages of galaxy collapse and star formation, the
interpretation of the kinematics of high-velocity stars, and the shape

of the underlying dark matter that generates the gravitational potential in which these stars move.

3.1 The shape of the metal-poor spheroid

The high-velocity, metal-poor field stars in the solar neighborhood have an anisotropic velocity-dispersion tensor, with $\sigma_r^2 : \sigma_\theta^2 : \sigma_\phi^2 \sim 2 : 1 : 1$. Since the velocity-dispersion tensor behaves as an anisotropic stress tensor in the equations governing stellar dynamics, one may expect this anisotropic 'pressure' to result in an anisotropic shape, i.e. a flattened metal-poor spheroid (see §2 above for the relevant equations). Binney & May (1986) investigated this idea in more detail, and concluded that in the locally non-spherical potential felt by the subdwarfs, due to the presence of the disk, the observed velocity dispersion anisotropy implies a substantially flattened spheroid, with shape $\sim$ E7 or axis ratio $\sim$1:4.

The kinematic data of Ratnatunga & Freeman (1985; 1988) for distant metal-poor K giants can also most easily be explained by allowing these stars to form a flattened distribution. The most important feature of the data is the fact that the line-of-sight velocity dispersion in the SGP does not increase with distance despite the increasing contribution of the radial (relative to the Galactic centre) component of the velocity dispersion (σ_r) to the observed stellar radial (relative to the sun) speed. The assumption behind the expectation of a rising line-of-sight dispersion with distance is that the distant metal-poor K giants trace the same population as the local metal-poor K giants, the high ΔS RR Lyraes, and the local subdwarfs, and hence should have the same radially-biased velocity-dispersion tensor. To model this, one can depress the observed velocity dispersion at large distances by allowing suitable discontinuities in the stellar distribution function; essentially we are assuming that all stars beyond a given galactocentric radius are on circular orbits. This approach allows a fit which retains a spherical spatial structure for these stars (Sommer-Larsen 1987; Sommer-Larsen & Christensen 1989; Dejonghe & de Zeeuw 1988). Alternatively, adopting a global form of the distribution function in either a spherical (White 1985, 1989) or in an oblate (Levison and Richstone 1986) potential requires a flattened spatial distribution for the spheroid stars, again with axis ratio $\sim$1:4.

In the light of this kinematic and dynamical evidence, it is mildly puzzling that direct star-count studies suggest the subdwarf stellar system is approximately round (Freeman 1987). The most-quoted evidence for a spherical distribution of field spheroid stars comes from the modelling by Bahcall & Soneira (1980, 1984) (hereafter BS) of the faint star counts of Koo & Kron (1982) in two fields; BS conclude that the axis ratio of the spheroid stars is $c/a = 0.80^{+0.20}_{-0.05}$. Their technique is based on the fact that fields in the $l = 90°, 270°$ plane are at equal galactocentric distances if at equal distances from the solar neighborhood (i.e. us) and hence a spherical distribution of stars will contribute equally to all fields in this plane. Thus if one compares magnitude-limited samples in fields at high and low Galactic latitude one should obtain equal numbers of spheroid stars in the two fields. A flattened distribution of stars will yield lower counts in the higher-latitude field.

BS complicate their analysis somewhat by adopting different color-magnitude relations for the two fields, and thus they do not predict equal numbers of stars for a spherical distribution; they are forced to do this to obtain an acceptable fit for their model in each of the two fields, due to a combination of inadequacies in the model, such as lack of the thick disk component, and in the data, discussed below. There is no physical basis for such a variation of color-magnitude relation (metallicity gradients are not relevant since the fields are supposed to be at the same galactocentric distance) and it is a potential source of uncertainty. Adoption of a metal-poor color-magnitude relation has the effect of assigning a low intrinsic luminosity to stars of a given color. Hence, in an apparent-magnitude limited sample one will be comparing lower luminosity, less distant stars in the 'metal-poor' field with higher luminosity, more distant stars in the other field. The predictions of relative star-counts are therefore sensitive to the shape of the subdwarf luminosity function as well as to the shape and density profile of the stellar tracer population, so that it is possible to produce predictions for the ratio of counts that can exceed unity in a spherical distribution, as BS derived.

A new study of this problem, utilising a larger dataset and using a more general model-Galaxy program which requires internally consistent properties for a given stellar population in different fields, and which allows the inclusion of a thick disk is described by Wyse & Gilmore (1989). Following BS they counted stars blueward of a colour limit (B-V = 0.6) chosen to minimise contamination of the tracer sample by nearby old disk stars, with the precise value of this limit not being critical. The two fields used were ($l = 0°$, $b = 90°$; area surveyed = 0.75 square degrees) and ($l = 272°, b = -44°$; area surveyed = 0.75 square degrees). The observed ratio of blue stars in the two fields was 0.59. This disagrees strongly with Koo & Kron's counts in two fields at similar Galactic latitudes but at much fainter magnitudes, which yield a ratio of 1.09 for blue stars with $20 \lesssim V \lesssim 22$; the more recent calibration of the same data by Koo, Kron & Cudworth (1986) gives 1.3 (note that these numbers are based on a somewhat uncertain color cut, but this should not matter provided one is blue enough to have isolated the metal-poor spheroid stars). We suspect that this apparent disagreement reflects the uncertainty in the Koo and Kron counts, due to the difficulty of reliable star-galaxy discrimination at faint magnitudes. This suspicion is based on the results of the Koo, Kron & Cudworth (1986) 'subdwarf' category counts for their north Galactic pole field, together with predictions from the WG model and from the BS model, each model with an assumed spheroid axis ratio of 0.8. There is an obvious disagreement between the data and the models; the data fail to increase towards fainter magnitudes, contrary to both of the models, and contrary to intuition.

The BS model predictions (their Table 3) combined with our low value of the relative observed counts would imply that the spheroid had an axis ratio $c/a \leqslant 0.5$. When one considers the presence of the thick disk, and also models the observed **total** counts as well as their ratio, the best estimate for the axis ratio of the metal-poor subdwarf stellar population within a few kpc of the sun is $c/a \sim 0.6$.

One can also utilise direct counts of other spheroid tracers, such as RR Lyrae stars, to derive the density profile of spheroid light. Early work based on RR Lyrae stars in the Palomar-Groningen and Lick surveys, which were towards the Galactic center (Kinman, Wirtanen & Janes 1966; Oort & Plaut 1975) concluded that these stars were distributed in a nearly spherical system. These results have now been superseded by better photometric data (Wesselink et al. 1987); the more modern analysis finds in contrast that the RR Lyrae stars towards the Galactic center have a rather flattened distribution, with axis ratio $\leqslant 0.6$, in excellent agreement with the star-count result above. A possible complication in this picture is due to the work of Hartwick (1987), who suggests that the RR Lyraes form a two-component system, with the more metal-rich stars being part of the thick disk. However, Hartwick analysed the available data for **metal-poor** RR Lyrae stars separately, and concluded that the axis ratio of the RR Lyrae system varies with galactocentric radius, being flattened (axis ratio $c/a \sim 0.6$ and scale height ~ 1.5 kpc, i.e. rather similar to the parameters of the more metal-rich thick disk RR Lyraes) interior to the solar circle. At very large Galactocentric distances the RR Lyrae data somewhat favour a more spherical distribution.

Hartwick finds a similar two-component structure for the metal-poor ([Fe/H] < -1) globular clusters from the distribution of their projected positions on the sky; this should not be confused with the two well-established distinct components in the globular cluster system (Zinn 1985; Thomas 1988) one metal-rich and one metal-poor. However, the small number of clusters involved gives this small statistical weight. The kinematics of the metal-poor globular cluster system was found by Frenk & White (1980) to consist of negligible net rotation and isotropic velocity-dispersion tensor. These properties would suggest a spherical spatial distribution, given a spherical potential, and a flattened distribution given a flattened potential. However, Norris (1986) found there to be no statistically significant difference between the 'isotropic' velocity-dispersion tensor of the globular clusters and the markedly **anisotropic** velocity dispersions of the local subdwarfs, while Thomas (1988) has shown that one cannot in general draw **any** strong conclusions about the kinematics of the globular cluster system, due to the effects of distance errors. While it is interesting that the kinematic parameters derived for the globular clusters are so similar to those of the field stars, there is no compelling theoretical or observational evidence that the globular cluster system is intimately related to the field star system, so such similarities should not be over-interpreted.

In summary, the available evidence on the shape of the metal-poor field stars which make up the Galactic spheroid suggests that these stars form a rather non-spherical system, whose flattening may vary with radius, but is c/a ~ 0.5 within a few kpc of the sun, and within a few kpc of the Galactic centre.

4 THE THICK DISK

Confirmation that the Intermediate Population II stellar system defined at the Vatican Conference is indeed characterised by a vertical scale height of ~1 to 1.5 kpc, a vertical velocity dispersion of ~45 km·sec^{-1}, a typical stellar chemical abundance of ~-0.75 dex, and a mean asymmetric drift of ~30 to 50 km·sec^{-1} has been provided by a very large number of photometric and spectroscopic surveys in the last few years (cf. Gilmore & Reid 1983, Freeman 1987). The detailed values of the descriptive parameters remain poorly determined however, primarily because the offset in the mean values characterising the thick disk distribution function over age, metallicity, and kinematics from those characterising the oldest thin disk stars is much less than the dispersions in these quantities. Reliable determination of the parameters of the distribution function is important since it may allow a discrimination between the several currently viable models of the formation of the thick disk.

Possible formation mechanisms for the thick disk include:
1) A slow pressure-supported collapse phase following formation of the extreme Population II system;
2) Violent dynamical heating of the early thin disk by, for example, satellite accretion or violent relaxion of the Galactic potential;
3) Accretion of the thick disk material directly;
4) A period of enhanced kinematic diffusion of stars formed in the thin disk to high energy orbits. This might be due perhaps to a transient bar, or to a large population of high velocity black holes in the Galactic halo.

Discrimination amongst these several types of model is possible from appropriate age-abundance data. The first type of model noted above will lead to an old system, with an abundance gradient. The second will have a small internal age range, but is unlikely to have an extant abundance gradient (modulo the details of the merger process). The third has a wide variety of allowed combinations of age and abundance, while the fourth will have a range of ages, but a kinematic discontinuity between the old disk and the thick disk. In view of this possibility to determine the evolutionary history of the thick disk, an extensive debate is underway to determine a reliable description of the kinematic, abundance and age structure of the thick disk. This is well reviewed in Sandage (1987) and Norris (1987).

Here we summarise the data on the determination of the age range of thick disk stars, and discuss the difficult question of the kinematic relationship between the thick disk and the old disk near the sun.

4.1 The age of the thick disk near the sun

An age determination for samples of thick disk stars is an extremely difficult observational problem. In part this is due to the usual difficulty in assigning a reliable age to anything in astronomy, but in this case the situation is complicated by the point noted above that there is no obvious a priori way to define a sample of purely 'thick disk' stars. Any sample selected by abundance, kinematics, or chemistry will inevitably include old disk and/or extreme population II stars in addition to the thick disk. Thus determination of the age of the **youngest** or the **oldest** star in a sample, while tractable, is not an obviously clever way to answer the question of interest. Some information may be derived from Figure 2, in §2.5 above. It is evident from this figure that the age of the oldest stars with [Fe/H] $\lesssim$ -0.8 is comparable with that of the globular cluster system. This abundance range probably is dominated by thick disk stars for -1 $\lesssim$ [Fe/H] $\lesssim$ -0.8, suggesting that the most metal-poor thick disk stars are among the oldest in the Galaxy. More metal-rich thick disk stars may or may not be the same age. It is impossible to tell from comparison with diagrams like Figure 2, as some old disk stars will contaminate the sample. This point is worth emphasising, as it removes the rigorous basis for the conclusion of Norris & Green (1989) that the thick disk is several Gyr younger than the subdwarfs near the sun. Their conclusion **may** be correct, but it cannot be derived reliably from photometric data alone. This point is discussed in more detail by Sandage (1989).

However, the horizontal branch, as studied by Norris & Green (1989), does provide important information on the age of the thick disk stars. The observational evidence that metal-rich RR Lyrae stars form a thick disk system (e.g. Hartwick 1987) is an important clue. The **youngest** dated population of RR Lyrae stars is that in NGC 121, which has an age of ~12 Gyr. Thus, if the small ΔS RR Lyrae stars do indeed belong to the thick disk, 12 Gyr is a lower limit on the age of at least some of the thick disk. Further studies of these stars is of considerable interest.

Similarly, while the relationship of the globular clusters to field stars is very non-obvious, **if** the disk globular cluster system studied so well by Zinn and collaborators (cf. Armandroff 1989 for the most recent analysis) is indeed part of the thick disk, then the antiquity of the thick disk is reliably established. All these arguments however leave open the possibility of a large age **range** in the thick disk. There is no reliable information yet available on this point.

4.2 Is the thick disk kinematically discrete?

The velocity ellipsoid for the extreme population II subdwarf system is reliably determined to be $\sigma_{rr}:\sigma_{\phi\phi}:\sigma_{zz} = 128^2:96^2:93^2$ (Carney & Latham 1986). The vertical velocity dispersion of the thick disk is ~45 km·sec^{-1} (cf. Figure 12a of Gilmore & Wyse 1987). The number of stars with abundances and kinematics such that they might plausibly be assigned to either the low velocity tail of the extreme population II or to the high velocity tail of the thick disk (i.e. those stars with [Fe/H] ~ -1) is very small (cf. Carney, Latham, & Laird 1989; Paper VIII in their current series). Thus the thick disk is apparently kinematically discrete from the subdwarf system to an adequate approximation. This means simply that the rate of dissipation in the vertical direction was relatively high, compared to the star formation rate, as the proto-disk collapsed.

The relationship of the thick disk to the high velocity tail of the old disk is more problematic, and has been discussed extensively by Sandage (1987) and Norris (1987). To oversimplify the point at issue, is there a continuous relationship of vertical velocity dispersion with metallicity extending all the way to the ~45 km·sec^{-1} vertical velocity dispersion of the thick disk (Norris 1987, Figure 7), or does the old

Figure 4. The distribution of vertical (W) velocities of stars in the Gliese catalogue, excluding stars with $\delta_{0.6} > 0.15$. The two lines illustrate models with a kinematically discrete thick disk (dashed lines) and with a continuous kinematic relationship between the old disk and the thick disk (solid line). Distinction between these models on the basis of these data is clearly not possible.

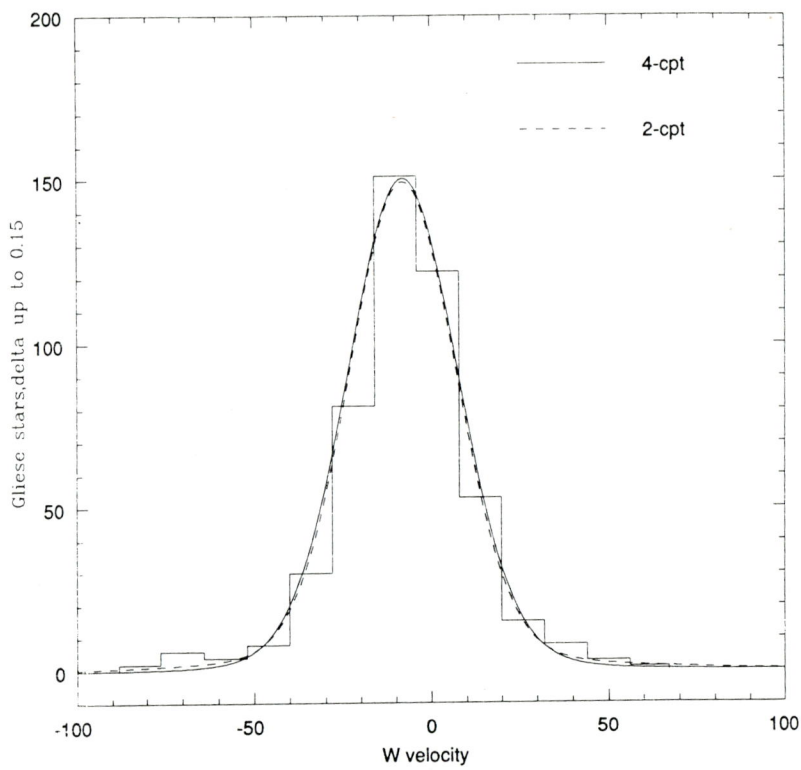

disk velocity dispersion become asymptotically constant at the value of ~ 22 km·sec^{-1} appropriate for spectroscopically selected samples of old dwarfs near the sun (Fuchs & Wielen 1987; Sandage 1987)? The difficulty in deciding this question is illustrated in Figure 4. This shows the vertical velocity distribution of those stars in the Gliese catalogue with photometric abundance parameter $\delta_{0.6} < 0.15$ (this abundance range was chosen so as to exclude high velocity subdwarfs). The two models overlaying the data histogram are a two-component model, with discrete old disk and thick disk, and the four-component approximation to a continuous relation between the old disk and the thick disk fitted by Norris. The two models are clearly both an excellent description of the data, and are equally clearly indistinguishable.

It is evident from Figure 4 that available **local** data are incapable of determining if the old disk and the thick disk are kinematically discrete. Resolution of this uncertainty, with its important implications for the formation history of the Galaxy, must await completion of the several extant **in situ** surveys of the stellar distribution several kpc from the Galactic plane. The available data marginally favour a model in which the thick disk is a kinematically discrete component of the Galaxy, but the issue remains to be decided by observational test.

ACKNOWLEDGEMENT

RFGW acknowledges partial support from the NSF, through grant AST-88-07799.

REFERENCES

Armandroff, T. (1989). A.J., in press.
Bahcall, J.N. & Soneira, R.M. (1980). Ap.J.Suppl.,44, 73.
Bahcall, J.N. & Soneira, R.M. (1984). Ap.J.Suppl.,55, 67.
Binney, J. & May, A. (1986). M.N.R.A.S.,218, 743.
Carney, B. & Latham, D.W. (1986). A.J.,92, 60.
Dejonghe, H. & de Zeeuw, P.T. (1988). Ap.J.,329, 720.
Feast, M. (1987). In The Galaxy, eds. G. Gilmore & B. Carswell, p. 1. Dordrecht: Reidel.
Freeman, K.C. (1987). Ann. Rev. Astron. Astrophys.,25, 603.
Fuchs, B. & Wielen, R. (1987). In The Galaxy, eds. G. Gilmore & B. Carswell, p. 375. Dordrecht: Reidel.
Gilmore, G. & Reid, I.N., (1983). M.N.R.A.S.,202, 1025.
Gilmore, G. & Wyse, R.F.G. (1986). A.J.,91, 855.
Gilmore, G. & Wyse, R.F.G. (1987). In The Galaxy, eds. G. Gilmore & B. Carswell, p. 247. Dordrecht: Reidel.
Gilmore, G., Wyse, R.F.G., & Kuijken, K. (1989). Ann. Rev. Astr. Astrophys.,27, in press.
Hartwick, F.D.A. (1987). In The Galaxy, eds. G. Gilmore & B. Carswell, p. 281. Dordrecht: Reidel.
Frenk, C.S. & White, S.D.M. (1980). M.N.R.A.S.,193, 295.

Iben, I. (1986). In Cosmogonical Processes, eds W.D. Arnett, C.J. Hansen, J.W. Truran & S. Tsuruta, p. 155. Utrecht: VNU Science Press.
Kinman, T.D., Wirtanen, C.A. & Janes, K.A. (1966). Ap.J.Suppl.,13, 379.
Koo, D.C. & Kron R.G. (1982). Astron. & Astrophys.,105, 107.
Koo, D.C., Kron, R.G. & Cudworth, K. (1986). P.A.S.P.,98, 285.
Kuijken, K. & Gilmore, G. (1989). M.N.R.A.S., in press.
Laird, J.B., Carney. B.W., & Latham, D.W. (1988). A.J.,95, 1843.
Levison, H.F. & Richstone, D.O. (1986). Ap.J.,308, 627.
Matteucci, F. & Greggio, L. (1986). Astron. & Astrophys.,154, 279.
Norris, J. (1986). Ap.J.Suppl.,61, 667.
Norris, J. (1987). In The Galaxy, eds. G. Gilmore & B. Carswell, p. 297. Dordrecht: Reidel.
Norris, J., & Green, E.M. (1988). Preprint.
Oort, J.H. & Plaut, L. (1975). Astron. & Astrophys.,41, 71.
O'Connell, D.J.K. (1965). Stellar Populations. Amsterdam: North Holland.
Ratnatunga, K.U. & Freeman, K.C. (1985). Ap.J.,291, 260.
Ratnatunga, K.U. & Freeman, K.C. (1988). Ap.J., in press.
Sandage, A. (1987). In The Galaxy, eds. G. Gilmore & B. Carswell, p. 321. Dordrecht: Reidel.
Sandage, A. (1989). In The Calibration of Stellar Ages, in press. Middletown, CN: Van Vleck Observatory.
Sneden, C., Wheeler, J.C. & Truran, J. (1989). Ann. Rev. Astr. Astrophys.,27, in press.
Sommer-Larsen, J. (1987). M.N.R.A.S.,227, 21P.
Sommer-Larsen, J. & Christensen, P.R. (1989). Preprint.
Thomas, P. (1988). Preprint.
Tinsley, B.M. (1979). Ap.J.,229, 1046.
Vandenbergh, D., & Bell, R.A. (1985). Ap.J.Suppl.,58, 711.
Wesselink, T.H., Le Poole, R.S. & Lub J. (1987). In Stellar Evolution and Dynamics in the Outer Halo of the Galaxy, eds. M. Azzopardi & F. Matteucci, p. 185. Garching: ESO.
White, S.D.M. (1985). Ap.J.,294, L99.
White, S.D.M (1989). Preprint.
Wielen, R. (1974). In Highlights of Astronomy, Vol. 3, ed. G. Contopoulos, p 395. Dordrecht: Reidel.
Wyse, R.F.G. & Gilmore,G. (1988). A.J.,95, 1404.
Zinn, R. (1985). Ap.J.,293, 424.

HORIZONTAL BRANCH EVOLUTION

R.T. Rood
University of Virginia, Charlottesville, VA

D.A. Crocker
University of North Carolina, Chapel Hill, NC

Abstract. In 1973 the outstanding problems confronting the theory of horizontal branch evolution were the "second parameter" problem and the Oosterhoff Effect. Despite significant progress, particularly in the observations and in the observation/theory interface, they remain as the outstanding problems of 1988. The Oosterhoff Effect is now discussed primarily in the guise of the Sandage Period Shift Effect. The morphology of the HB seems more complicated than ever. E.g., many clusters show bimodal distributions along the HB. Here we will tentatively consider those to be manifestations of the second parameter problem. We will indicate why we feel that all previously suggested solutions have all been chimeras.

1 A RETROSPECTIVE—RTR

By the early 1970's the basic theory of horizontal branch (HB) stars seemed on firm ground. Faulkner and Iben had convincingly identified the HB with the core helium burning phase of globular cluster stars. The solution to the "first parameter" problem—the HB gets redder as metalicity increases—followed naturally from this identification and thus never became a problem. The major oversight in the first models, the presence of first overshooting and then semiconvection, at the boundary of the convective core had been demonstrated by Castellani, Giannone, and Renzini. Models incorporating this result had been constructed by, e.g., Demarque, Mengel, and Sweigart. The HB had been shown not to be an evolutionary sequence, but most probably a sequence of stars with equal core masses and slightly different total masses due to some differential mass loss process on the red giant branch. I was gratified when Monte Carlo simulations constructed along these lines bore a striking similarity to observed HBs.

At that time there seemed to two remaining problems. The "second parameter" problem—at a given metalicity some clusters have HBs that are too red or too blue—was particularly intriguing. If, for example, the second parameter could be identified with age, one would have direct observational evidence on the collapse time of the galaxy to a precision of 1 Gyr or perhaps even better. The other was the Oosterhoff effect—on the basis of the properties of their RR Lyrae stars the clusters fell into two distinct groups. For Oo I clusters about 75% of the RR Ly were pulsating in the fundamental mode (Type ab) with mean period 0.55 day. For the Oo II clusters the corresponding numbers were 50% and 0.65 day. The trouble was that the second parameter problem was too easy to solve—variations in age, helium abundance, CNO abundances all worked, with little but aesthetics to chose among them. On the other hand nothing worked for the Oosterhoff effect. (I had started the investigation convinced that solution of the Oosterhoff problem would identify *the true* second parameter.)

One could go no further at this time. It was obvious that mass loss was extremely important, but there was no mechanism for producing the mass loss, much less knowing how it scaled with parameters like the metal abundance. The abundances themselves were poorly known. The conversions from theoretically determined $\log L$, $\log T_{\text{eff}}$ to observables were suspect. The photometric samples were too small and often incomplete. There was uncertainty whether the evolutionary history of an HB star could affect whether it pulsated in the first harmonic or fundamental mode (hysteresis). I punted.

Surprisingly, by the end of the 1970's my wish list of 1973 was complete. For mass loss, we had both the model of Fusi Pecci and Renzini (now unfortunately ruled out by X-ray observations) and the empirical Reimer's formula. At the least, these provided a whipping boy against which to test hypotheses. The atmospheres were in much better shape due to the work of Bell, Gustafsson, Kurucz, and others. The work of Stellingwerf provided a concrete way of dealing with the mode of the variables. Many observers (bless them) had increased both the quantity and quality of the photometry and spectroscopy. The first suggestions of variations in [CNO/Fe] and in star-to-star abundance variations were appearing. Photometry extending below the main sequence turnoff point was becoming commonplace. I approached the problem with new optimism. With the help of first Pat Seitzer and then Debe Crocker, the simulation program was updated, expanded, and made fully interactive along the lines of the data reduction programs then becoming available at NRAO. It was clear that the profusion of data required a far more efficient approach.

As with a decade earlier my optimism was misplaced. For the second parameter problem things got worse. Perhaps it was only due to the efficiency that we could now produce models, but the sensitivity of the results to assumptions about mass loss seemed worse than ever. When one finds oneself twiddling with the Reimer's efficiency η_R in the third significant figure, how can much significance possibly be placed on the results? One could argue that η_R was a function of [Fe/H], or not, depending (sensitively) on how [CNO/Fe] varies with [Fe/H]. Peterson found significant rotation in HB stars. The number of possible second parameters was actually increasing with time! The problem with the Oosterhoff effect, likewise, was resistant to solution. At least it acquired a new name—the Sandage Period Shift Effect.

The following sections informally summarize some of the results obtained in the last few years. For a more detailed review and complete references I direct you to the excellent review of Renzini and Fusi Pecci (1988).

2 THE STANDARD MODEL

Many lines of evidence suggest that globular cluster red giants lose 0.1–$0.2\,M_\odot$ of their mass either in the form of a stellar wind or at helium flash. The standard model of the HB suggests that for some reason there is a dispersion of order 10–20% in the amount of mass lost. Since the core mass at flash is insensitive to total mass, stars begin quiescent core helium burning along a line of constant core mass and varying total mass (the ZAHB). The observed HB is the result of the superposition of evolutionary tracks of stars of differing mass which arrived on the ZAHB at sometime in the past. While this model is widely accepted at the moment, the possibility remains that some parameter other than, or in addition to, total mass varies along the HB. Firmly established are: (1) The lower envelope of the observed HB is close to the ZAHB except possibly for the hottest stars. (2) The HB is not an

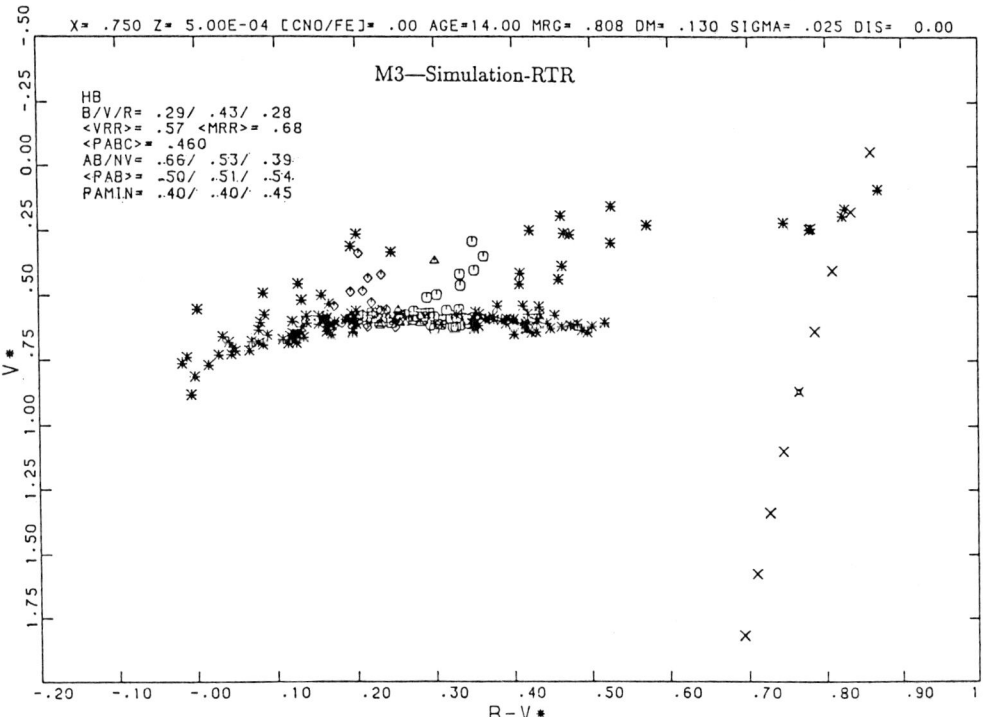

Fig. 1.—Simulation for M3 using the Rood (1973) fits to evolutionary tracks.

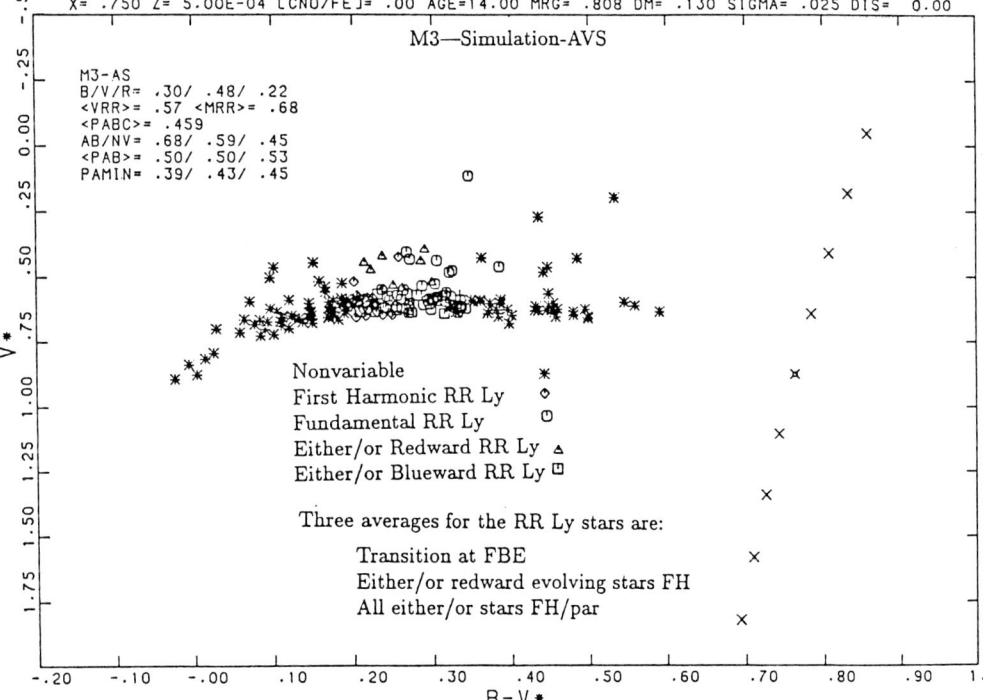

Fig. 2.—Simulation for M3 using Sweigart (1987) tracks

evolutionary sequence like the RGB. I.e., something besides age varies along the HB. Otherwise the HB would have "sharp" ends rather than the observed trickle out. The true skeptic should bear these points in mind.

Because the HB is a superposition of evolutionary tracks most of the details of the tracks are lost. For this reason we have not worried too much about our representation of the evolution away from the ZAHB. To illustrate this point we compare a simulation for M3 using the old Rood (1973) approximation (Figure 1) to a fit (admittedly crude) to the much newer Sweigart (1987) tracks (Figure 2. (Note that an "*" on the axis labels as in $B - V*$ denotes that " observational errors" have been added.) As expected, the results are indistinguishable. We anticipate the same for the ongoing calculations of Van den Berg which use updated input physics. The observed HB for M3 (Buonanno et al. 1988) is shown in Figure 3. The agreement is fairly good except for the observed blue extension. The adopted mass distribution, which is purely *ad hoc*, could easily be fudged to give such a tail in the simulations.

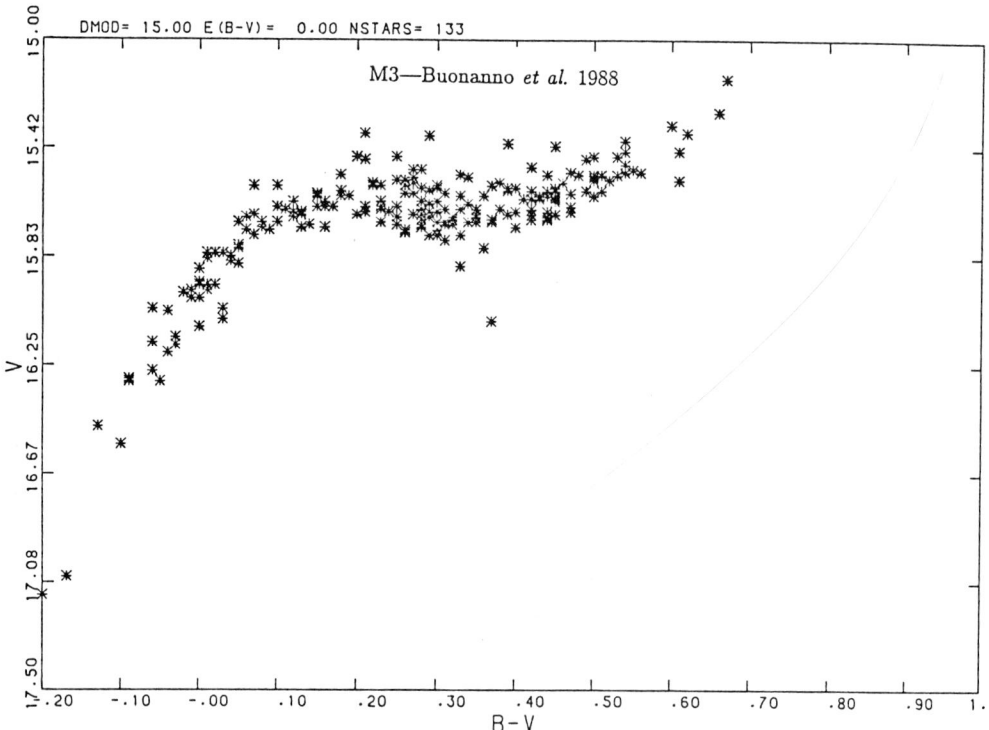

Fig 3.—M3 HB for stars with $B - V \geq -0.20$ as observed by Buonanno *et al.* 1988.

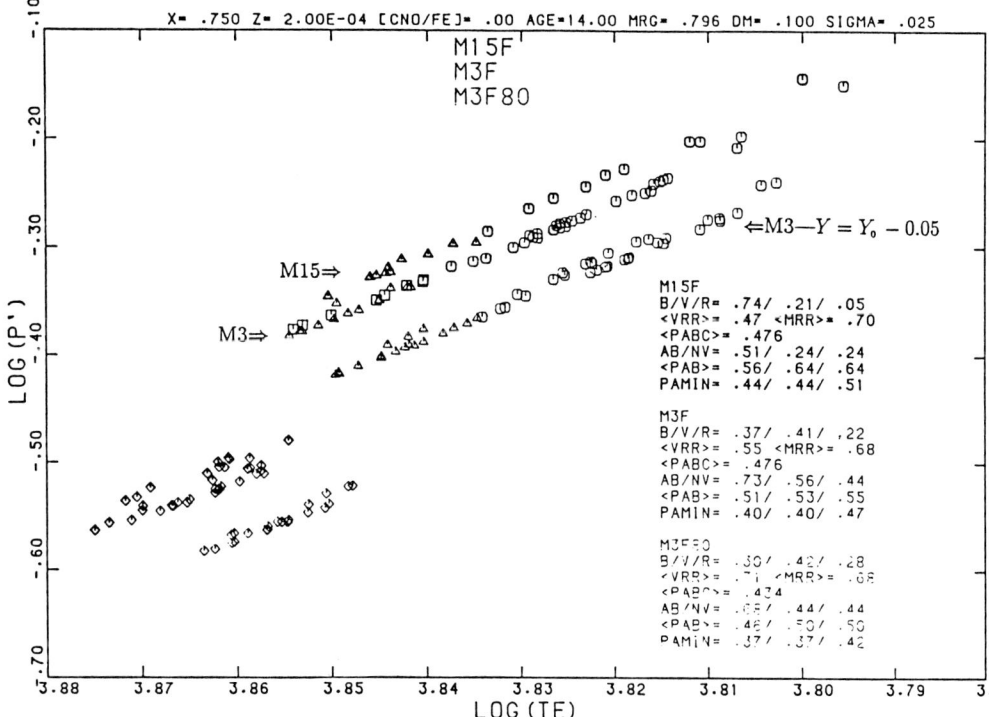

Fig. 4.—Model period shifts for a standard M3, M15, and helium depleted M3.

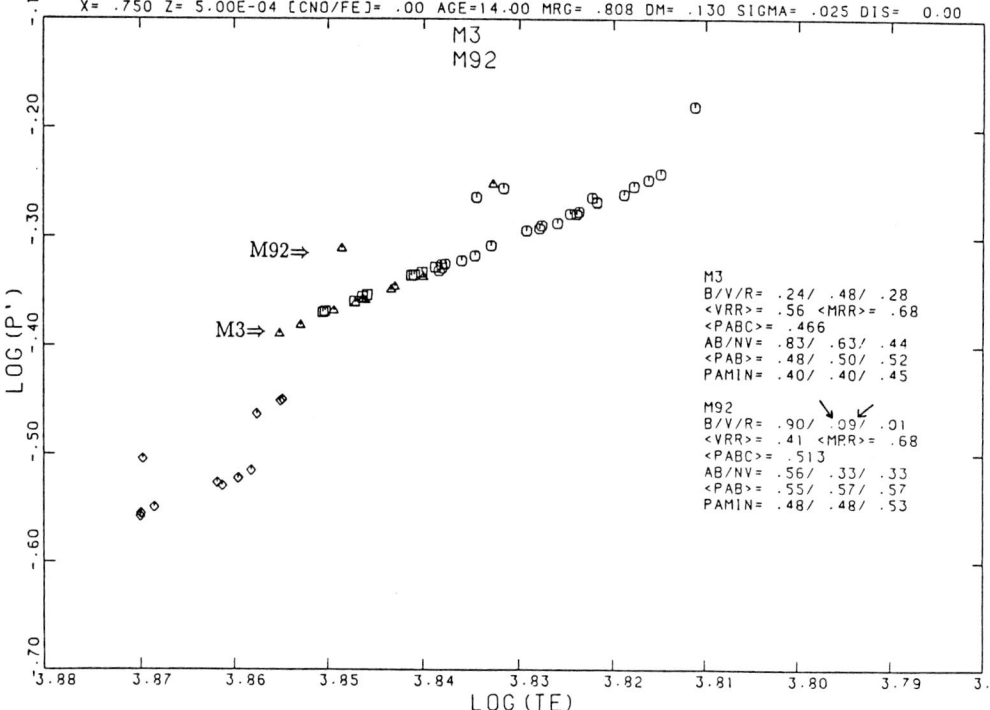

Fig. 5.—Models for M92 and M3.

One particular improvement we can point to in our understanding of the standard model is the solid evidence that has been obtained on HB lifetimes and thus the degree of mixing during core helium burning. Buzzoni et al. (1983) and Buonnano et al. (1985) use the ratio of asymptotic giant branch stars to HB stars to show that the standard treatment of overshooting and semiconvection lead to about the correct amount of mixing.

3 THE OOSTERHOFF EFFECT OR SANDAGE PERIOD SHIFT

In the early 1980's Sandage first argued that the periods of RR Lyraes decreased as cluster metallicity increased. In particular, he noted "...at **every temperature** the shifts exist ..." (emphasis ours) The shift was most prominent in the log P'-log T_{eff} diagram. The use of log P' $[= \log P + 0.336(m_{\text{bol}} - \langle m_{\text{bol}} \rangle)]$ removes much of the scatter due to the intrinsic dispersion of luminosity among the variables. This period shift was much larger than predicted by the standard models. The only explanation Sandage could find was an **anti**correlation of helium with metallicity. This was a most unpalatable explanation (we think for Sandage as well). It was, however, most important because it showed that the difference between the periods between the Oosterhoff classes was due to some systematic difference between the clusters. The Oosterhoff dichotomy was not present just in the average properties. It could not be blamed on some quirky distribution or on hysteresis.

While the case for the shift was fairly convincing, there was some cause for concern. In particular, note that in Fig. 11 of Sandage (1982), for M15 the slope in the (log P-log T_{eff})-diagram is much less than that for M3 which agrees quite well with theory. Now if there is any thing we should get right, its this slope. Hence, we from the outset were suspicious of the "observed" temperatures for M15. (Sandage addresses this point in his paper in this volume.) After much maneuvering with many parameters (Rood 1983; Rood and Crocker 1985a) we reached the same conclusion as Sandage— only the anticorrelation of Y and Z worked as shown in Figure 4. A similar conclusion was reached in a far more extensive study by Sweigart, Renzini, and Tormambè (1987). As a simple explanation was not forthcoming, we would have blamed the observations were it not for the paper of Bingham et al. (1984) which presented a convincing case that the shift exists between M15 and M3. Since that time we have considered the observational basis of the Sandage period shift to lie primarily in the M15, M3 pair.

Recently Lee, Zinn, and Demarque (1987, 1988) have claimed to have explained the Sandage shift. Lee et al. argue that their result arises, in part, from the inclusion of HB evolution all the way to central helium exhaustion and the use of an unusually fine grid of models. We believe these are not important new additions. Instead, their results arise partially from extreme parameter values which exaggerate the theoretical shift (variable strip wider than standard), and partially from "measuring" the shift a single (too cool) temperature in the log P-log T_{eff} plane rather than shift at *every temperature* found by Sandage and Bingham et al. in the log P'-log T_{eff} plane. Further, by their own admission, they cannot account for the crucial M15, M3 pair.

Lee et al. do indeed find a significant shift between M92 and M3. We have never had any difficulty with M92 either as shown in Figure 5. We never considered this significant because of the small number of variables in M92. More than 90% of the M92 HB is blueward of the variable strip. For such a cluster **all** of the RR Lyrae and red HB can be near helium core exhaustion. In Figure 6 we show log L as a function of time. Roughly 10% of the HB life is spent at large enough $\log L$ to lead to $\delta \log P' \sim 0.04 - 0.07$. If this works for M92, could "improved" tracks work with M15? Since only about 70% of M15 is blue HB, roughly 30% of the HB lifetime would have to be spent $\delta \log L \sim 0.08$ or ~ 0.2 mag "above" the ZAHB. These stars would not be confined to the variable strip in M15; in other clusters, like M3, they would lie above the HB. There is no evidence for such a large number of pre-exhaustion stars in any metal poor cluster.

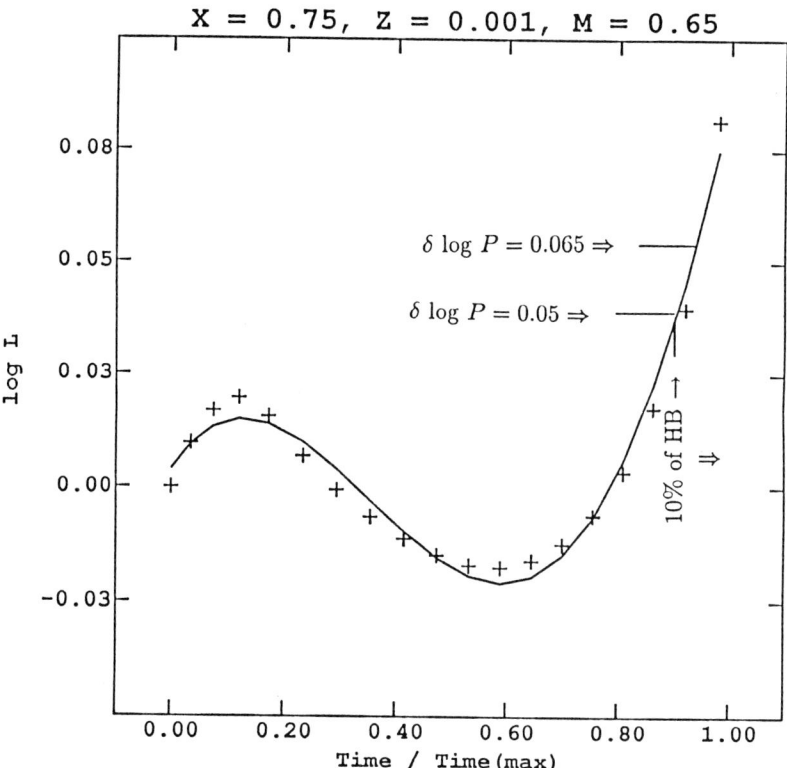

Figure 6. Sweigart's evolution of $\log L$ vs. time (crosses) extrapolated to $Y_c = 0$ and our fit (line).

In part, the Lee et al. "explanation," as well as that of Caputo (1988), rests on the absence of a shift in ω Cen and its implications for the amplitude and rise time vs. temperature diagrams. Basically, could part of the "shift" be a temperature shift rather than a period shift? Dickens, elsewhere in this volume, presents evidence concerning the role of ω Cen. (He also notes that ω Cen is also more than 90% blue HB. It may well be that ω Cen is like M92—"no problem." Indeed, the standard theory may predict no shift within the cluster.)

We would like here to provide some additional information (confusion?) on the matter of the temperatures of the RR Lyrae. It has always been possible to interpret the shift

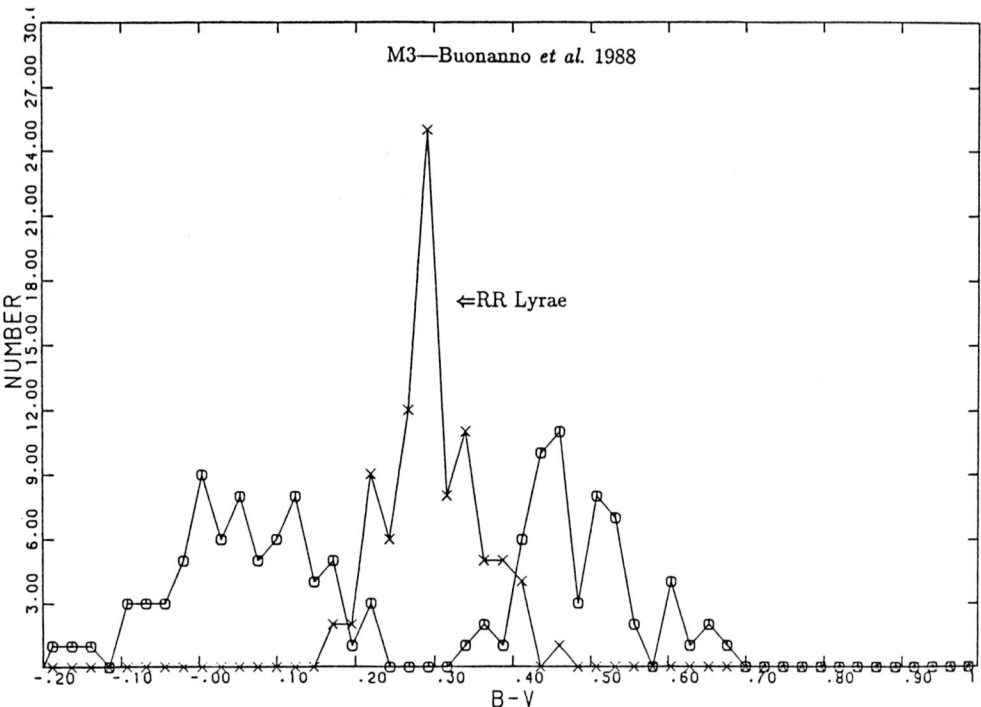

Fig. 7.—The observed distribution in color of the HB stars for M3.

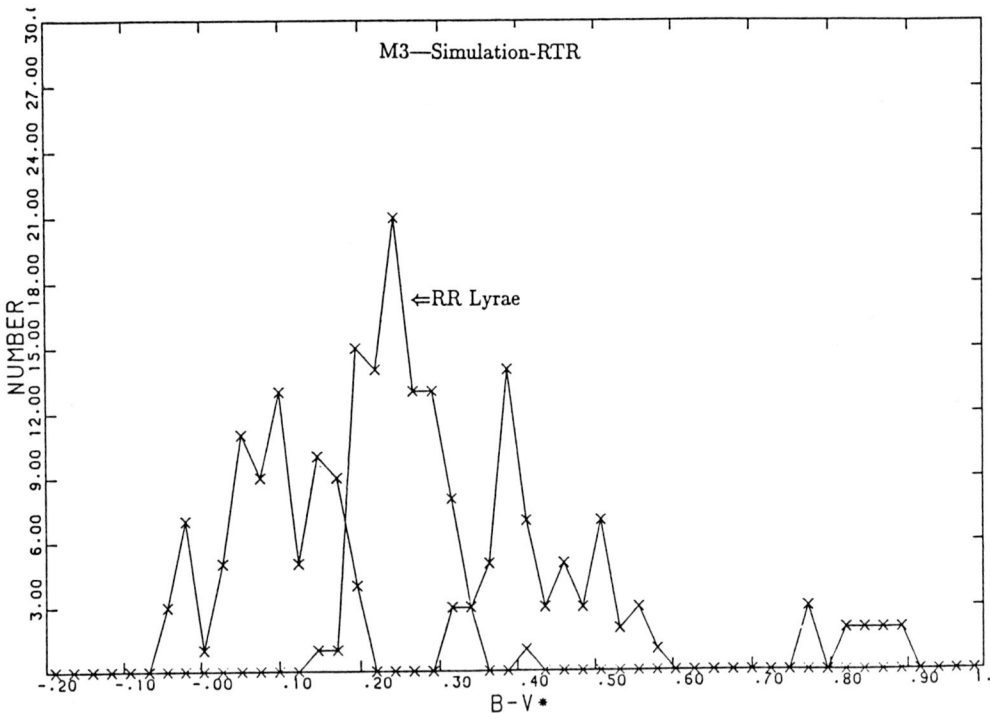

Fig. 8.—Simulation of the color distribution in M3.

between M15 and M3 *either* as a period shift, $\delta \log P' \approx -0.065$, *or* a temperature shift, $\delta \log T_{\text{eff}} \approx -0.02$ ($\delta(B-V) \approx 0.06$). Such an interpretation is not without difficulties, but considering the intractability of the period shift, we feel that it deserves some consideration. Indeed, there is a hint of difficulty with the temperatures which can be inferred from our models.

When comparing HB simulations with the observations, the pertinent temperatures, periods, etc. are those the star would have if it were not pulsating. Unfortunately the observers cannot stop the stars from pulsating, but must use some sort of averaging. We always have been uneasy with this process. In particular, it seems that zero point jumps in "$B-V$" might enter at the boundaries of the variable strip and between the fundamental and first harmonic. There could easily be amplitude related factors (again, see Sandage in this volume). We have often been told that there is no overlap between the variables and non variables. The largest complete sample to quantify such a statement is that of Buonanno et al. for M3. The color histogram (Figure 7) shows virtually no overlap. In addition there is no overlap between the RRab's and RRc's for M3 (Bingham et al., 1984). Even if these boundaries are "theoretically sharp," our simulations show that there should be significant overlap observed. Our simulation for M3 is shown in Figure 8. There is significantly more overlap, particularly at the blue edge, than observed. We have taken $\sigma(B-V) = 0.02$, the samples are the same size, and the BHB/RR/RHB ratio is about that observed. Likewise, there is significant overlap between the RRab and RRc in the simulated ($\log P'$, $\log T_{\text{eff}}$)-diagram. Although we doubt that the results are statistically significant we feel there is a strong hint at a jump in the temperature scale at both the BHB/RRc boundary and RRc/RRab boundary. These are rather nitpicking points which have until this time been quite justifyably neglected by the observers, but they are of the magnitude that they could be important for the Sandage Effect.

It is possible that assorted temperature errors could explain, or at least make more manageable, the Sandage Effect problem. Before one rejoices too much at this possibility, one should recall that this will still leave us with the Oosterhoff problem almost as it existed in 1973. The difference between the mean periods of Oo I and Oo II clusters will be without explanation. The one definite bit of progress is that the fraction of RR Lyrae in the fundamental mode follows quite naturally from nonuniform distribution of stars along the HB. The blue HBs of Oo II clusters give about 50% RRc so long as the transition takes near the fundamental blue edge. At least hysteresis and a wide either/or strip seemed to have been ruled out.

We should also mention two other failures of the standard theory as it applies to the RR Lyrae. Alert to the possibility of systematic temperature errors, one might be concerned by the extremely peaked observed color distribution for M3 shown in Fig. 7. Could this be an indication of an amplitude related error in averaging $B-V$? Suggesting otherwise, is the period distribution as shown from the catalog of Cacciari and Renzini (1976) in Figure 9. There is a pronounced peak at the blue edge of the RRab's. This cannot be due to some sort of averaging error—the stars know their true colors even if we don't. There appears to be a real pile-up at this point. The same is typical for other clusters. Simulations, as shown in Figure 9 on the right, are always much flatter. If simulations could produce peaked distributions, in other than

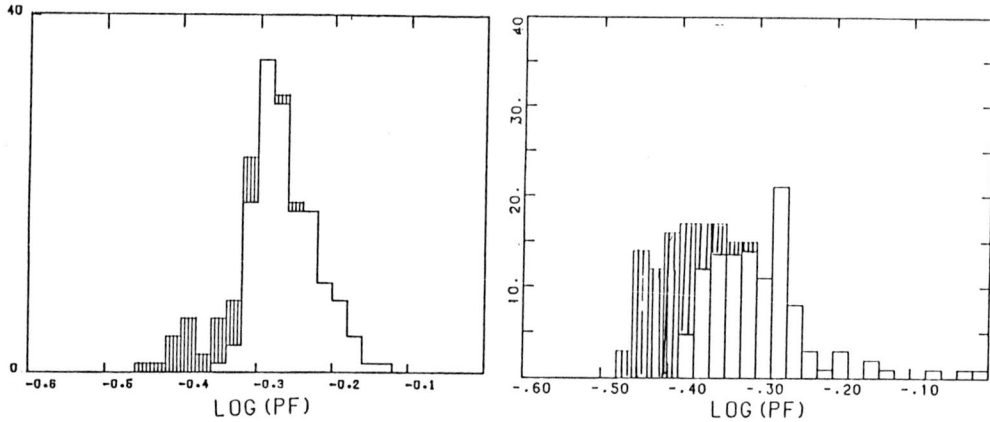

Fig. 7.—On the left is the period distribution for M3 from the catalog of Cacciari and Renzini (1976). On the right is a simulated period distribution for M3.

a purely *ad hoc* way, there could be some relevance for the mean properties of the RR Lyrae and thus the old Oosterhoff problem.

In discussions at this meeting Art Cox and Jim Nemec have again pointed out that the masses derived for double mode RR Lyrae in Oo I and Oo II clusters differ significantly. Again the standard model fails, predicting essentially equal masses. Even the $\Delta Y \propto -\Delta Z$ solution gives essentially the same masses for the two groups. We may well not have given adequate consideration to this point. Perhaps the solution when it is found will give the correct mass differences.

4 THE SECOND PARAMETER PROBLEM—1988

For some time the second parameter problem in its most obvious form—the distribution along the HB—has been caught in the quagmire of mass loss. Variations in cluster age, Y, and [CNO/Fe] affect the HB directly. The HB is indirectly affected by anything which can affect mass loss. Rotation delays He flash increasing the duration of the strongest winds; rotation could determine the size of magnetic fields and thus the magnitude of MHD winds. Magnetic fields might vary in some way independent of rotation. The list has only been limited by our imaginations. Various pieces of observational evidence at times seem to favor one or the other. Yet no convincing identification has been made. At the very least, to approach the problem in a brute force way, one has to know how mass loss depends on [Fe/H] to at least two significant figures. In reality perhaps even the sign of this variation is uncertain. One can, of course, parameterize things, e.g., making η_R a function of [Fe/H], but things rapidly degenerate into parameter twiddling. Recently our interest has been directed toward clusters with bimodal HBs (Rood and Crocker 1985b; Crocker and Rood 1985; Crocker, Rood, and O'Connell 1988, hereafter CRO).

The HB of NGC 6752 was rather a shock when it was first shown by Cannon and Lee at a Frascati workshop in 1973. It wasn't horizontal, but rather a blue droop stretching down to magnitudes as faint as the turnoff. Even worse—right in the middle was a big ugly gap. Since that time many other clusters have been found with HBs with gaps or somewhat less pronounced bimodal distribution. Even such familiar clusters as M15 have turned out to have a gap in the BHB. Suggestions of less extreme bimodality are commonplace. We have adopted as a working hypothesis, that bimodal HBs are a special case of, or extreme example of, the second parameter problem. The bimodality could be the result of some second parameter operating within a single cluster. In some cases like NGC 6752, this parameter might be taking on an extreme value and thus be easier to identify.

One must be careful in defining bimodality. The distribution in $B-V$ is inappropriate. An clump is produced in the blue as $B-V$ saturates as a temperature measure. A clump is produced in the red HB as the mass dependence of ZAHB $\log T_{\mathrm{eff}}$ decreases with increasing mass. Neither of these clumps are in some sense "real." We feel HB structure is best examined by "straightening out" the HB. To do so we define one coordinate X_{HB} measured "along" the HB and another Y_{HB} perpendicular to the HB (Figure 10). The (X_{HB}, Y_{HB})-diagram depends on many factors both observational and theoretical—the photometry, $E(B-V)$, $(m-M)_V$, the theoretical ZAHB and thus assumed composition, the color temperature relation and bolometric corrections, etc. Despite the complexity things are not as bad

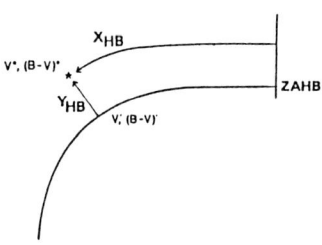

Fig. 10.—Definition of X_{HB} and Y_{HB}.

as one might fear. Some things are not particularly important, e.g., the assumed composition of the ZAHB. Others produce characteristic and easily recognized distortions. E.g., based on a downward bend at the red end of the HB (Figure 11), we conclude that for M92 $(m-M)_V = 14.8$ rather than the 14.5 reported by Harris and Racine (1979). This is in accord with our discussion above concerning the period shift for M92. Other uncertainties are reduced by considering only differential studies, i.e., one cluster as compared to another.

Because X_{HB} is defined for each star relative to the ZAHB it leads directly to a "temperature" $\log T_{\mathrm{eff}}(X_{HB})$ and "mass" $M(X_{HB})$. $M(X_{HB})(MASSD$ in the figures) differs slightly from the true mass because of evolution away from the ZAHB. As shown in Figure 12 the difference between the "real" and $M(X_{HB})$ distributions in a simulation for M3 is small. The observed $M(X_{HB})$ distribution for M3 is shown in Figure 13. One is able to define HB morphology in a much more quantitative way using the $M(X_{HB})$ distribution. The bimodality of clusters such as NGC 6752 and M15 stands out (CRO). Bimodality appears in less obvious cases like M92. Clusters, such as M5, which superficially appear bimodal in $B-V$, prove to unimodal.

The distribution in $M(X_{HB})$ will of course be just as subject to the problems of mass loss as is the morpholgy in the color-magnitude diagram. Ad hoc bimodal distributions in your favorite second parameter will produce bimodal $M(X_{HB})$ distributions. Maybe thats all there is to clusters like NGC 6752. Nature has provided precedence for such a solution in dromedaries and Bactrian camels. We hope this isn't the case—we certainly don't need more free parameters.

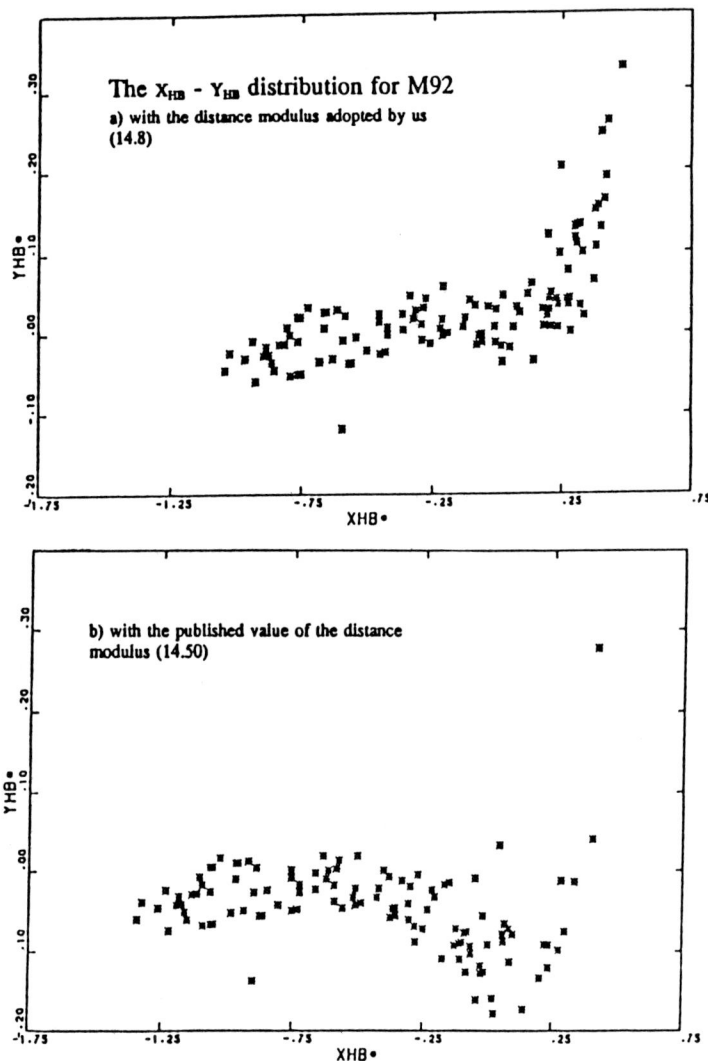

Fig. 11.—The (X_{HB}, Y_{HB})-diagrams for M92 with different assumptions for $(m - M)_V$.

The solution we suspect lies in another direction—Y_{HB}. The distribution in Y_{HB} should be much less affected by the details of mass loss. One requires quite good photometry to do this in the (X_{HB}, Y_{HB})-diagram. A simulation using Sweigart's new tracks is show in Figure 14. The real data is shown in Figure 15. We have not analyzed this result in detail, but there is a suggestion that the simulation errors $(\sigma(V) = 0.01; \sigma(B - V) = 0.02$ are too small in particular for the variables. However the general agreement is not bad. It is quite obvious that far less than 30% of the HB

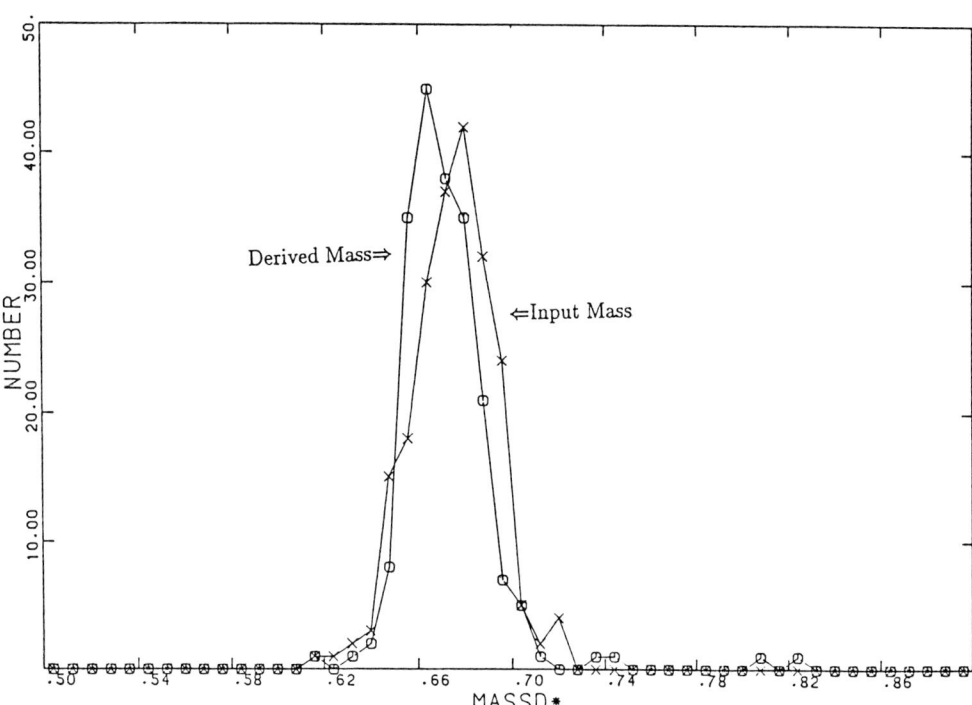

Fig. 12.—Distribution in $M(X_{HB})$ as compared to input mass distribution.

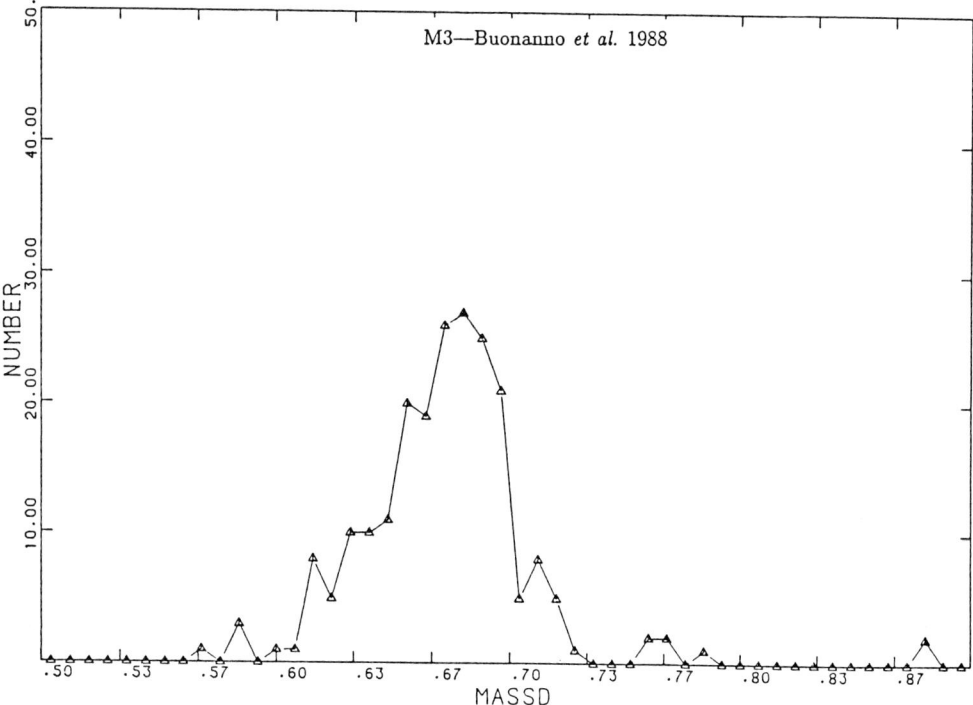

Fig. 13.—Observed "mass" distribution for M3.

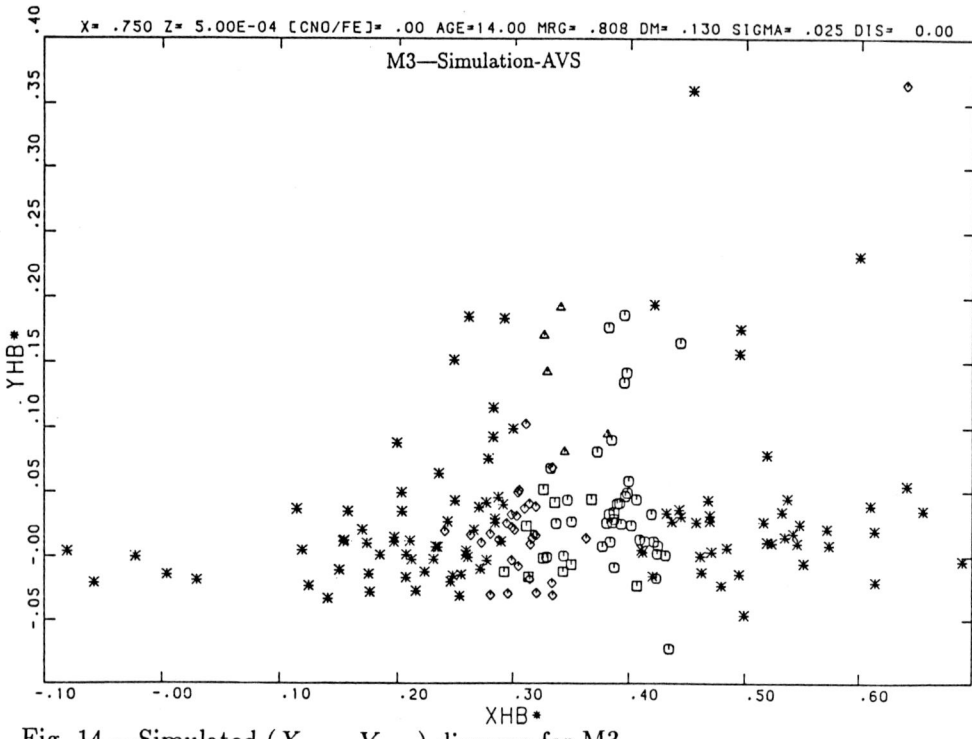

Fig. 14.—Simulated (X_{HB}, Y_{HB})-diagram for M3.

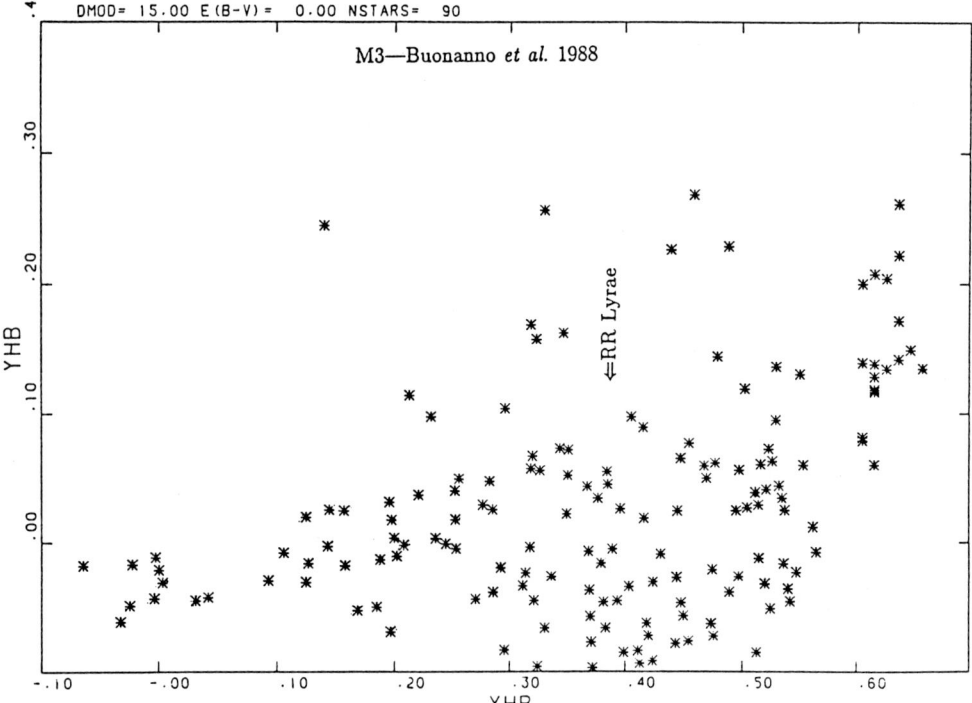

Fig. 15.—The observed (X_{HB}, Y_{HB})-diagram for M3.

lies at $Y_{HB} \gtrsim 0.20$, which would be the case if the solution of the M15, M3 Sandage shift were the inclusion of the helium core exhaustion phase of the HB.

To this point our investigation of the Y_{HB} distribution has been primarily confined to the blue HB stars in the (log g, log T_{eff})-diagrams of gap clusters. The possibilities we were searching for are summarized in Figure 16. Our results which have just appeared in CRO do not appear to be consistent with any of our hypotheses. Qualitatively the stars blueward of the gaps are displaced as they would be if rapidly rotating. However, the displacement appears much larger than we would expect. Such nice solutions as [O/Fe] (Rood and Crocker 1985b) appear to be ruled out.

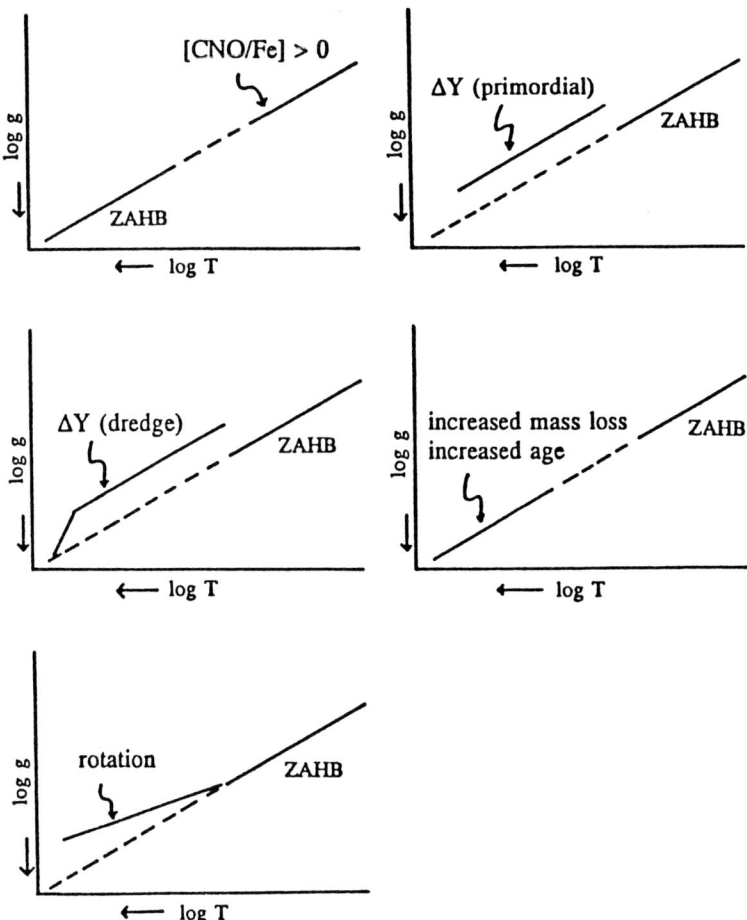

Figure 16. (log g, log T_{eff})-diagrams for assorted "solutions" to BHB gap clusters.

The first detailed investigation of Y_{HB} distributions was, in fact, by Sandage—the period shift can be mapped into Y_{HB}. The $(\log P, \log T_{\rm eff})$-diagram is equivalent to the $(\log g, \log T_{\rm eff})$-diagram. The expected shift is $\delta \log g \approx 1.19 \delta \log P$. The observations in CRO were designed to detect this "Sandage Shift" in the nonvariable stars. Unfortunately, what appears to be an intrinsic dispersion in our results hides any shift if present.

5 CONCLUSIONS

The main difficulties facing the theory of HB stars are the second parameter problem and the Sandage Period Shift. There appears to be a slight chance that the Sandage Period Shift will "go away," becoming a temperature shift. Then we are still left with the Oosterhoff Effect and we are back to 1973 (theoretically). We suspect that the solutions to these problems will arise from new kinds of observations and new ways of "looking at" the old. The observational status of HB stars is much better than in 1973 and promises in many ways to get much better. Unfortunately, some kinds of crucial observations do not appear to be stylish. It sure would be nice to have some more old fashioned RR Lyrae work like that of Bingham et al. (1984) for M15 for **large** samples in other clusters.

Its hard to know whether to be optimistic or not. Still we must keep chipping away. The solutions to these problems have interest far beyond those of the globular cluster and variable star aficionados and stellar structure theorists who wish to check off another obscure success. In understanding these stars we have probably our most direct evidence of what was going on at the formation of the Milky Way and one ofour most important probes of the early universe.

We wish to thank Flavio Fusio Pecci for providing the data M3 data in machine readable form.

REFERENCES

Bingham, E. A., Cacciari, C., Dickens, R. J., and Fusi Pecci, F. 1984, *M. N. R. A. S.*, **209**, 765.
Buonanno, R., Corsi, C., and Fusi Pecci, F. 1985, *Astr. Ap.*, **145**, 97.
Buonanno, R., Buzzoni, A., Corsi, C., Fusi Pecci, F., Sandage, A. R. 1988 , in *IAU Symposium 126, Globular Cluster Systems in Galaxies*, ed. J. Grindley and A.G.D. Philip (Dordrect: Reidel), p. 621.
Buzzoni, A., Fusi Pecci, F., Buonanno, R., and Corsi, C. 1983, *Astr. Ap.*, **123**, 94.
Cacciari, C., and Renzini, A. 1976, *Astr. Ap. Suppl.*, **25**, 303.
Caputo, F. 1988, *Astr. Ap.*, **189**, 70.
Crocker, D. A. and Rood, R. T. 1985, in *Hot Population II Stars*, ed. A. G. D. Philip (Schenectady: L. Davis Press), p. 107.
Crocker, D. A., Rood, R. T., and O'Connell, R. W. 1988, *Ap. J.*, **332**, 236.
Harris, W. E., and Racine, R. 1979, *Ann. Rev. Astr. Ap.*, **17**, 241.
Lee, Y.-W., Demarque, P., Zinn, R. 1988, in *Faint Blue Stars*, ed. A. G. D. Philip (Schenectady: L. Davis Press), p. 000.
Lee, Y.-W., Demarque, P., Zinn, R. 1988, in *The Calibration of Stellar Ages*, ed. A. G. D. Philip (Schenectady: L. Davis Press), p. 000.
Renzini, A., and Fusi Pecci, F. 1988, *Ann. Rev. Astr. Ap.*, **26**, 199.
Rood, R. T. 1973, *Ap. J.*, **184**, 815.

Rood, R. T. and Crocker, D. A. 1985b, in *Hot Population II Stars*, ed. A. G. D. Philip (Schenectady: L. Davis Press), p. 99.
Rood, R. T. and Crocker, D. A. 1985a, in *ESO Workshop on Production and Distribution of C, N, O Elements*, ed. I. J. Danziger, F. Matteucci, and K. Kjär (Garching: ESO), p. 61.
Sandage, A. R. 1982, *Ap. J.*, **252**, 553.
Sweigart, A. V. 1987, *Ap. J. Suppl.*, **65**, 95.
Sweigart, A. V., Renzini, A., Tornambè, A. 1987, *Ap. J.*, **312**, 762.

THE ABSOLUTE MAGNITUDES OF RR LYRAE STARS AND THE AGE OF THE GALAXY

Allan Sandage
Mount Wilson and Las Campanas Observatories
of the Carnegie Institution of Washington

Abstract. It is shown that the intrinsic spread in the absolute magnitudes of the RR Lyrae variables in a given globular cluster can reach 0.5 magnitudes at a given period or at a given color, due to luminosity evolution away from the zero age horizontal (ZAHB). The size of this intrinsic luminosity spread is largest in clusters of the highest metallicity.
The absolute magnitude of the ZAHB itself also differs from cluster to cluster as a function of metallicity, being brightest in clusters of the lowest metallicity. Three independent methods of calibrating the ZAHB RR Lyrae luminosities each show a strong variation of $M_V(RR)$ with [Fe/H]. The pulsation equation of $P<\rho>^{0.5} = Q(M,T_e,L)$ used with the observed periods, temperatures, and masses of field and of cluster RR Lyraes gives the very steep luminosity-metallicity dependence of $dM_V(RR)/d[Fe/H] = 0.42$. Main sequence fitting of the color-magnitude diagrams of clusters which have modern main-sequence photometry gives a confirming steep slope of 0.39. A summary of Baade-Wesselink $M_V(RR)$ values for field stars determined in four independent recent studies also shows a luminosity-metallicity dependence, but less steep with a slope of $dM_V(RR)/d[Fe/H] = 0.21$.
Observations show that the magnitude difference between the main sequence turn-off point and the ZAHB in a number of well observed globular clusters is independent of [Fe/H], and has a stable value of dV = 3.54 with a disperion of only 0.1 magnitudes. Using this fact, the absolute magnitude of the main sequence turn-off is determined in any given globular cluster from the observed apparent magnitude of the ZAHB by adopting any particular $M_V(RR) = f([Fe/H])$ calibration.
Ages of the clusters are shown to vary with [Fe/H] by amounts that depend upon the slopes of the $M_V(RR) = f([Fe/H])$ calibrations. The calibrations show that there would be a steep dependence of the age on [Fe/H] if $M_V(RR)$ does not depend on [Fe/H]. No dependence of age on metallicity exists if the RR Lyrae luminosities depend on [Fe/H] as $dM_V(RR)/d[Fe/H] = 0.37$. If Oxygen is not enhanced as [Fe/H] decreases, the absolute average age of the globular cluster system is 16 Gyr, independent of [Fe/H], using the steep $M_V(RR)/[Fe/H]$ calibration that is favored. If Oxygen is enhanced by [O/Fe] = - 0.14 [Fe/H] + 0.40 for

[Fe/H] < -1.0, as suggested from the observations of field subdwarfs, then the age of the globular cluster system decreases to 13 Gyr, again independent of [Fe/H], if the RR Lyrae ZAHB luminosities have a metallicity dependence of $dM_V(RR)/d[Fe/H] = 0.37$.

1 INTRODUCTION

Knowledge of the absolute magnitudes of RR Lyrae variables has been traditionally important in finding the distances of globular clusters and eventually to the nearest galaxies. Ages of individual clusters and of the Galactic globular cluster system can be determined when the absolute luminosities of the main sequence turn-offs are known by combining the apparent magnitudes of the turn-off points with the adopted distances.

Until recently, $M_V(RR)$ for the RR Lyraes had generally been assumed to be a fixed number of small intrinsic dispersion, and to be independent of the metal abundance. These assumptions were originally justified from (1) the observed near horizontal nature of the horizontal branch in the V pass band in most globular clusters, (2) the small observed scatter in V magnitudes of the variables in a given cluster, and (3) the approximate equality of the observationally determined $M_V(RR)$ values in clusters of different metallicity obtained from the early data on main sequence fittings in clusters studied from 1950 to 1970.

Based on modern data, the two assumptions of (1) small intrinsic dispersion in $M_V(RR)$ in a given globular cluster, and (2) independence of the absolute luminosity of the zero age horizontal branch (ZAHB) on metallicity are almost certainly incorrect. The purpose of this report is to show the importance of the changes of these views for the question of the ages of globular clusters as a function of [Fe/H]. Details have been set out in three papers submitted to the Astrophysical Journal. The summary given here is an extension of an earlier discussion on the vertical structure of the globular cluster horizontal branch (Sandage 1987).

2 INTRINSIC DISPERSION OF $M_V(RR)$ DUE TO POST ZAHB EVOLUTION

The observed width of the horizontal branch (HB) in many globular clusters is real rather than due to observational photometric errors. The evidence is that the RR Lyrae variables that are measured to be brighter than others of the same color have measured longer periods. This observed period-magnitude correlation at constant temperature has the slope close to $dV/d(\log P) = 3$ required (at constant mass) by the pulsation equation of $P<\rho>^{0.5} = Q$ that have been calculated by van Albada & Baker (1971), Iben (1971), Cox (1987) and others.

The observed HB data for M3 and M15, based on measurements of many photographic plates in each cluster, are shown in Figure 1. The variables and the nonvariables blueward and redward of the RR Lyrae variable star instability region are shown. If the observed spread in V magnitudes is real, the periods of the brighter variables must be longer than for the fainter variables at a given color. Furthermore, the correlation of magnitude and period shift must have a slope close to 3, as derived from the pulsation equation (assuming constant mass). This test for such a correlation is shown in Figure 2 for M3 and Figure 3 for M15. The slopes of the correlations in the bottom panels of each diagram are, in fact, close to 3.

Figure 4 shows the data for M4. The observed spread in V of the HB is large at dV = 0.5 magnitudes. The good correlation of magnitude and period shift in the bottom panel shows again that the observed spread in

Figure 1. Horizontal branch photometry for M3 and M15 based on many plates in each color such that the photometric measuring errors are less than 0.01 magnitude in each color. RR Lyrae stars are shown as open circles for c types, and crosses for ab types. The colors have been corrected for the adopted reddenings shown in the code.

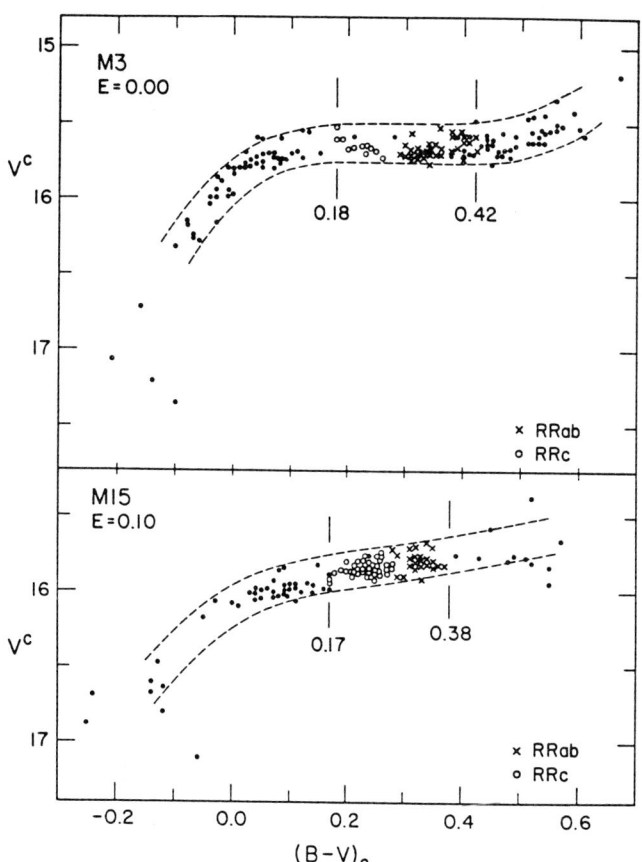

Figure 2. (top): The HR diagram for type ab RR Lyraes in M3 where the data in Figure 1 (top) have been converted to temperatures. (bottom): Correlation of the observed apparent magnitudes for M3 RR Lyraes with the period ratio of observed period to period on the ZAHB at the same temperature. The correlation shows that the observed dispersion in V in the top panel is real.

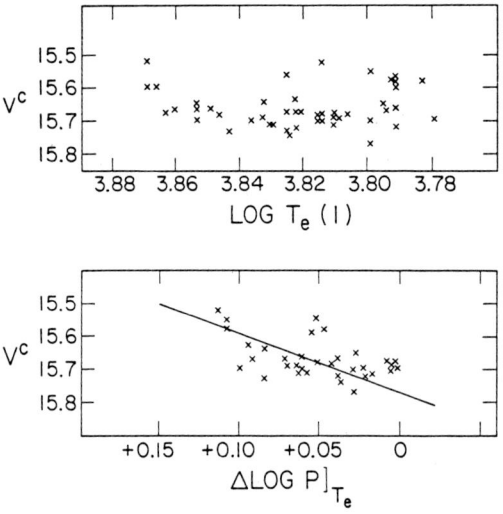

Figure 3. Same as Figure 2, but for the RR Lyrae data in M15. The sense of the correlation in the bottom panel is that the brighter observed stars have longer periods than fainter variables at the same temperature.

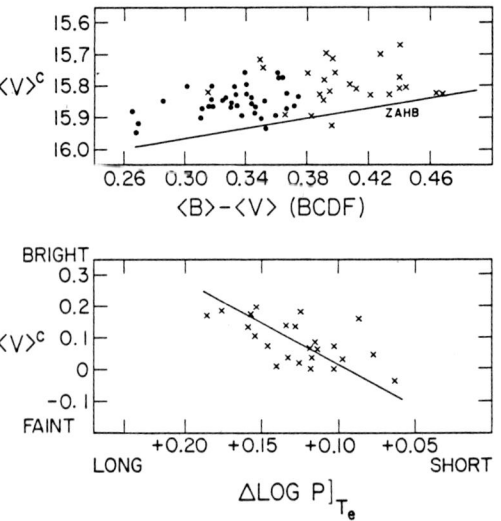

Figure 4. Same as Figures 2 and 3 for the HB stars in M4. The edges of the RR Lyrae region are shown in the top panel.

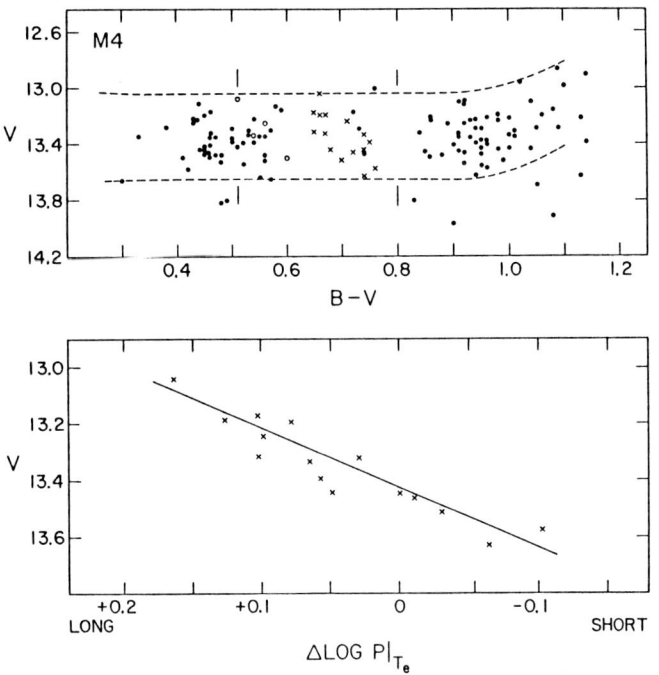

Figure 5. Summary of the intinsic width of the HB as a function of metallicity.

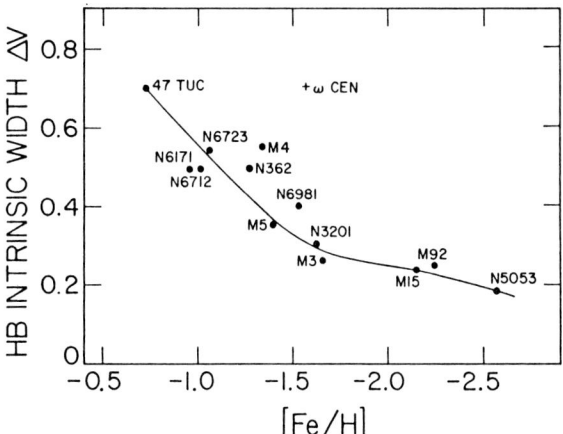

RR Lyrae magnitudes in M4 is real.

Similar data for other clusters, summarized in Figure 5, show that this intrinsic dispersion in M_V(RR) for the RR Lyrae variables in a given cluster is correlated with [Fe/H]. For high metallicity clusters, the spread in RR Lyrae absolute magnitudes at constant [Fe/H] reaches 0.5 magnitudes. Gratton, Tornambé & Ortolani (1986) suggest that the deviation of Omega Cen from the good correlation for the other clusters is due to this post ZAHB evolution, combined with the variation of the luminosity level of the ZAHB as a function of [Fe/H], discussed next.

3 VARIATION OF THE ZAHB RR LYRAE ABSOLUTE MAGNITUDES WITH METALLICITY DERIVED FROM THE PULSATION EQUATION

The pulsation equation derived by van Albada & Baker (1967) by a proper integration over that part of the stellar envelope that controls the period of pulsation (similar to the equations derived also by Iben, 1971, and by Cox, 1987) is

$$\log (L/M^{0.81}) = (\log P + 3.48 \log T_e - 11.497)/0.84. \qquad (1)$$

The effective temperature can be obtained from observed colors from a calibration of the color-temperature relation as a function of surface gravity and metal abundance (Bell, as summarized by Butler, Dickens & Epps 1978). Temperatures, and the observed periods, as reduced to the ZAHB values that would apply in the absence of post ZAHB evolution, permit ZAHB values of $L/M^{0.81}$ to be calculated from equation (1). These mass-to-luminosity ratios have been found to be correlated with metallicity (Sandage 1989a,b) with the result shown in Figure 6.

Figure 6. Correlation of metallicity and the luminosity-to-mass ratio for the ZAHBs of 10 clusters as determined from the pulsation properties of their RR Lyraes.

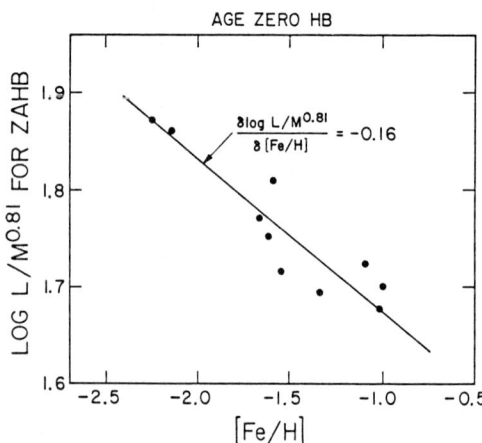

Similar data for the field RR Lyraes from the measurements of Lub (1977, 1987) are shown in Figure 7 over the larger metallicity range of [Fe/H] from 0 to -2.2. The slope of the correlation is $d\log(L/M^{0.81})/d[Fe/H] = -0.11$. The equation of the line in Figure 7 is

$$\log (L/M^{0.81}) = -0.11 [Fe/H] + 1.71, \qquad (2)$$

where L and M are in solar units. The mass can be eliminated from equation (2) using the RR Lyrae masses determined from the period ratios of the fundamental to first overtone in the double mode RR Lyrae stars in M15, M3, and IC 4499 by Cox, Hudson, & Clancy (1983), and by Clement et al. (1986), following Stellingwerf (1975) and Petersen (1978, 1979). The result (Sandage 1989b) is

$$\log M = -0.1 [Fe/H] - 0.41. \qquad (3)$$

Equation (3) put into equation (2), and changed into magnitudes by assuming $M_{bol} = 4.75$ for the sun, gives

$$M_{bol}(RR) = 0.48 [Fe/H] + 1.30. \qquad (4)$$

From stellar models by Kurucz and others, the bolometric correction at a surface gravity of $\log g = 3.0$ to 3.5 and a temperature of $\log T_e = 3.85$ appropriate for the RR Lyrae stars is found to be

$$B.C. = 0.06 [Fe/H] + 0.06, \qquad (5)$$

which, when put into equation (4) gives

$$M_V(RR) = 0.42 [Fe/H] + 1.24, \qquad (6)$$

as the relation we are seeking. Note that equation (6) has been derived entirely from the pulsation condition of equation (1), together with the empirical correlations of equations (2) and (3). No appeal has been made to theories of the ZAHB. Equation (6) is then an empirical determination based only on the pulsation properties of mechanical oscillators, governed simply by gravity. For this reason, the RR Lyrae luminosities obtained by this method is given the highest weight in what follows.

4 RR LYRAE ABSOLUTE MAGNITUDES DERIVED FROM GLOBULAR CLUSTER MAIN SEQUENCE FITTINGS

Buonanno, Corsi, & Fusi Pecci (1988), in an important paper where the consequences of the HB models of Sweigart & Gross (1976) and Sweigart, Renzini, & Tornambé (1987) are discussed concerning the observed period-shift, metallicity effect, again determine RR Lyrae star luminosities in globular clusters by the method of main sequence fitting. Their result has high weight because of the new precision in their main sequence photometry for their program clusters. Using a particular adopted series of fiducial main sequences for various metallicities, their result is

$$M_V(RR) = 0.37 [Fe/H] + 1.29, \qquad (7)$$

similar to equation (6) in the last section. However, both equations differ in their slope dependences from the results found by four independent research groups that have used the Baade-Wesselink method on field RR Lyrae field variables, as discussed in the next section. To determine how sensitive the main sequence fitting results are to the adopted positions of the fiducial main sequences for different [Fe/H] values, Cacciari & Sandage (1989) have carried out a series of main sequence fittings, using the Buonanno et al. basic main sequence observational data. They began with a different set of assumptions for the positions of the various fiducial main sequence positions than were used by Buonanno et al. with the following result.

Consider first the position of the fiducial main sequence to be used for stars with [Fe/H] = 0. It has often been assumed that the distance to the Hyades is required to establish this ZAMS relation. However, the Hyades distance can be circumvented by using trigonometric parallax stars of high weight, such that the error in absolute magnitude of any given MS star is less than say 0.15 magnitudes (one sigma value). There are 34 such high weight trigonometric parallax stars with metallicities between [Fe/H] of +0.05 and -0.08 in the first Yale Parallax Catalog. The data are shown in Figure 8, where the adopted ZAMS line is drawn as given originally for a similar purpose (Sandage & Eggen 1959, Table III). In an independent study, based on theoretical unevolved main

Figure 7. Same as Figure 6 but for the field variables studied by Lub. The temperatures required to calculate $L/M^{0.81}$ from equation (1), and the metallicities plotted on the abscissa have been determined by Lub from his Walraven photometry.

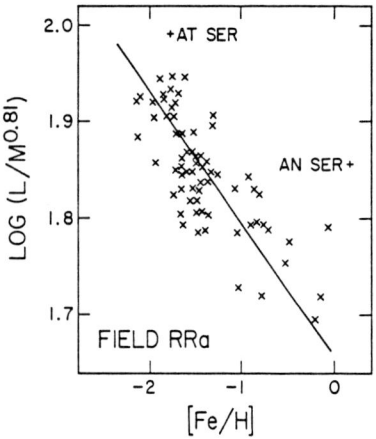

sequence stellar models, VandenBerg (1987) has derived an interpolation equation for the solar abundance ZAMS as

$$M_V = 2.837 - 6.796(B-V) + 31.77(B-V)^2 - 31.6(B-V)^3 + 10.57(B-V)^4. \tag{8}$$

This equation produces the empirical line that is drawn in Figure 8 to better than 0.01 magnitudes at all colors, and, to be emphasized again, is independent of the Hyades distance.

As the metallicity is decreased, the position of the MS in B-V becomes fainter, due to the combined effects of (1) a fundamental change in the stellar models in the M_{bol}, log T_e plane and (2) to the blanketing effects on the B-V colors. Ideally, the main sequence depression could be calibrated empirically from trigonometric parallax stars (e.g. Sandage & Eggen 1959, Eggen & Sandage 1962, Cameron 1985), but the parallax data are not yet of adequate quality to give a definitive solution. Nevertheless, Figure 9 shows that the MS depression does, in fact, occur with decreasing metallicity, but because the scatter of the data is too large to be of quantitative use, we have relied on calculated models of the effect in the M_V, B-V plane.

Figure 8. Adopted fiducial main sequence for stars with [Fe/H] = 0. The points are trigonometric parallax stars with parallax values larger than 0.067 arc sec, whose parallax errors are smaller than 7%, giving magnitude errors (one sigma) of less than 0.15 magnitudes. The line is the ZAMS adopted by Sandage & Eggen (1959), and is closely defined by equation (8) due to Vandenberg. The fiducial main sequence defined in this way is independent of the distance to the Hyades.

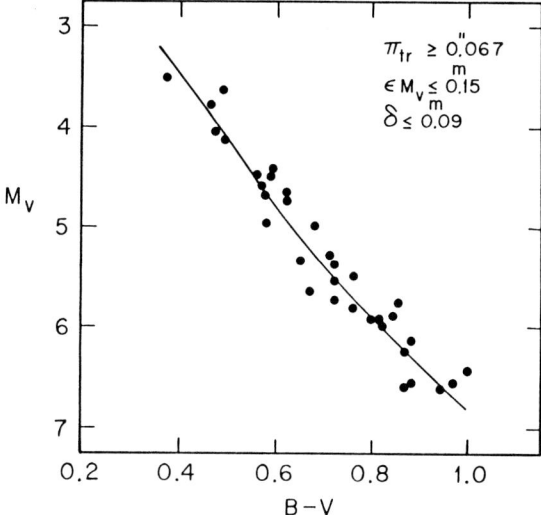

Figure 10 shows our adopted calibration of the depression of M_V below the fiducial main sequence. The solid line in Figure 10 has been read from Figure 3 of VandenBerg & Bridges (1984) and Figure 4 of VandenBerg & Bell (1985), both at B-V = 0.6. The four plotted points are the adopted mean values from trigonometric parallax star data summarized elsewhere (Sandage 1970, Table 11) based on the earlier discussion by Eggen & Sandage (1962). Use of the solid line relation to be discussed in the next paragraph, gives a steeper slope to the resulting M_V(RR) calibration than does the dashed line which gives the slope of $dM_V(RR)/d[Fe/H] = 0.21$ discussed later.

Combining Figures 8 and 10 gives the adopted main sequence positions for different metallicities shown in Figure 11, using the same coding as in Figure 10. Note the very high sensitivity of the slope of the M_V(RR)/[Fe/H] relation to the small changes in the main sequence positions, shown by the dashed lines.

The solid lines in Figures 10 and 11 are well fit by the equation for the chemical composition correction to equation (8) as

$$\Delta M_V = 2.6(Y-0.27) - \{[Fe/H]\}(1.444 + 0.362[Fe/H]) \qquad (9)$$

Figure 9. Trigonometric parallax stars with parallaxes larger than 0.045 arc sec whose ultraviolet excess values are larger than 0.16 magnitudes, corresponding to [Fe/H] values smaller than - 1.

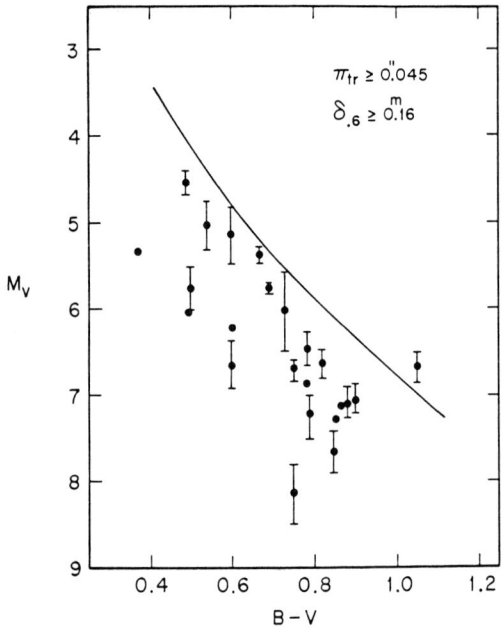

Figure 10. The depression in V magnitudes below the fiducial main sequence as a function of metallicity. The solid line is read from Figure 4 of Vandenberg and Bell (1985) at B-V = 0.6. It is given closely by equation (9). The open circles are from trigonometric parallax stars as discussed by Eggen and Sandage (1962) and Sandage (1970). The dashed line is that required to give the indicated $M_V(RR)$ relation using MS fits for the program globular clusters.

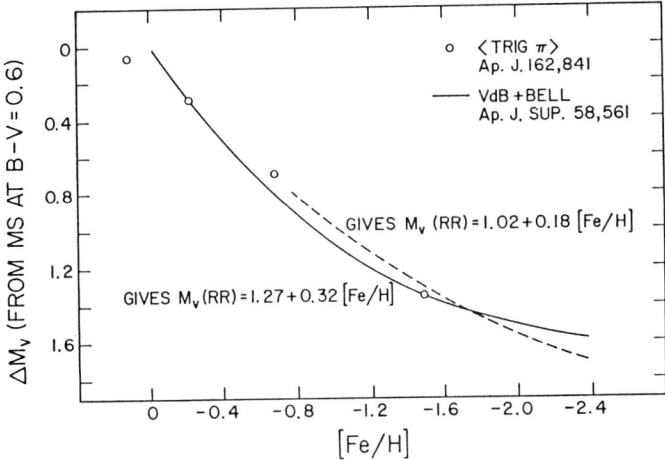

Figure 11. The adopted CMD for different metallicities, obtained by combining Figures 8 and 10.

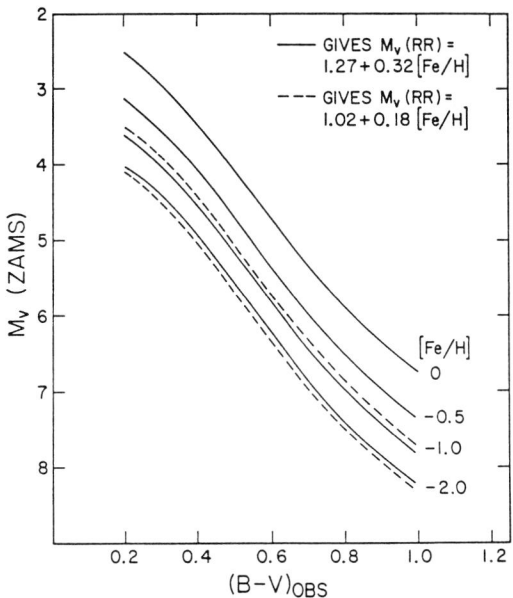

for [Fe/H] > -1.5 derived by VandenBerg (1987). The final result of fitting the MS photometric data of Buonanno, Corsi & Fusi Pecci (1988) to the solid curves in Figures 10 and 11 is shown in Figure 12. The impartial least squares line (averaging the two solutions by interchanging M_V and [Fe/H] as the independent variable) is

$$M_V(RR) = 0.39 \ [Fe/H] + 1.39. \tag{10}$$

If only $M_V(RR)$ is used as the independent variable, then the equation is

$$M_V(RR) = 0.32 \ [Fe/H] + 1.27, \tag{11}$$

as marked in Figures 10 and 11.

5 SUMMARY OF VARIOUS RECENT DETERMINATIONS OF M_V(RR) USING THE BAADE-WESSELINK METHOD

Combining the Baade-Wesselink results of Cacciari, Clementini & Buser (1988), Liu & Janes (1988), Jameson, Fernley & Longmore (1987), and Jones, Carney & Latham (1988), gives the correlation shown in Figure 13, which has the regression equation of

$$M_V(RR) = 0.21 \ [Fe/H] + 1.05. \tag{12}$$

The consistency of the independent data from the four investigations that are combined in Figure 13 would seem, of course, to give considerable weight to equation (12). However, the B-W method is much more complicated (e.g. Gautschy 1987) than the quite direct determination from pulsation data that gives equation (6). In addition, each of the four analyses of the B-W data, although independent, use the same basic assumptions in their central details of the computations. They also use the same Kurucz model atmospheres, which, although the best currently available, are not yet definitive (Kurucz, comment at the Van Fleck Age Dating Workshop, 1988). It might then be that all present B-W results could contain a common systematic error, if such exists. The large slope difference between equation (12) and equations (6), (7), (10), and (11) might be explained in this way.

6 SUMMARY OF VARIOUS DETERMINATIONS OF RR LYRAE LUMINOSITIES

Figure 14 shows four assumptions for the variation of RR Lyrae star luminosities with metallicity. The line labeled "pulsation" is from equation (6) of §3. The line labeled MS is equation (11) of §4. The result from equation (12) is labeled B-W. Also shown is the assumption of a constant absolute magnitude of $M_V = 0.70$, based on the statistical parallax solution of Hawley et al. (1986), as slightly modified by Barnes & Hawley (1986). Ages based on these four assumptions are discussed in the next two sections, with and without Oxygen enhancement.

Various other determinations from the literature are compared in Figure 15. The line marked <BCF>(MS) is the calibration obtained by Buonanno, Corsi & Fusi Pecci (1988), mentioned previously, from their direct main sequence fits, giving equation (7) of §4. The dashed line marked Lub (field) is from Lub's (1987) determination of the $L/M^{0.81}$ ratio using

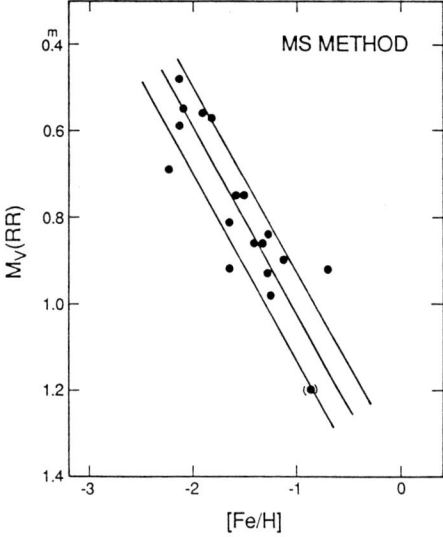

Figure 12. The calibration of $M_V(RR)$ as a function of [Fe/H] obtained by main sequence fitting the program globular clusters studied by Buonanno, Corsi, & Fusi Pecci (1988) to the grid of solid lines in Figure 11, defined by equations (8) and (9).

Figure 13. Summary of the calibration of $M_V(RR)$ as a function of [Fe/H] obtained from the Baade-Wesselink method obtained by the four research groups discussed in the text.

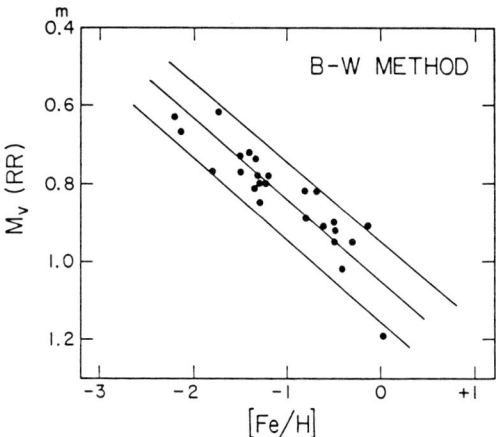

equation (1), applied to his field star data, assuming a fixed value of the RR Lyrae mass. The dashed line marked ND is by Noble & Dickens (1988) from their direct fitting of globular cluster CMDs to an adopted fiducial main sequence defined by subdwarfs with moderaltely large trigonometric parallaxes. The two lines marked <u>theory</u> are from the HB models of Sweigart, Renzini & Tornambé (1987, SRT) for two values of the Helium abundance. From their Figure 1, and by applying the bolometric correction of equation (5), one can read the predicted calibrations from their theory to be

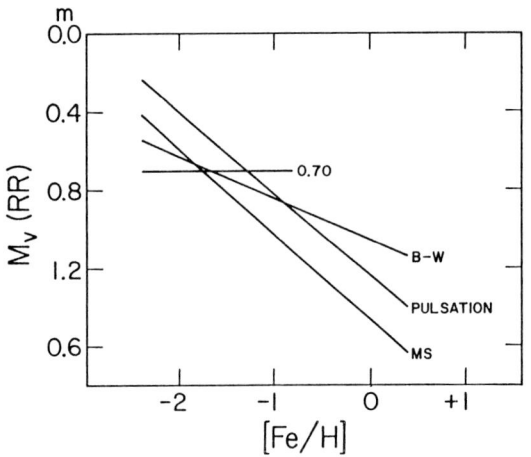

Figure 14. Comparison of four calibrations of $M_V(RR) = f([Fe/H])$.

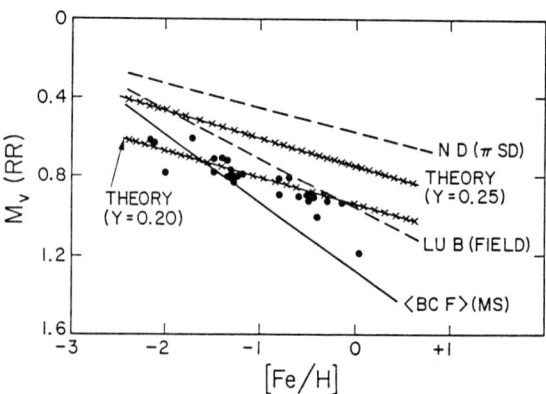

Figure 15. Comparison of additional calibrations of RR Lyrae absolute magnitudes. The plotted points are the same as shown in Figure 13, obtained from the Baade-Wesselink method.

$$M_V(RR) = 0.15 \, [Fe/H] + 0.95, \quad (13)$$

for Y = 0.20,

$$M_V(RR) = 0.13 \, [Fe/H] + 0.74, \quad (14)$$

for Y = 0.25,

$$M_V(RR) = 0.10 \, [Fe/H] + 0.54, \quad (15)$$

for Y = 0.30 where the zero points are based on the theoretical calculations. These model values for $M_V(RR)$ have consistently been brighter than those favored by the observers by about 0.2 magnitudes for any reasonable He abundance such as Y = 0.23. Concerning the He abundance, Figure 15 shows the well known fact that the Sweigart and Gross, and the SRT HB models require an anticorreleation of Y and [Fe/H] if the steep slope of the the <BCF> calibration of equation (7), or the pulsation calibration of equation (6), is adopted. The steeper lines labeled <BCF> and Lub in Figure 15 cut across the two theory lines, showing that higher Y is required for lower [Fe/H] if these ZAHB models are correct.

7 CLUSTER AGES USING THE VARIOUS $M_V(RR)$ CALIBRATIONS ASSUMING NO OXYGEN ENHANCEMENT

The age, T, of a cluster whose bolometric luminosity at the main sequence turn-off is M_{bol} (TO) is

$$\log T = 8.319 + 0.41 M_{bol} (TO) - 0.15 [Fe/H] - 0.43(Y - 0.24), \quad (16)$$

as interpolated (Sandage, Katem & and Sandage 1981, equation (18) and Figure 14) from the extensive Yale tables of Ciardullo & Demarque (1977, CD). The results are very similar to those obtained earlier by Simoda & Iben (1968, 1970), and Iben & Rood (1970), and later by VandenBerg (1983) and by VandenBerg & Bell (1985). The accuracy of equation (16) in reproducing the CD tables is better than 5% over the range of Y, [Fe/H], and T of interest for the globular system.

Equation (16) can be related to $M_V(RR)$ by noting that the observed magnitude difference between the ZAHB and the main sequence turn-off is 3.54 magnitudes in V, and that the bolometric correction for MS stars at the turn-off is 0.1 magnitudes to give, from equation (16) with Y = 0.24,

$$\log T = 9.734 + 0.41 \, M_V(RR) - 0.15 \, [Fe/H]. \quad (17)$$

Substituting $M_V(RR) = a \, [Fe/H] + b$ for the RR luminosity in equation (17) shows directly how the age T should vary with metallicity. Clearly, the condition for no variation of T with [Fe/H] is

$$a = 0.15/0.41 = 0.37,$$

which is close to the calibration values determined from the pulsation data [equation (6)] and from the MS fitting (equations (7), (10), and (11)].

The ages calculated from equation (17), using three assumptions for how $M_V(RR)$ varies with metallicity, are shown in Figure 16. If the observed Fe abundance tracks the total metal abundance which determines the interior opacity, the resulting ages, calculated from equation (17), are on the left. Age determinations for each of the program clusters, based on the individual magnitude differences between the ZAHB and the MS turn-off are shown as dots. The trend, assuming that each cluster has a 3.54 magnitudes difference (in V) between the ZAHB and the MSTO, is shown as the solid lines. Our preferred calibration of $M_V(RR)$, which lies between the bottom two panels, gives a mean age of 16 Byr with no measurable variation with [Fe/H]. Ages with an Oxygen enhancement of [O/Fe] = +0.6, discussed in the next section, along the right hand ordinate, are smaller by a factor of 1.2.

Figure 16. Summary of the age determinations for the program clusters using three different assumptions for the RR Lyrae absloute magnitudes as a function of metallicity. Left hand ordinate is with no Oxygen enhancement. Right hand ordinate assumes [O/Fe] = + 0.6 for all relevant [Fe/H] values.

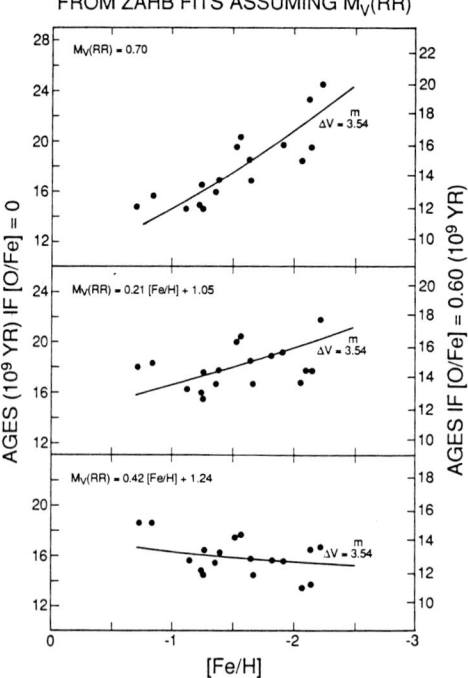

8 CLUSTER AGES FROM $M_V(RR)$ ASSUMING OXYGEN ENHANCEMENT

Strong evidence exists from field subdwarfs that [O/Fe] > 0 for [Fe/H] less than -1. Figure 17 from Sneden's (1985) review shows that for [Fe/H] less than -1, the [O/H] enhancement is at least = 0.5. Because Oxygen is more abundant than Fe in any standard adopted chemical cosmic mixture by about a factor of 250, the opacity in a mix that has an O enhancement is not as low at a given [Fe/H] value as it would have been otherwise. The approximate effect on the ages can be estimated from equation (17) by replacing [Fe/H] by the true metal abundance [M/H] relative to the sun.

Suppose the curves in Figure 17 are represented by

$$[O/Fe] = c [Fe/H] + d. \tag{18}$$

To a good approximation, [M/H] = [O/H] = (c + 1) [Fe/H] = d. This, put into equation (17) by replacing the [Fe/H] term, gives

$$\log T = 9.734 + 0.41 M_V(RR) - 0.15 \{(c + 1) [Fe/H]\} - 0.15 d. \tag{19}$$

As a compromise between curves A and B in Figure 17 we adopt

$$[O/Fe] = -0.14 [Fe/H] + 0.40 \tag{20}$$

for the Oxygen enhancement. This, put into equation (19) gives the age dating results shown in Figure 18. Again, our preferred solution, based on the most reliable $M_V(RR)$ calibration between the last two panels in Figure 18, gives the average age of the globular cluster system to be 13 Byr.

Figure 17. Summary of the observed Oxygen enhancement given by Sneden (1985). The adopted variation of [O/Fe] = -0.14 [Fe/H] + 0.40 for [Fe/H] smaller than -1 is between curves A and B.

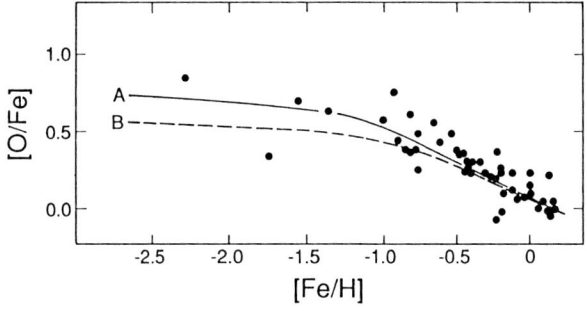

Figure 18. Same as Figure 16 but assuming an Oxygen enhancement given by equation (20).

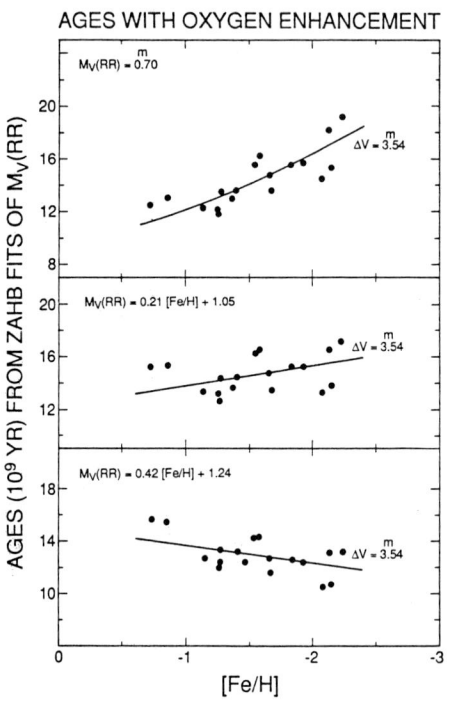

Figure 19. Ages with and without Oxygen enhancement using direct main sequence fits to determine $M_V(TO)$. The ages are larger than in Figures 16 and 18 because of the 0.2 magnitudes offset shown in Figure 14 between the lines labeled MS and pulsation.

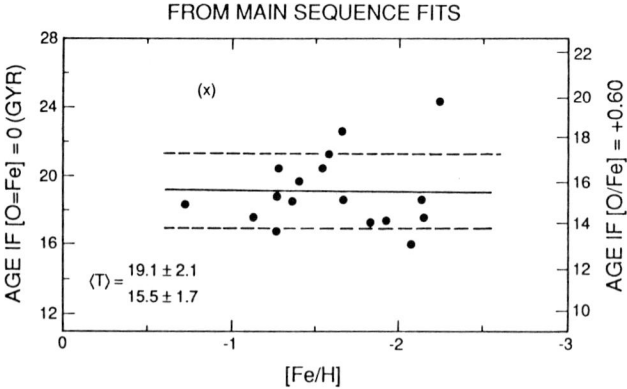

9 CLUSTER AGES FROM M_{bol} (MSTO) VALUES USING DIRECT MS FITTING

The M_V values of the main sequence turn-offs for each of the program clusters can be used directly in equation (16), (adding the bolometric correction of -0.1 magnitudes) to obtain ages, independent of any assumption about the RR Lyrae calibration. The result is shown in Figure 19, with and without [O/Fe] enhancement. The ages of 19.1 Byr and 15.5 Byr are longer than those given in Figures 16 and 18 because the main sequence fitting data for M_V(RR) are fainter in the mean by about 0.2 magnitudes from those given either by the pulsation or the B-W M_V(RR) calibrations. This is shown by the off-set of the MS curve from the others by this amount in Figure 14. As with our preferred RR Lyrae calibrations in Figures 16 and 18, there is no measurable variation of cluster ages with metallicity in Figure 19 within the scatter of about 10% on either side of the mean horizontal line. This independence of age on metallicity is consistent with the rapid formation of the galactic halo in about the free fall time (no dissipation, i.e. no pressure support in the halo formation phase) envisaged in the model suggested by Eggen et al. (1962).

REFERENCES

Barnes, T.G. & Hawley, S.L. (1986). Ap.J.,307, L9.
Buonanno, R., Corsi, C.E. & Fusi Pecci, F. (1988). Astron. & Astrophys., submitted, (ESO preprint 594).
Butler, D., Dickens, R.J. & Epps, E. (1978). Ap.J.,225, 148.
Cacciari, C., Clementini, G. & Buser, R. (1988). STScI preprint 268.
Cameron, L.M. (1985). Astron. & Astrophys.,146, 59.
Ciardullo, R.B. & Demarque, P. (1977). Yale Trans., 33.
Clement, C.M., Nemec, J.M., Wells, N.R., Wells, T., Dickens, R.J. & Bingham, E.A. (1986). A.J.,92, 825.
Cox, A.N. (1987). In Second Conference on Faint Blue Stars, IAU Colloquium No.95, eds. A.G. Davis Philip, D.S. Hayes & J.W. Liebert, p. 161. Schenectady, NY: L. Davis Press.
Cox, A.N., Hodson, S.W. & Clancy, S.P. (1983). Ap.J.,266, 94.
Eggen, O.J., Lynden-Bell, D. & Sandage, A. (1962). Ap.J.,136, 748.
Eggen, O.J. & Sandage, A. (1962). Ap.J.,136, 735.
Gautschy, A. (1987). Vistas in Astronomy,30, 197.
Gratton,R.G., Tornambé, A. & Ortolani, S. (1986). Astron. & Astrophys.,169, 111.
Hawley, S.L., Jeffreys, W.H., Barnes, T.G. & Lai, W. (1986). Ap.J.,302, 626.
Iben, I. (1971). P.A.S.P.,83, 697.
Iben, I. & Rood, R.T. (1970). Ap.J.,159, 605.
Jameson, R.F., Fernley, J.A. & Longmore, A.J. (1987). M.N.R.A.S., in press.
Jones, R.V., Carney, B.W. & Latham, D.W. (1988). Ap.J.,332, 206.
Liu, T. & Janes, K.A. (1988). In Calibrating Stellar Ages, IAU Colloquium, ed. A.G. Davis Philip. Schenctady, NY: L. Davis Press.
Lub, J. (1977). Thesis, University of Leiden.

Lub, J. (1987). In Stellar Pulsation, Lecture Notes in Physics, No. 274, eds. A.N. Cox, W.M. Sparks & S.G. Starrfield, p. 218. Berlin: Springer-Verlag.
Noble, R. & Dickens, R.J. (1988). In New Ideas in Astronomy, ed. F. Bertola, J.W. Sulentic & B.F. Madore, p. 59. Cambridge: Cambridge University Press.
Peterson, J.O. (1978). Astron. & Astrophys., 62, 205.
Peterson, J.O. (1979). Astron. & Astrophys., 80, 53.
Sandage, A. (1970). Ap.J., 162, 841.
Sandage, A. (1987). In Second Conference on Faint Blue Stars, IAU Colloq. 95, eds. A.G. Davis Philip, D.S. Hayes & J.W. Liebert, p. 41. Schenectady, NY: L. Davis Press.
Sandage, A. (1989a). Ap.J., in press.
Sandage, A. (1989b). Ap.J., in press.
Sandage, A. & Cacciari, C. (1989). Ap.J., in press.
Sandage, A. & Eggen, O.J. (1959). M.N.R.A.S., 119, 278.
Sandage, A., Katem, B.N. & Sandage, M. (1981). Ap.J. Suppl., 46, 41.
Simoda, J. & Iben, I. (1968). Ap.J., 152, 509.
Simoda, J. & Iben, I. (1970). Ap.J. Suppl., 22, 81.
Sneden, C. (1985). In Production and Distribution of C,N,O Elements, eds. I.J. Danziger, F. Matteucci & K, Kjar, p. 1. Garching: ESO.
Stellingwerf, R.F. (1975). Ap.J., 199, 705.
Sweigart, A. V. & Gross, P. G. (1976). Ap.J. Suppl., 32, 367.
Sweigart, A.V., Renzini, A. & Tornambé, A. (1987). Ap.J., 312, 762.
van Albada, T.S. & Baker, N. (1971). Ap.J., 169, 311.
VandenBerg, D.A. (1983). Ap.J. Suppl., 51, 29.
VandenBerg, D.A. (1987). Colloquium at Space Science Science Institute, Baltimore.
VandenBerg, D.A. & Bell, R.A. (1985). Ap.J. Suppl., 58, 561.
VandenBerg, D.A. & Bridges, T. (1984). Ap.J, 278, 679.

THE GLOBULAR CLUSTER ωCENTAURI AND ITS RR LYRAE VARIABLES

R.J. Dickens
Rutherford Appleton Laboratory, Chilton, Didcot,
Oxfordshire OX11 0QX, UK.

Abstract. The significance of some of the unusual characteristics of the globular cluster ωCentauri in various fundamental problems is explored. Interest is centred on the properties of the cluster RR Lyraes, and what they can contribute to studies of early cluster chemical enrichment, stellar pulsation, the distance scale, stellar evolution, stellar ages and the Oosterhoff period-shift problem. This article, which is intended to highlight problems and progress rather than give a comprehensive review, includes new results based on photometry of the RR Lyraes, red giants, subgiants, horizontal-branch and main sequence stars in the cluster.

1 INTRODUCTION
1.1 The cluster

The globular cluster ωCentauri has been the subject of intensive study over the last two decades. The motivation for much of this work has been the realisation that the cluster possesses unusual properties, first hinted at in its peculiar HR diagram (Woolley 1966) which showed a large intrinsic scatter in the red giant branch. One of the more notable of its peculiarities is the spread in chemical composition, either measured or inferred among all samples of stars studied (red giants, subgiants, RR Lyraes and main-sequence stars).

ωCen is also the most luminous and massive of the galactic globular clusters, properties which might well be related to its apparent peculiarity. This fact leads one immediately to the first major question posed by the cluster's properties, is the cluster unique in some fundamental way or is it just an extreme example of a galactic globular cluster, its "peculiarities" showing up primarily because of the richness of the stellar sample it provides? However, because the cluster is so massive, this in itself might be responsible in a fundamental way (e.g. in the conditions at formation) for the cluster's apparent uniqueness. This question assumes great importance if we are to apply what we learn about the physical causes of the properties of the "sub-populations" in ωCen to interpret the properties of other clusters.

In many ways, ωCen resembles a small galaxy. It is noticeably elliptical in shape, as expected from its observed rotation about the minor axis (Harding 1966). The existence of identifiably different stellar populations within it provides a perhaps unique opportunity to study differential effects between them, especially valuable in view of the close proximity of the cluster as compared to any galaxy, even the nearby dwarf spheroidal systems which its stellar content appears to resemble.

1.2 The RR Lyrae variables

Since the turn of the century, the RR Lyrae variable stars in ωCen have attracted the attention of astronomers. Pioneering work by Bailey (1902) was followed later by a major study of Martin (1938), a classic paper which has become the fundamental reference work for subsequent studies. More recent papers include a study of the period changes (Belserene 1964), a photometric study (Dickens & Saunders 1965) and spectroscopic studies by Freeman & Rogers (FR, 1975), Butler Dickens & Epps (BDE, 1978) and Gratton, Tornambé & Ortolani (GTO, 1986).

These spectroscopic studies confirmed that the range in chemical composition expected from the intrinsic spread in colour on the giant branch occurred also among the RR Lyrae population. The spread in abundance is now firmly established from spectroscopic studies of individual giant stars (e.g. Cohen 1981; Caldwell & Dickens 1988), and has demonstrated that both light and heavy elements are involved, the spread arising most probably from both primordial and mixing origins (see Smith 1987 for a comprehensive review).

To set the scene for what follows, Figure 1 shows a composite colour-magnitude diagram (CMD) for ωCen, in which stars brighter than about V = 16 are taken from Dickens et al. (1988) and those fainter from Noble (1987). Note the huge colour spread amongst the giants, and the more modest spread on the main sequence, both being much larger than can be accounted for if all the stars had the same heavy-element abundance.

2 CLUSTER FORMATION AND EARLY CHEMICAL ENRICHMENT

Evidence bearing on conditions at the time the cluster was formed comes from the observed range in heavy-element abundance, its shape and the existence or otherwise of a radial gradient. We look first at the RR Lyraes, which provide the only **direct** evidence of the full range and shape of the abundance distribution.

Figure 2 shows a histogram of the distribution with [Fe/H] for all RR Lyraes for which [Ca/H] has been determined by the ΔS method. Most values come from BDE, which incorporated FR, but stars in common with GTO have new values obtained from averaging with theirs. The filled part of the histogram represents the Bailey c-type variables. The overall distribution is clearly asymmetric, tailing off slowly towards the more metal-rich end. The c-types alone show a similar distribution, but there is a deficiency of metal-rich stars of this type. The ab-types alone show less marked asymmetry. Although there may appear to be no immediately obvious reason to separate the sample according to type, and therefore pulsation mode (ab being fundamental and c first harmonic), the stars do occur in different parts of the horizontal branch (HB), and this would be influenced in part by their abundance,

Figure 1. A composite CMD for ω Cen, in which stars brighter than V=16 comprise photographic photometry from an annular region, and those fainter from CCD photometry of several fields located about 20 arcmin west of the cluster centre.

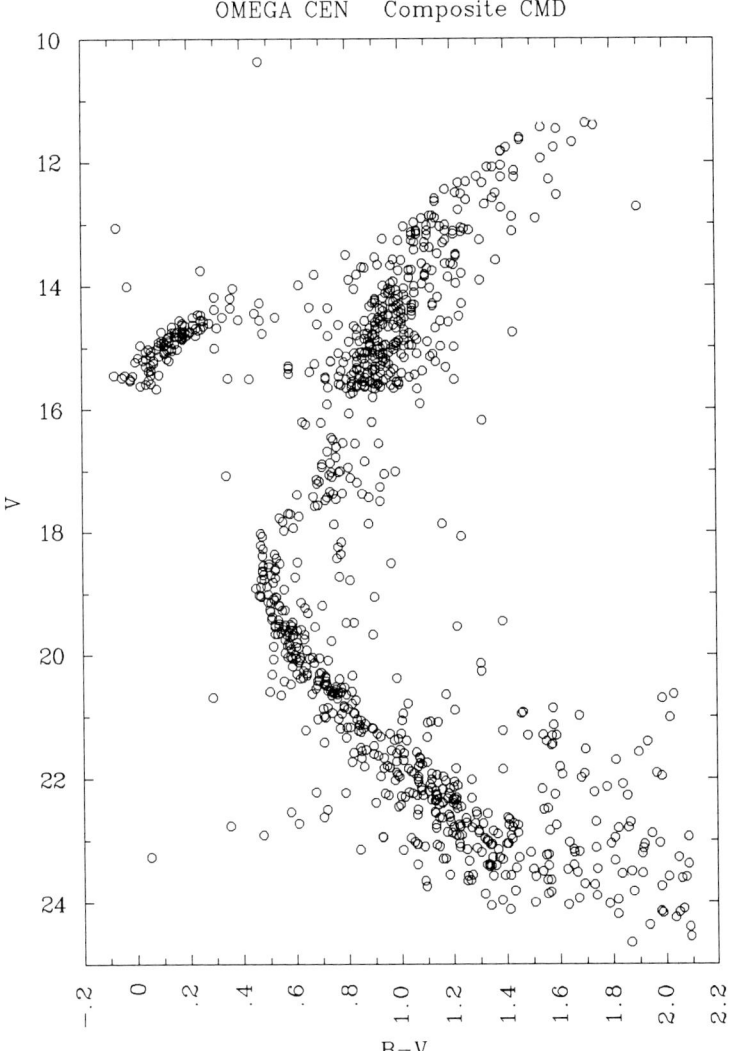

especially if they lay near the zero-age HB (ZAHB). On the basis of HB models, one might expect a tendency for the (bluer) c-types to be on average more metal-deficient, as is indeed suggested by their distributions. Another interesting property of these distributions is the fairly strong indication of bimodality. Although difficult to prove conclusively, we have investigated this possibility further, and show in Figure 3 a Maximum Likelihood fit of a double-Gaussian distribution to the full sample (the binning is slightly different in this diagram). The parameters derived for the two underlying populations are listed in Table 1. From Monte-Carlo simulations, we find that the probability (in a Kolmogorov-Smirnov test) that this double-Gaussian is a true representation of the data is only about 5%, i.e. on the borderline of being a satisfactory fit.

Thus we are teased with the possibility that there could be two underlying populations of stars in ωCen, as sampled in the instability strip, which might have arisen in two discrete bursts of star formation, or even in a merger of two separate systems (Searle 1977). However we

Figure 2. The distribution of [Fe/H] amongst the ω Cen RR Lyraes, as derived from [Ca/H] measurements. The filled part of the histogram represents the c-type variables alone. The distribution is clearly asymmetrical and also looks bimodal.

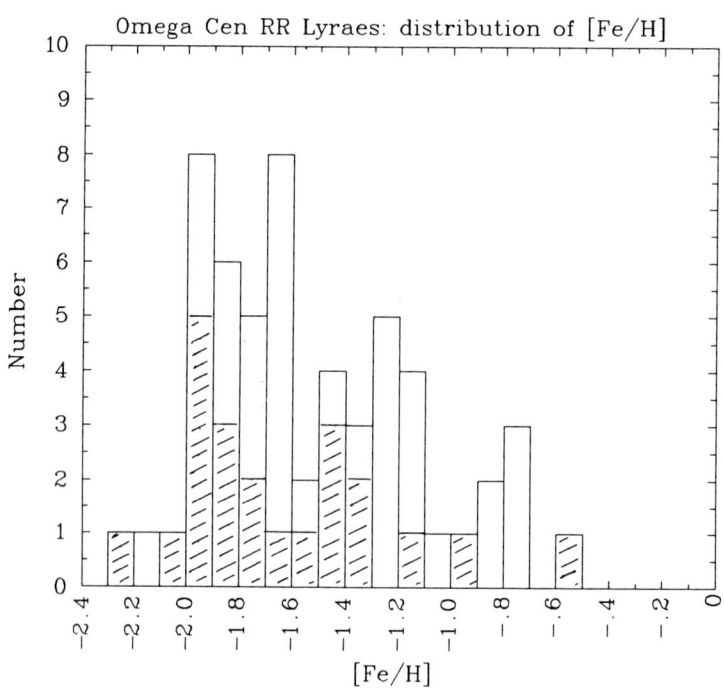

are not yet able to rule out a more steady enrichment process in which an initial burst of star formation forming the metal-poor peak was followed by a gradual decline in the rate as the interstellar material became steadily enriched.

Turning now to the question of a radial gradient in calcium abundance, as has been claimed by Freeman (1985) and earlier discussed by Smith (1981), we show in Figure 4 abundance histograms for three radial groupings. These show marginal evidence for a bias towards more metal-poor stars at radii greater than 10 arcmin. This effect is perhaps more marked in the c-types, as shown in Figure 5 where there appears to be a deficiency in metal-richer c-types at the larger radii

Table 1.
Parameters for underlying populations

Sample	No.	<[Fe/H]>	dispersion
metal-poor	26	-1.84	+/-0.17
metal-rich	30	-1.26	+/-0.35

Figure 3. The Maximum Likelihood fit of a double-Gaussian to the RR Lyrae abundance distribution.

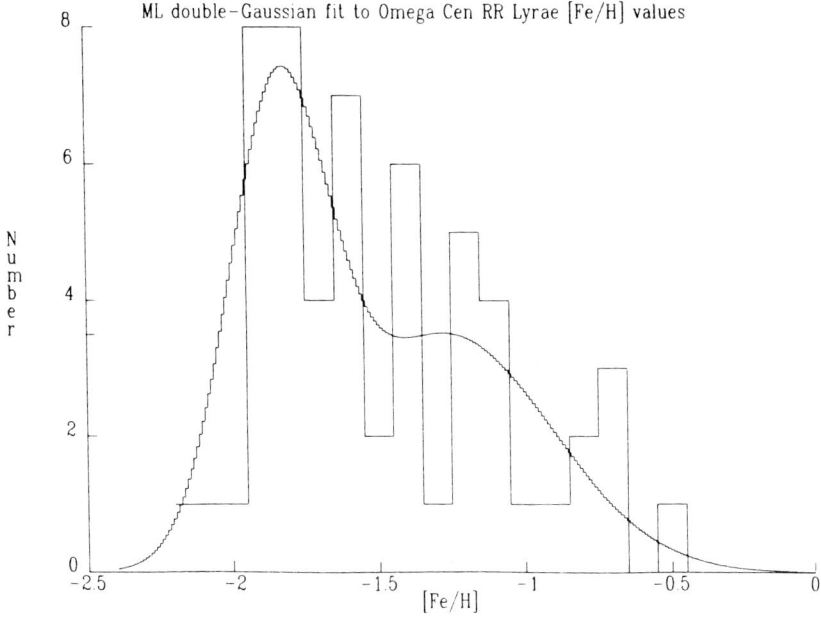

Figure 4. RR Lyrae abundance distributions for three radial groupings. The outer group show a less uniform distribution, with an excess of metal-poor RR Lyraes as compared to the inner groups.

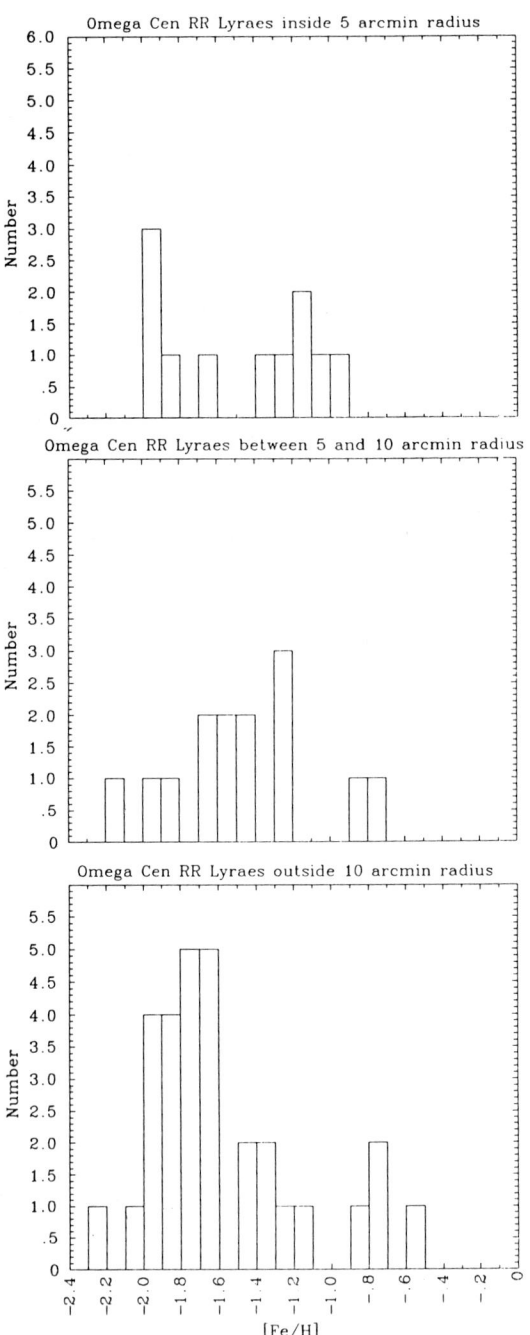

in contrast to the ab-types. None of these trends appear to have formal statistical significance, but it must be kept in mind that even moderately strong gradients would be obscured in projection. A substantially larger sample of stars with abundance measurements would clearly be of interest in order to follow up these tantalising trends, which are similar in kind to those found in galaxies, and if real, must have been laid down at the epoch of star formation in the cluster.

Indirect evidence on the abundance distribution comes from stellar colours in various parts of the CMD, on the assumption that the intrinsic width (corrected for observational error) is largely due to a variation in abundance. This is indeed the most likely explanation for giants, subgiants and main sequence stars; in the turnoff region, a range in age would also contribute to the spread. Figure 6 shows a distribution with colour for a half-magnitude slice of the giant branch, between radii of 10 and 25 arcmin, taken from the data of Dickens et al. (1988). This is a typical distribution for stars on the giant branch, with a suggestion of a red (higher metal content) tail as found in the RR Lyraes. A similar (asymmetrical) distribution in residual (V-K) colour (an abundance parameter) was obtained for all stars studied

Figure 5. [Fe/H] versus radius for RR Lyraes in ωCen; open circles are c-types, filled circles ab-types. The most metal-rich variable is V84, formerly designated an ab-type but now thought to be an overtone pulsator, and hence a c-type. In spite of this, there appears to be a deficiency of metal-rich c-types outside a radius of 0.1 degree.

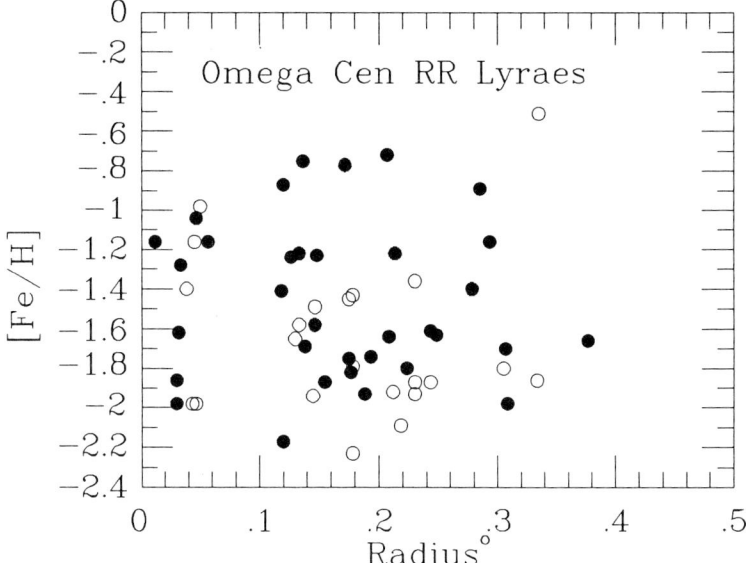

by Persson et al. (1980, Figure 4). Their photometrically unbiased sample showed no asymmetry, but the large spread in colour, and hence abundance, remained.

An intrinsic spread in abundance is also implied for subgiants, on the basis of the distribution with colour found by Da Costa & Villumsen (1981, Figure 2). The observations again show a red (presumed high abundance) tail, but no detailed analysis of these data appears to have been published.

Observations of lower subgiant and main-sequence stars are illustrated in Figure 7, taken from Noble (1987). Superposed on the data are 16 Gyr isochrones for four values of [Fe/H]. Since the photometric accuracy on the upper main sequence is of order 0.01 to 0.02 magnitude, it is clear that a large range in [Fe/H] amongst the stars can account for the observed range in colour, the extreme isochrones of [Fe/H] = -2.23 and -0.77 roughly delineating the extremes. Although there is some contamination by field stars, this does not materially affect this conclusion, which therefore suggests a "primordial" spread comparable to that shown by the RR Lyraes. These results are further quantified in

Figure 6. Distribution with colour in a sample of red giants in ⍵ Cen. The blue side of the distribution is contaminated with AGB stars. Allowing for this reinforces the apparent asymmetry, which has a red, presumed more metal-rich tail.

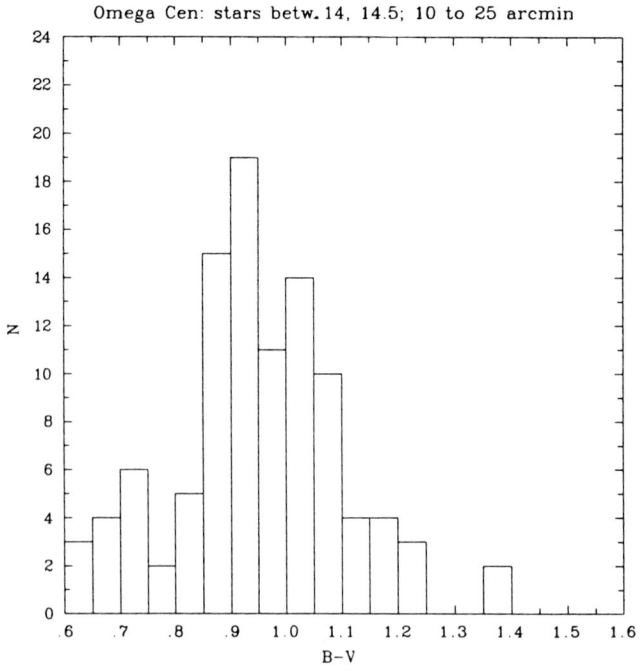

Figure 8, which shows histograms of the observed colour residuals about a fiducial line close to the isochrone for [Fe/H] = -1.27 in Figure 8, for one magnitude slices. The simple Gaussian fits to the residuals shown are not good representations in general, and the true shape of these distributions is yet to be investigated. Note that the isochrones indicate a non-linear relationship between colour and metallicity on the main sequence and subgiant branches. These results will be fully discussed elsewhere (Noble et al. 1989).

In summary, all evidence points to a large intrinsic spread in [Fe/H], from less than or of order -2.0 to about -0.5. Evolved stars indicate a skew distribution having a high metal-abundance tail. There could be at least two underlying distributions. Further work, some already in hand, promises to sharpen these results considerably.

Figure 7. CMD of the main sequence and turnoff region of ω Cen. Four isochrones of differing [Fe/H] are shown, the extreme ones embracing virtually all stars on the upper main sequence and subgiant branch.

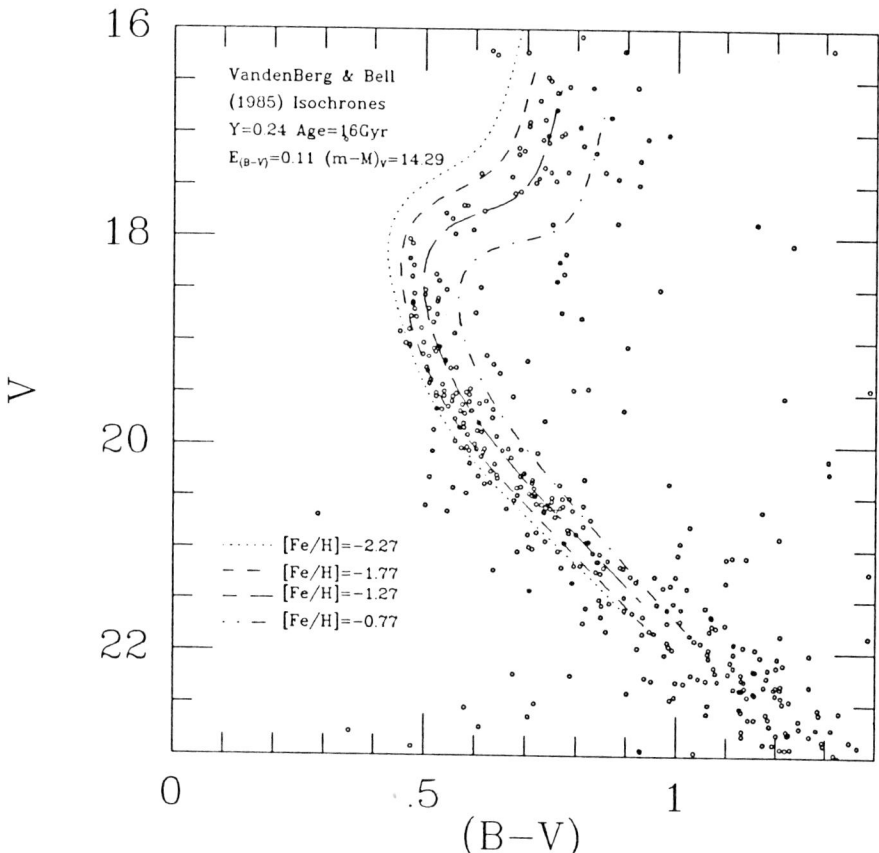

Figure 8. Histograms of colour residuals measured from a fiducial mean main-sequence line, for one magnitude slices. Fitted Gaussians are shown, with indicated dispersions much larger than the observational errors.

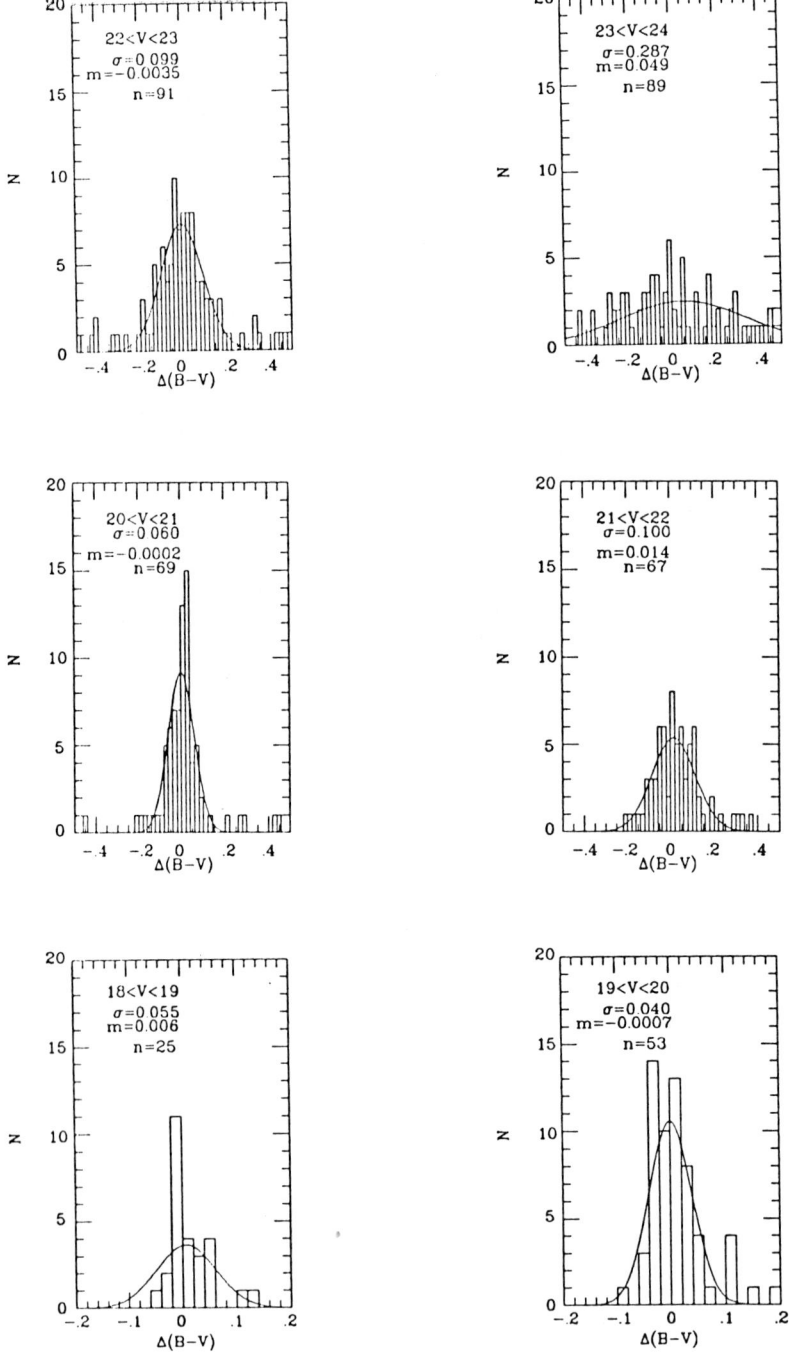

3 STELLAR PULSATION

The pulsational properties of the RR Lyraes of course play a central role in our understanding of the stars themselves, but also in understanding the peculiarities of ωCen, and in most of the wider questions being addressed here. Accurate light curves in several wavebands enable comparisons with sophisticated modelling, and provide the data necessary to derive important physical parameters that can be used to test stellar evolution theory (including some assessment of the correctness of the input physics to models; e.g. see Sweigart et al. 1987). The data to be discussed for ωCen are primarily B,V photometry by Dickens & Bingham (1989), but some reference will be made to Walraven photometry by de Bruijn & Lub (1987) and infrared K photometry by Longmore et al. (1989).

The position of the RR Lyraes in the CMD is shown in Figure 9. An equilibrium colour, $(B-V)_{eq} = 2/3 \langle B-V \rangle_{mag} + 1/3 (\langle B \rangle_{int} - \langle V \rangle_{int})$, has been calculated following Lub (1977), which is seen to combine both

Figure 9. The positions in the CMD of all the RR Lyraes with B,V photometry are given. The location of the blue edge of the instability strip (HBE) is shown as a vertical line. In this and subsequent Figures, open circles are c-types, filled circles ab-types. Most stars occupy a well-defined region about 0.1 magnitude deep, with only a few stars (flagged) lying off. It is believed that the fainter stars lie near the ZAHB, whereas the bulk are more evolved (see text).

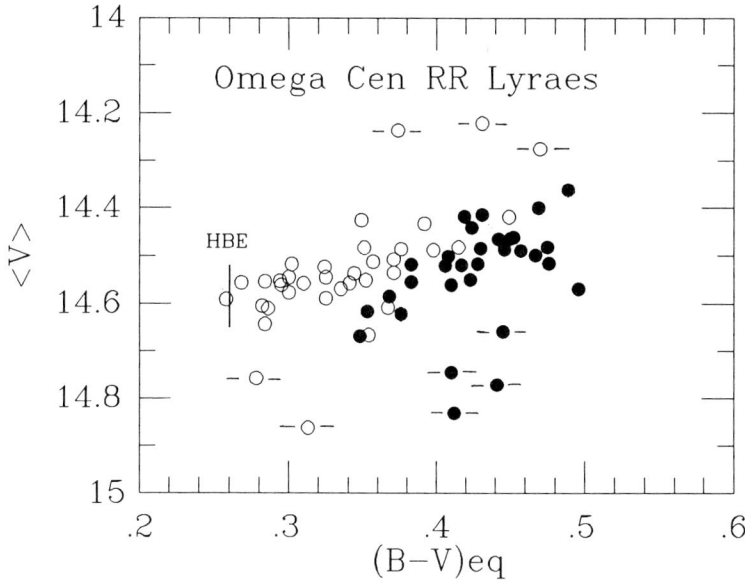

Figure 10. The correlation of blue amplitude with a) log P and b) log P' = log P + 0.33 (m_{bol} - $<m_{bol}>$) where $<m_{bol}>$ is the mean value at the observed colour, and P' the "corrected" period. The smaller scatter in b) shows that some of the luminosity spread is real.

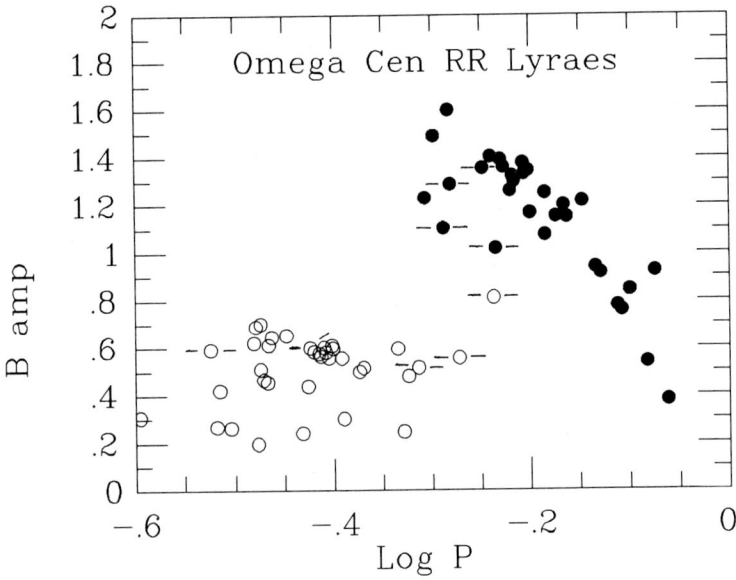

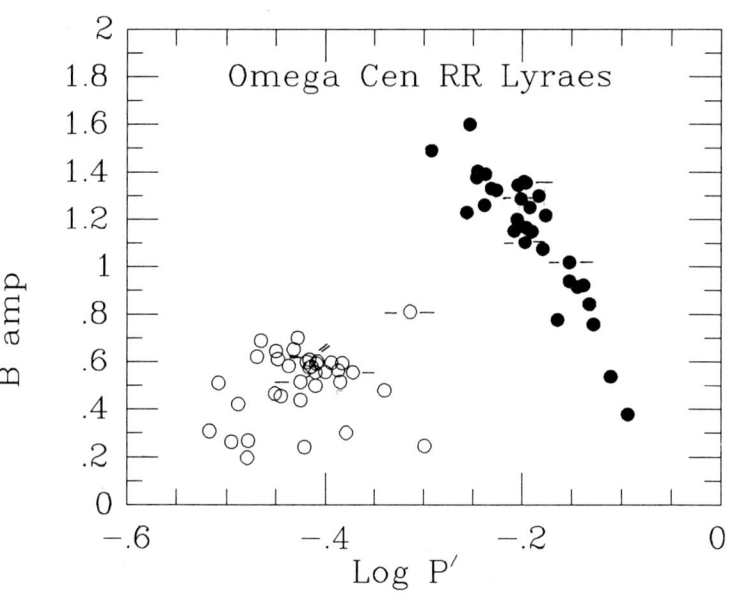

magnitude and intensity means. Although most variables occupy a well-defined region on the HB, as in other clusters, about 10% of the sample lie well away from the HB, and are flagged in this and the subsequent three figures. Only one of these stars is a non-cluster member. These stars tend to lie off other correlations between the light-curve parameters, and must differ in some important property from the majority of HB stars (see, for instance the period-amplitude diagram of Figure 10(a)). The individual periods may be corrected for the differing luminosity within the cluster, by normalising to the mean relationship followed by the stars in Figure 9. This is convenient when comparing mean properties between clusters, as for example in the (corrected) period-amplitude diagram shown in Figure 10(b). The location of the mean line of the ab-types in this diagram has often been used to define the Oosterhoff period-shift between clusters. The tightening of the period-amplitude relation is evidence that some of the luminosity spread is indeed real, as shown by van Albada & Baker (1973).

Figure 11 shows how the amplitude varies across the instability strip. Note the occurrence of a number of small-amplitude c-types close to the

Figure 11. The correlation of blue amplitude with equilibrium colour for the complete B,V sample. Note the occurrence of a number of very small-amplitude c-types near the blue edge. These stars appear to be absent in M15 (see Bingham et al. 1984), whereas it is the multimode variables occurring at the transition colour in M15 that are entirely absent in ⍵ Cen. These properties are thought to be related to the track morphology (see text).

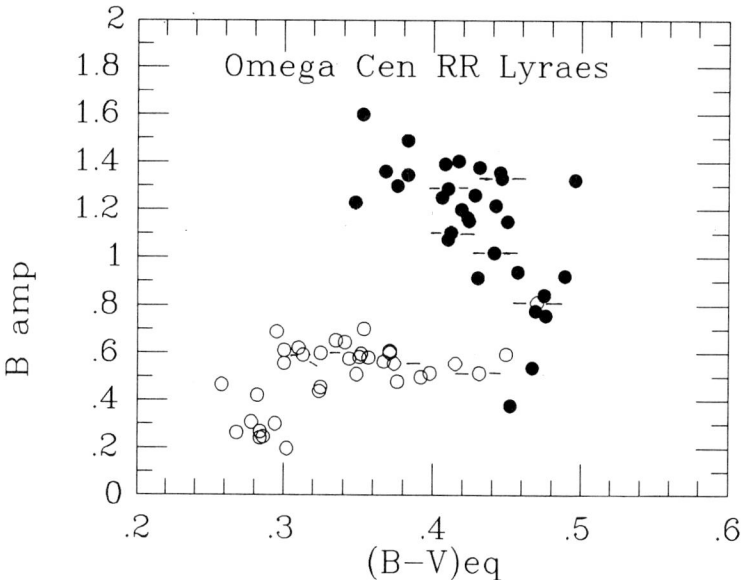

blue edge; these stars do not appear to exist in M15 (Bingham et al. 1984, Figure 16). This could be a selection effect, but might also relate to differences in track morphology between the clusters. Another notable difference between these two clusters is the presence of many multimode variables in M15, right at the transition temperature, and the complete absence of any such stars in ωCen (Stellingwerf & Dickens 1983, Nemec et al. 1986). Again this could be related to the details of track morphology, as has been discussed by Bingham et al. (1984) and will be discussed later. Finally, the correlation of period with colour for all the sample is shown in Figure 12. There is a considerable scatter within each sequence (ab or c) in this diagram, due both to the stars lying away from the HB (flagged) and to the inclusion of data of poorer quality. The complete data set has been used here, and in the three previous figures to illustrate the overall characteristics. Selection of only first quality photometry leads to improved correlations as shown in Figures 13 to 16. In the CMD of Figure 13, non-variable blue HB (BHB) stars have been included to delineate the blue edge of the strip. The B amplitude - log P' relationship in Figure 14 also shows schematically the mean locations of the RR Lyraes in M3

Figure 12. The period-colour relation for the complete B,V sample. The c-types (open circles) and ab-types (filled circles) define two distinct relations, separated by $\Delta \log P = 0.125$ because of their different mode of pulsation. Lines with the expected theoretical slope are added. Note how the flagged stars (Figure 9) lie off the mean relation, which can be understood because of their different luminosity.

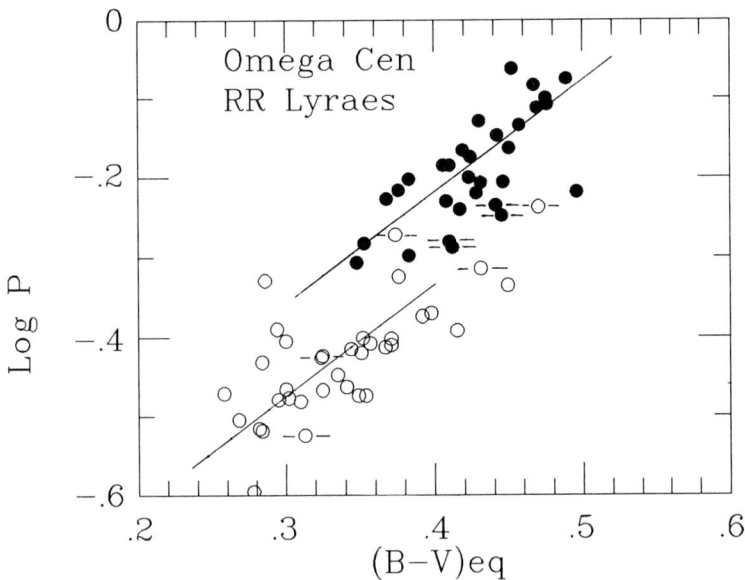

Figure 13. RR Lyraes with the best (quality 1) photometry in the CMD, together with some of the non-variable BHB stars measured with high internal accuracy along with the variables. This pinpoints very precisely the blue edge of the instability strip.

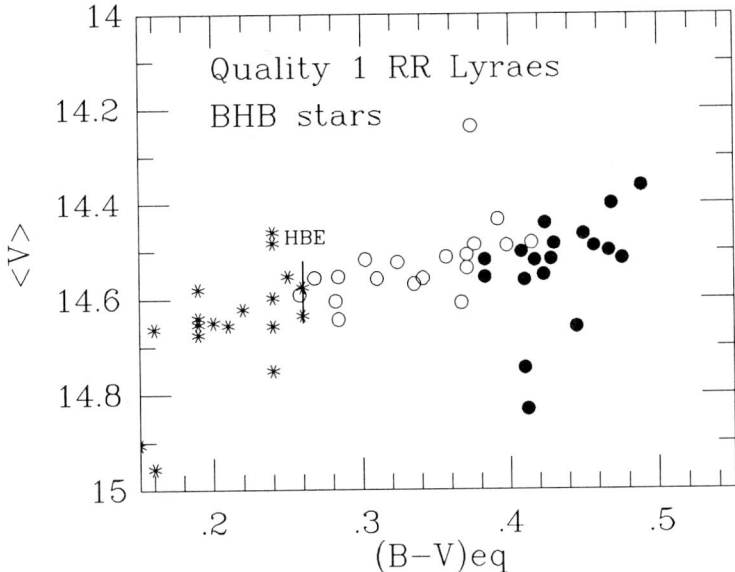

Figure 14. The blue amplitude-log P′ correlation for quality 1 data. The mean lines for M3 (Oosterhoff group I) and M15 (Oosterhoff group II) are indicated, showing that ωCen (group III?) is shifted further towards longer periods. The position of ωCen can be understood, at least in part, if the stars are more evolved, and therefore over-luminous with respect to the ZAHB (see text).

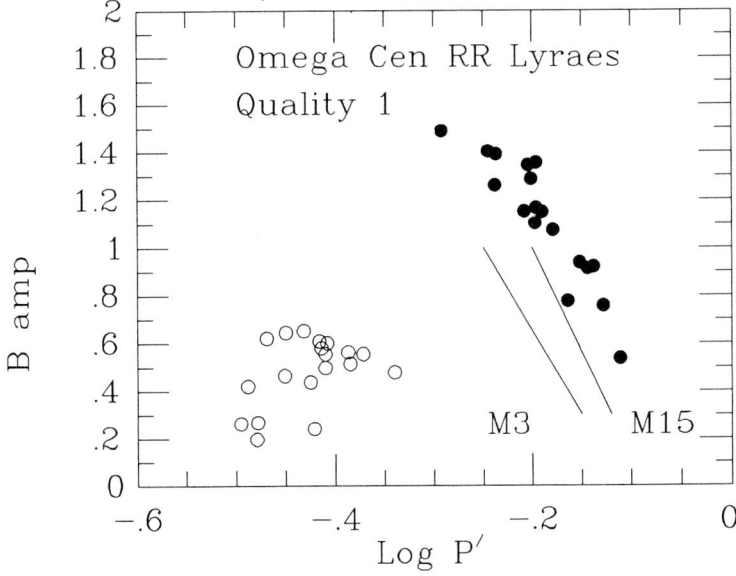

Figure 15. The blue amplitude-colour relationship for stars with quality 1 photometry. The large range in amplitude among the ab-types persists, suggesting a non-unique relationship of amplitude with colour.

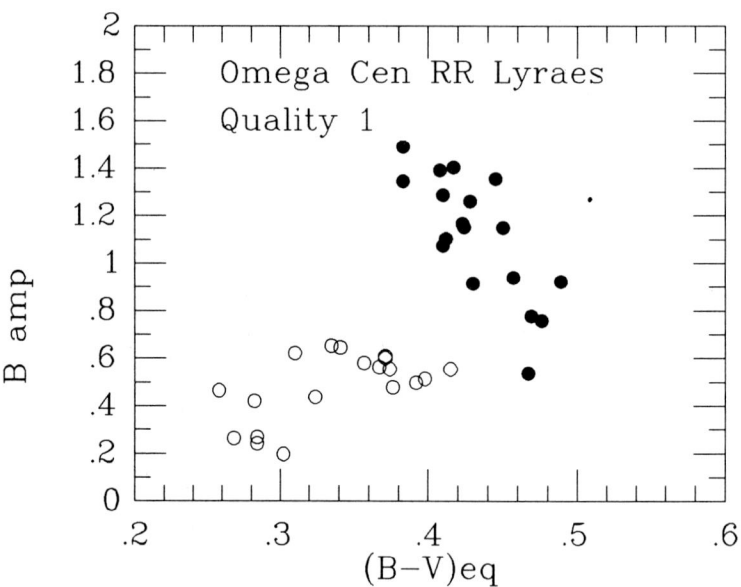

Figure 16. The (corrected, fundamental-mode) period-colour relationship for stars with quality 1 data. There is a hint that the c-types have a shallower slope than the ab-types. Most of the remaining spread in this diagram must be observational error, since the photolectric Walraven photometry, and (V-K) colours show a substantially smaller scatter.

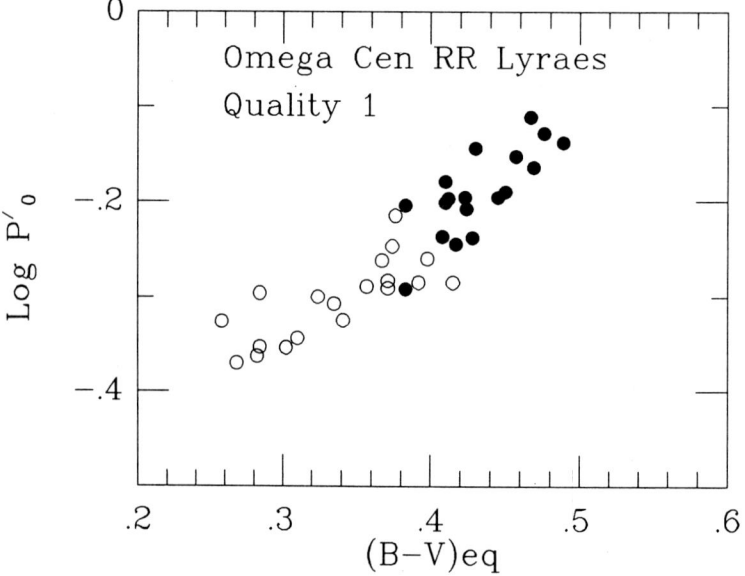

and M15 (Oosterhoff Groups I and II respectively). The B
amplitude-colour, and period-colour relations of Figures 15 and 16
respectively retain the main characteristics of the full sample, but
with somewhat reduced but still appreciable scatter. There is some
suggestion of a difference in slope between ab and c types.

The next step required for the study of abundance groups in ⍵Cen,
comparison with other clusters, and in the derivation of physical
parameters, is the conversion of colour to temperature. BDE used models
calculated by Manduca using Bell-Gustafsson programs, and with the
Doppler Broadening velocity, DBV = 3.6 km/s following Butler (1975).
More recent model colours (but not specifically for RR Lyraes) are
available by Kurucz (1981, see Lester et al. 1986) and VandenBerg & Bell
(VdBB, 1985). A detailed comparison cannot be entered into here; for
the BV data, we have used VdBB scales, which give temperatures in good
agreement with those derived from the Walraven photometry (based on
Kurucz 1981), but are about 150 to 200K cooler than BDE temperatures.
Temperatures derived from infrared (V-K) colours (Longmore et al. 1989)
are cooler still by about 100K.

The CMD for stars of different abundance groupings is shown in Figure
17. Most, but not all of the stars lying away from the main HB belong

Figure 17. This shows the position of those ⍵Cen RR Lyraes
with abundance data in the theoretical m_{bol} -log T_e diagram.
Note that four of the five most metal-rich stars lie away
from the bulk of the stars. There are otherwise no strong
abundance-related effects apparent in the diagram.

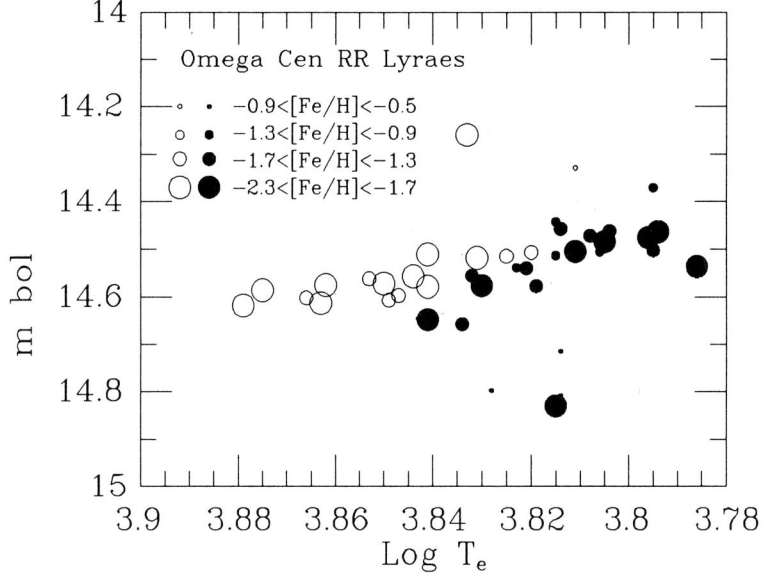

to the highest metallicity group. There is a small metallicity effect exhibited by the most metal-rich group in the period-colour diagram, at least among the ab-types, as shown in the log P-log T_e relationship of Figure 18. This effect is much smaller than that found in the period-temperature relationships of Oosterhoff I and II clusters covering a similar range of metallicity (e.g. see Bingham et al. 1984, Figure 24). In fact the effect may be even smaller, as a smaller spread in this diagram is found using temperatures from infrared colours (Longmore et al. 1989), and from the Walraven photometry (de Bruijn & Lub 1987). Clearly, as has been pointed out previously (e.g. Dickens 1982) the RR Lyraes in ωCen differ in some important property from those of comparable metallicity in other clusters. We shall return to this point later when discussing their evolutionary status, and the physical parameters (in particular the mass-to-light ratio) deduced from the period-temperature relation.

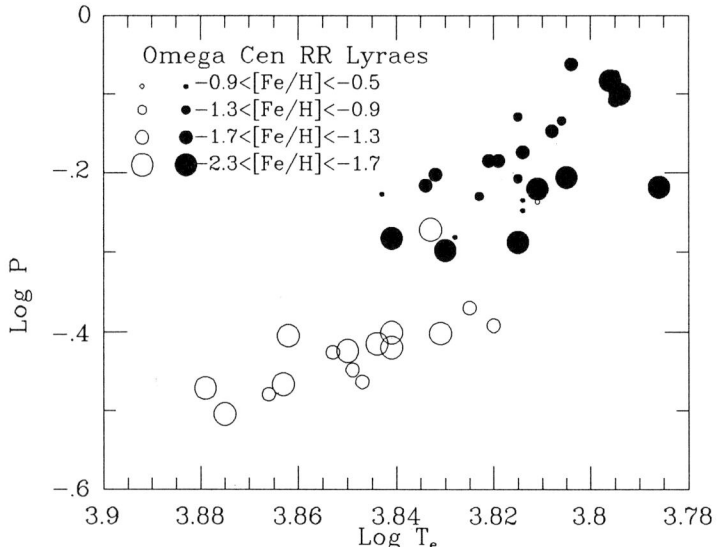

Figure 18. The log P - log T_e relationship for the abundance sample. Although some effect of the differing luminosity of the "extreme" stars (see Figure 9) can be detected, there is little if any "Oosterhoff effect" separating different abundance groups apparent for most of the sample. It is possible that some effect (less than that shown between clusters of different Oosterhoff groups) could be present, but masked by observational error. Note also the possibility of a smaller slope amongst the c-types.

4 THE DISTANCE SCALE

The apparently small range in luminosity found amongst RR Lyrae variables has led to their extensive use as distance indicators in our own and nearby galaxies. Studies of their pulsation properties, and evolutionary status within clusters has opened up the prospect of a considerably greater precision in their use as distance calibrators. Similarly, refinement of the Baade-Wesselink and related methods of radius determination in field RR Lyraes has led to great improvements (see Moffett, this volume). Much interest now centres on understanding the dependence of luminosity on metallicity, as manifest in the Oosterhoff, or Sandage period-shift effect between clusters. Using a simple pulsation model, it is easy to show that a logarithmic relationship between period and temperature for a group of RR Lyraes (such as in a cluster) gives information about the mean mass-to-light ratio, M/L, so that a shift in such a relationship between clusters implies a difference in M/L. A full understanding of the pulsational and evolutionary properties of the RR Lyraes therefore holds the key to accurate distance calibration via M/L. This has received considerable attention in recent years, and is of great importance as it can in principle lead to improved cluster ages, via the more precise distance measurement. We concentrate here on the role the RR Lyraes in ωCen can play in these fundamental problems.

A classical, and essentially empirical method for finding distances to globular clusters comes from fitting their observed main sequences to nearby subdwarfs whose distances are known from their trigonometric parallaxes. This has been carried out by Noble & Dickens (1988) for a sample of clusters for which high accuracy CCD CMDs were then available. Although there have been more recent treatments of a greater number of clusters now available (e.g. Buonanno et al. 1988), the former study serves to illustrate the way in which RR Lyrae properties enter the problem, and how those in ωCen fit in to the picture.

Figure 19 shows the fit of three isochrones of different age, having roughly the mean metallicity of ωCen, and with the distance determined from the subdwarfs, to the ωCen data shown in Figure 7. The mean age around 16 Gyr agrees well with that found for the other clusters considered by Noble and Dickens. The subdwarf fit provides a calibration of the luminosity of the ωCen RR Lyraes, which is compared with the results for the other clusters in Figure 20. Note the weak correlation of M_V with [Fe/H], with ωCen not apparently conspicuous (within the errors indicated). I would like to make two remarks on this diagram.

Firstly, it provides a direct check (in principle) on the role of M_V in driving the Oosterhoff period shift between clusters, which Sandage (1982) demonstrated could be the dominant parameter correlating strongly with [Fe/H] over many clusters. Since it is M/L which must vary as the mean period-colour line shifts, our observed small range in M_V must imply a range of mass among the HB stars of the clusters in Figure 20 (i.e. assuming a similarly significant role in driving the period shifts). Although the estimated errors shown are too large to yet allow a definitive solution, it appears likely that some variation in mass exists amongst the clusters, in the sense of higher mass for the more metal-rich clusters. Note also that allowance for the effects of evolution away from the ZAHB, which will be more important for the metal-poorer clusters (see below) will tend to flatten the slope further.

Secondly, the variation of M_V with abundance **within** ωCen is also small, as illustrated in the CMD of Figure 17, with the same trend as shown in Figure 20. However the luminosity of any particular RR Lyrae is infuenced by its evolutionary history which needs to be considered before the argument can be developed further. This is discussed below in the context of searching for consistency with stellar evolution

Figure 19. The CMD of the main sequence and turnoff regions is shown, with three isochrones of different age superimposed. They have been fitted using the distance deduced from fitting to nearby subdwarfs, and provide a satisfactory mean representation of the data.

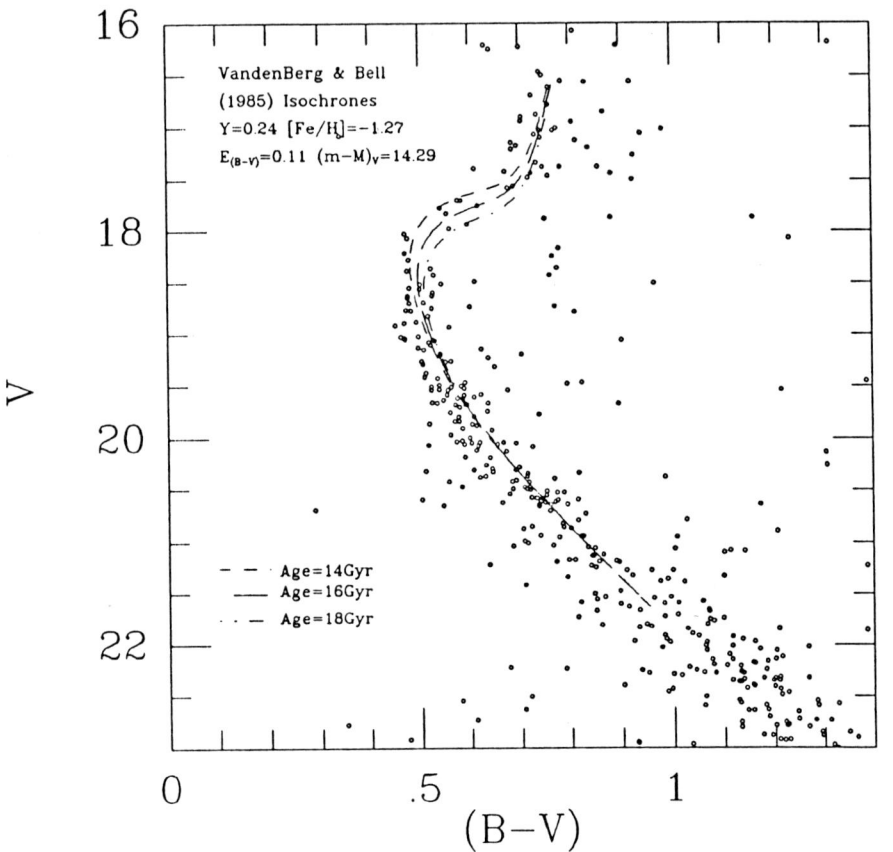

models. However we can already see from the above how a very precise knowledge of the intricacies of RR Lyrae behaviour would enable the crucial precise calibration needed to determine cluster ages.

5 RR LYRAES AND STELLAR EVOLUTION
5.1 Evolution

Figure 21 shows the correlation of period shift with metallicity found by Noble & Dickens (1988), which has a slope of +0.079 ± 0.011, in excellent agreement with that of +0.084 found by de Bruijn & Lub (1987). These results are essentially consistent with the earlier result by Sandage (1982). Why is this effect so much smaller, if not absent altogether **within** ⍵ Cen? Also, Figure 14 shows that ⍵ Cen is shifted towards longer periods still than the Oosterhoff II cluster, M15, yet its average metallicity is much higher. Do these properties provide any clues as to the cause of the Oosterhoff effect?

As discussed by many authors, one can account for a systematic period shift, as found in the period-temperature relation, in terms of a variation in M/L between the clusters of each Oosterhoff group (or on a

Figure 20 A plot of the RR Lyrae luminosity, M_V, as a function of [Fe/H] for 10 clusters. A least-squares linear fit to these is shown, together with the result from the fit to ⍵Cen. Note, however, that the four BHB clusters, shown as larger filled circles, are the most luminous, as would be expected if the stars populating the instability strip in these clusters are indeed more evolved, as discussed in the text. Allowance for this effect would give an even weaker correlation than shown.

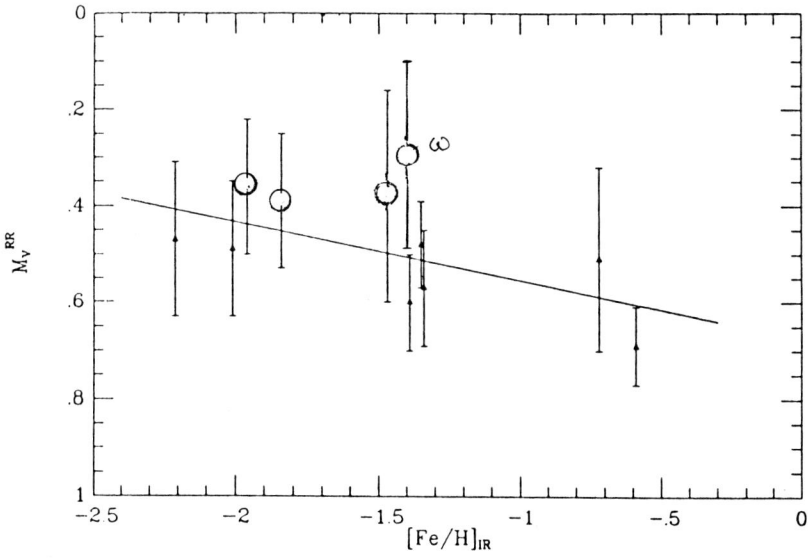

cluster-by-cluster basis). The "problem" is to be able to explain these differences, and now especially the correlation with metal abundance embodied in the "Sandage effect", in terms of stellar models. Furthermore, we should like to know how much of M and how much of L contribute to any given shift. (The current status of this entertaining pastime is discussed by Rood in this volume.) It turns out that the main contribution of ⍵Cen to this problem is, however, merely to "explain" the anomalous position of ⍵Cen itself!

A partial answer to the ⍵Cen anomaly was proposed by Gratton et al. (1986) who considered the behaviour of the luminosities of HB models within the instability strip. They drew attention to the much smaller dependence of luminosity on metal abundance for evolved HB models than for those on the ZAHB. In clusters with very blue HBs, like ⍵Cen and M92, most of the RR Lyraes are likely to be in a later stage of HB

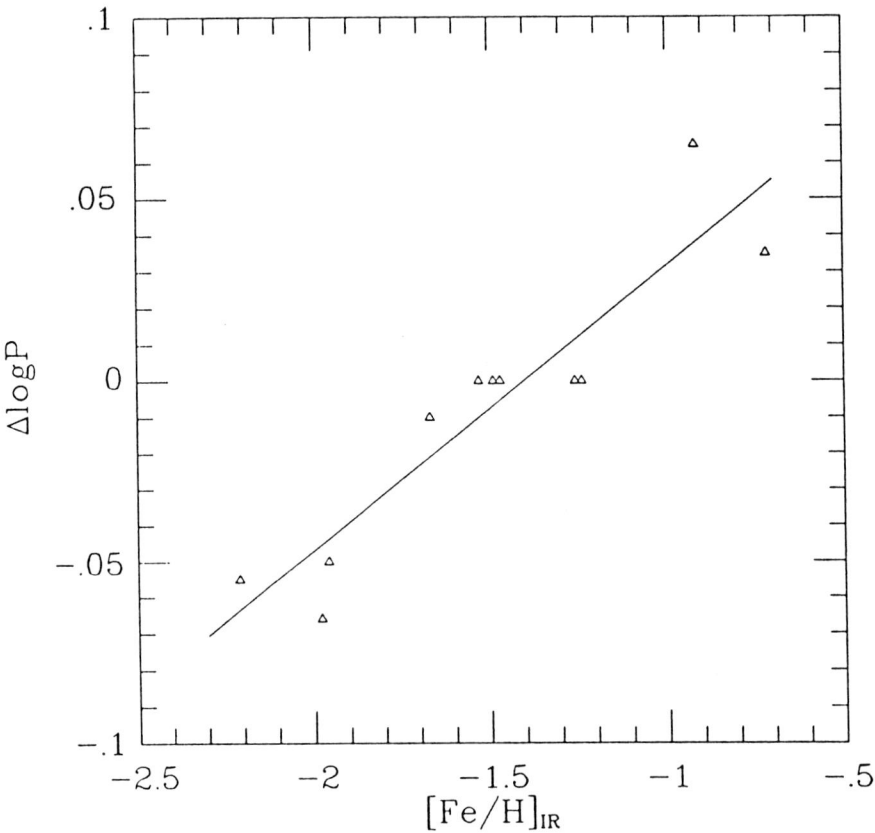

Figure 21. The period shift - metallicity relation as derived by Noble and Dickens. The metallicity scale comes from infrared photometry by Frogel et al. (1983). The line is a least-squares fit to the data (see text).

evolution than those in clusters with more red HBs (e.g. M15 and M3) and
according to the models should hence show less correlation of luminosity
with metals within the instability strip. This could account for the
absence of any strong metal-dependent luminosity differences among the
bulk of the RR Lyraes in ⍵Cen (see Figure 17). Also, this luminosity
excess would shift the bulk of the ⍵Cen RR Lyraes towards longer
periods as compared to ZAHB stars, of which there appear to be a
sprinkling in ⍵Cen (see the CMD of Figure 9). This could in principle
account for the positive shift of ⍵Cen with respect to M15, which must
be dominated by stars near the ZAHB. Note, however, that the ⍵Cen
stars, being more evolved, will tend to have slightly higher masses than
ZAHB stars, also affecting their position in the period-temperature
plane, in the sense of reducing the shift with respect to M15!

Nevertheless we believe this picture to be essentially correct. It is
indeed likely that the bulk of the RR Lyraes in ⍵Cen are in an advanced
stage of evolution. Figure 22 shows a representative CMD in which the
data accurately represent the true relative distributions on the HB.
Clearly the RR Lyraes account for only a few per cent of the HB, and
must therefore mostly be evolving away from the ZAHB locations on the
BHB. Comparison with other clusters, therefore, is inappropriate
without taking account of evolutionary effects, and this has now been
studied by Sandage elsewhere in this volume.

Thus ⍵Cen has demonstrated the importance of evolutionary effects at
least in clusters with extremely blue HBs. It is, however, likely that
the morphology of evolutionary tracks (rather than just the ZAHB
distribution) plays a significant role for all clusters in the details
of just how the variables are distributed in the instability strip, and
perhaps in their pulsational behaviour (e.g. double-mode pulsation,
location and width of either/or regions). This topic has been discussed
already in some detail by Bingham et al. (1984) in a comparison of the
properties of the Oosterhoff I and II clusters, M3 and M15 respectively.
The topic would, however, repay further study with detailed simulations
using the latest HB model tracks.

5.2 Physical parameters

We now discuss briefly the derivation of physical
parameters, derived from purely pulsational considerations, and their
comparison with those expected from evolutionary models, thereby
providing a powerful check on the theory. We include some preliminary
mean results from our data, and show how they bear on current stellar
evolutionary predictions. A full discussion of this topic is outside
the scope of this paper, and will be given elsewhere (Dickens & Bingham
1989).

Firstly, the mean mass-to-light ratio, M/L is derived by
fitting the de Bruijn and Lub slope of -3.48 to the data in Figure 18.
This slope is also the theoretical slope derived by van Albada & Baker
(1971) and provides an acceptable fit to our data. This gives a mean
value of

$$\langle \log(M^{0.81}/L) \rangle = -1.88 = A_{puls}$$

in exact agreement with that of -1.88 +/- 0.03 obtained by de Bruijn & Lub. A second result often derived from cluster RR Lyrae properties is the helium content, Y, using the temperature of the blue edge or the width of the instability strip. From Figures 13 and 18, we find $\log T_e(hbe) = 3.875$ at $\log P_1 = -0.525$, which from Tuggle & Iben (1972) implies $Y = 0.28$. More recent work by Stellingwerf (1984) has, however, cast doubt on this method. His convective blue edges show little sensitivity to Y, with $\log T_e(hbe) = 3.856$ ($Y = 0.2$) and $\log T_e(hbe) = 3.862$ ($Y = 0.3$). More theoretical work is needed to explore this further. However it is interesting to note that the cooler temperature scale from V-K would give better agreement with Stellingwerf's blue edges.

The above mass-to-light ratio may be combined with the luminosity of the RR Lyraes derived from the main-sequence distance calibration used for Figure 7 to derive a mean mass for the RR Lyraes. The distance modulus derived, using the original Lutz-Kelker corrections to the subdwarf parallaxes (see Carney 1979), is $(m-M) = 14.29$, or applying one half the corrections (e.g. see Hesser et al. 1987) gives $M_V(RR) = 0.30$ magnitude.

Figure 22. CMD of a statistically true representative sample of RR Lyraes and non-variable stars, showing the continuous distribution with colour expected if the RR Lyraes are evolving away from their ZAHB locations on the BHB. The RR Lyrae data have not been corrected for radial photometric error, which becomes important inside 10 arcmin radius, in order to match the colours of the non-variable stars.

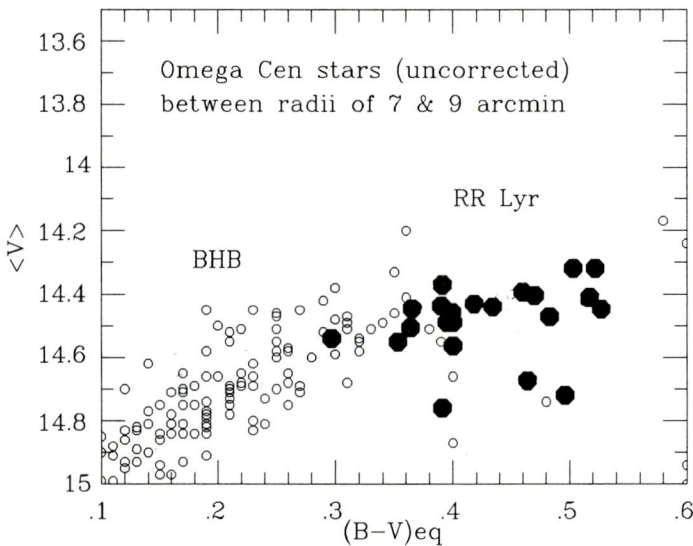

Thus $\log L_{RR}/L_\odot = 1.77$ which combined with the above mass-to-light ratio yields a mass of $M_{RR}/M_\odot = 0.73$. Are these values in accord with HB models? For Sweigart-Gross (1976, SG) models near the ZAHB, the mass-to-light ratio is given roughly by (see Bingham et al. 1984, equation 11)

$$\log(M^{0.81}/L)_{evol} = -1.40\,(Y - 0.3) - 1.87 = A_{evol}$$

which implies $Y = 0.3$ for $A_{puls} = -1.88$. The primordial value of $Y = 0.23$ requires $A_{evol} = -1.77$, implying a mass around $0.90\,M_\odot$ for $\log L = 1.77$ from the main-sequence fit, as compared to the pulsation mass of 0.73. If the ⍵ Cen RR Lyraes are in an advanced stage of evolution, as suggested above, then the appropriate $\log L_{evol}$ is about 0.1 fainter, and $M = 0.75\,M_\odot$ for $\log L_{zahb} = 1.67$, which agrees with that derived from pulsation considerations. Thus **only** by postulating that the ⍵ Cen RR Lyraes are in an advanced, more luminous stage of evolution can we reconcile pulsation theory, and the main sequence calibration with evolution theory, but this is a plausible and encouraging result.

Also note that about two-thirds of the discrepancy in luminosity (at constant mass) between pulsational and evolutionary values found by Bingham et al. for M15 (where the RR Lyraes must mainly be near the ZAHB) is removed with the revised B,V temperature scale used here, or entirely removed using the V-K calibration. This underlines the crucial importance accurate temperature scales have in this method of evaluating the validity of evolutionary models of the HB, and in the derivation of mass and luminosity from pulsation studies.

6 CONCLUDING REMARKS

So what have we learned from studying ⍵ Cen, and what further should be done? Although we seem to be no nearer understanding why ⍵ Cen is the way it is, the properties of its sub-populations over the whole HR diagram are broadly understandable in terms of an overall range in heavy-element abundance. We leave to one side the more complex role of mixing, especially in the red giant stars, which has recently been discussed by Dickens & Caldwell (1988).

Whilst we have no real understanding of the phenomenon, the size and shape of the abundance distribution appears to be similar for stars in the various stages of evolution observed, and could be caused by two dominant underlying populations of different mean metallicity. The observed pulsational properties (i.e. the period-amplitude-colour-luminosity correlations) of the bulk of the RR Lyraes are qualitatively the same as those in other clusters (as expected), but the sub-populations do not behave as those in other clusters of different metallicity, in showing a strong Oosterhoff effect. This is thought to be due to their advanced stage of evolution; the few RR Lyraes in clusters like M13, M92 are likely to be similarly over-luminous through advanced evolution, and this must be taken account of in comparative studies of the Oosterhoff effect. The mean mass and luminosity of the

RR Lyraes derived from pulsation theory and calibration of the cluster distance, are consistent with those expected from SG HB models, with preferred values of $M/M_\odot = 0.73$, $\log(L/L_\odot) = 1.77$ ($M_V = 0.30$), with estimated uncertainties of a few hundredths in each case. This is encouraging for the theory but greater precision is still needed for a definitive result, both to test the theory, and to provide better distance calibration.

Great effort is still needed to improve further the T_e-colour calibrations, and to extend the evolutionary tracks beyond core helium exhaustion to check the advanced evolution picture for ω Cen RR Lyraes - the lack of much luminosity difference as a function of abundance, if reflected in the models is a powerful qualitative check, and the pulsation parameters a powerful quantitative check. Spectra of good samples of subgiants, stars at turnoff and on the main sequence are needed to verify the abundance spread postulated from the photometry.

One disappointing aspect of the study of the ω Cen RR Lyraes concerns the solution of the Oosterhoff problem, since it was to be hoped that the occurrence of what appeared to be different Oosterhoff groups within one cluster would enable a direct check on their relative luminosities, and perhaps give some indication which are the important parameters driving the period shifts in other clusters. As demonstrated elsewhere in this volume, in spite of considering in detail the effects of evolution (Sandage), and recent model developments (Rood), this problem is yet to be solved. Although we have known for a long time that mass, luminosity, and track morphology must play a role, we still need to know precisely what the values are, and which other parameters also enter, so that we can use our knowledge effectively in the various fundamental applications that have been referred to.

REFERENCES

Bailey, S.I. (1902). Harvard Annals, 38.
Belserene, E.P. (1964). A.J., 69, 475.
Bingham, E.A., Cacciari, C., Dickens, R.J. & Fusi Pecci, F. (1984). M.N.R.A.S., 209, 765.
Buonanno, R., Corsi, C.E. & Fusi Pecci, F. (1988). ESO Preprint No.594.
Butler, D. (1975). Ap.J., 200, 68.
Butler, D., Dickens, R.J. & Epps, E.A. (1978). Ap.J., 225, 148.
Caldwell, S.P. & Dickens, R.J. (1988). M.N.R.A.S., 233, 367.
Carney, B. (1979). Ap.J., 233, 211.
Cohen, J.G. (1981). Ap.J., 246, 869.
Da Costa, G.S. & Villumsen, J.V. (1981). In IAU Colloquium 68, Astrophysical Parameters for Globular Clusters, eds. A.G.Davis Philip, D.L.Hayes, p. 527. Schenectady, N.Y.: L.Davis Press.
De Bruijn, J.W. & Lub, J. (1987). In Lecture Notes in Physics, Vol.274, eds. A.N. Cox, W.M. Sparks & S.G. Starrfield, p. 233. Berlin: Springer-Verlag.

Dickens, R.J. (1982). In Pulsations in Classical and Cataclysmic
    Variable Stars, eds. J.P. Cox & C.J. Hansen, p. 182.
    Boulder, CO: J.I.L.A.
Dickens, R.J. & Bingham, E.A. (1989). in preparation.
Dickens, R.J. Brodie, I.R., Bingham, E.A., & Caldwell, S.P. (1988).
    Rutherford Appleton Laboratory Report, No. RAL-88-004.
Dickens, R.J. & Caldwell, S.P. (1988). M.N.R.A.S.,233, 677.
Dickens, R.J. & Saunders, J. (1965). Roy. Obs. Bull., No. 101.
Freeman, K.C. (1985). In IAU Symposium 113,Dynamics of Star Clusters,ed.
    J.Goodman,P.Hut(Dordrecht:Reidel),p. 33.
Freeman, K.C. & Rodgers, A.W. (1975). Ap.J.Letts.,201, L71.
Frogel, J.A., Cohen, J.G. & Persson, S.E. (1983). Ap.J.,275, 773.
Gratton, R.G., Tornambé, A. & Ortolani, S. (1986). Astron. &
    Astrophys.,169, 111.
Harding, G.A. (1965). Roy.Obs.Bull.,No.99.
Hesser, J.E., Harris, W.E., VandenBerg, D.A., Allwright, J.W.B., Shott,
    P. & Stetson, P.B. (1987). P.A.S.P.,99, 739.
Lester, J.B.,Gray, R.O. & Kurucz, R.L. (1986). Ap.J.Suppl.,61, 509.
Longmore, A.J.,Dixon, R.I. & Skillen, I. (1989). in preparation.
Lub, J. (1977). Thesis, University of Leiden.
Martin, W.Chr. (1938). Leiden Annals,17.
Nemec, J.M., Nemec, A.F.L. & Norris, J. (1986). A.J.,92, 358.
Noble, R.G. (1987). Ph.D. Thesis, University of Leeds.
Noble, R.G. & Dickens, R.J. (1988). In New Ideas in Astronomy, p. 59.
    Cambridge: Cambridge University Press.
Noble, R.G., Dickens, R.J., Buttress, J. & Griffiths, W.K. (1989). In
    preparation.
Persson, S.E., Frogel, J.A., Cohen, J.G., Aaronson, M. & Matthews, K.
    (1980). Ap.J.,235, 452.
Sandage, A.R. (1982). Ap.J.,252, 553.
Searle, L. (1977). In The Evolution of Galaxies and Stellar Populations,
    eds. R. Larsen & B. Tinsley, p. 219. New Haven, CT: Yale
    University Observatory.
Smith, G.H. (1987). P.A.S.P.,99, 67.
Smith, H.A. (1981). A.J.,86, 538.
Stellingwerf, R.F. (1984). Ap.J.,277, 322.
Stellingwerf, R.F. & Dickens, R.J. (1983). unpublished.
Sweigart, A.V. & Gross, P.G. (1976). Ap.J.Suppl.,32, 367.
Sweigart, A.V.,Renzini, A. & Tornambé, A. (1987). Ap.J.,312, 762.
Tuggle, R.S. & Iben, I. (1972). Ap.J.,178, 455.
Van Albada, T.S. & Baker, N. (1971). Ap.J.,169, 311.
Van Albada, T.S. & Baker, N. (1973). Ap.J.,185, 477.
VandenBerg, D.A. & Bell, R.A. (1985). Ap.J.Suppl.,58, 56.
Woolley, R.v.d.R. (1966). Roy.Obs.Ann.,No.2.

VARIABLE STARS AND THE COSMIC DISTANCE SCALE

Jeremy Mould
Palomar Observatory, California Institute of Technology
Pasadena, California 91125

INTRODUCTION

For three quarters of a century pulsating variable stars have lain at the foundation of the extragalactic distance scale. The construction of larger telescopes, advances in detector technology, hard work by observers, and our understanding of stellar structure have all contributed to the expansion of the realm of the Cepheids to the distance of M101. Now, with the advent of Hubble Space Telescope (HST), we can look forward to the detection of Cepheids in the Virgo cluster and the removal of much of the remaining uncertainty in the Hubble constant.

It is an appropriate time, therefore, to think about some of the details of applying period luminosity (P-L) relations for variable stars to galaxies with necessarily different histories of star formation and chemical evolution.

EFFECT OF CHEMICAL COMPOSITION ON P-L RELATIONS

The first realization that P-L relations were different in different stellar populations had a major impact on the distance scale (Baade 1956). Nowadays, informed by knowledge of stellar evolution, we tend to talk about the masses and chemical composition of classes of variable stars, but we are still asking the same question: are there important second parameters in the P-L relation?

For model Cepheids Stothers (1988) has recently found that to first order at constant period:

$$\delta M_{bol} = 0.8 \, \delta Y - 1.8 \, \delta Z$$

and

$$\delta M_V = 0.5 \, \delta Y - 2.8 \, \delta Z$$

which is an almost insignificant composition dependence.

Several lines of argument concur that multiwavelength observations of Cepheids offer the best means of determining the distances of these stars. One can integrate the energy distribution given BVRIJHK photometry to measure bolometric magnitudes directly. Without bolometric corrections one is in a better position regarding the continuing choice between P-L and P-L-C relations (Feast and Walker 1987). And one can also determine the reddening to individual Cepheids (Freedman 1985).

An empirical test of the composition dependence of the P-L relation has not yet been successfully carried out. Freedman (1988c) has devised such a test, employing the metallicity gradients exhibited by disk galaxies, but the application to M 31 is currently incomplete. One can also look for discrepancies between the Cepheid distance moduli and the Population II distance moduli for galaxies of differing metallicity.

However, this raises the subject of the composition dependence of the absolute magnitude of RR Lyrae stars. In this case the theoretical expectation (Iben and Renzini 1984) is:

$$\log L = 1.73 + 1.40(Y - 0.3) - 0.073(3 + \log Z)$$

for RR Lyrae stars at a specific effective temperature. This metallicity dependence is shallower than that found by Sandage (1989) by considering the amplitudes of RR Lyrae light curves. A trend in the relation between M_{bol} and $\log Z$ is seen in the Baade-Wesselink measurements of RR Lyrae stars by Jones et al. (1988), Clementini and Cacchiari (1989), and Liu and Janes (1989). Carney (1988) indicates:

$$M_{bol} = 0.20[Fe/H] + 1.03$$

Böhm-Vitense et al. (1989) point out the sensitivity of Baade-Wesselink results to the transformation from color to effective temperature. The dependence of this transformation on metallicity because of line blanketing makes the coefficient of [Fe/H] in this equation problematical. A further opportunity to determine this coefficient exists with HST observations of the horizontal branches of M 31 globular clusters. Measurement of the composition dependence of the absolute magnitudes of RR Lyrae stars would greatly strengthen the Population II distance scale.

Table 1. Cepheid Distance Moduli for Local Group Dwarf Irregulars

Galaxy	$(m-M)_o$	O/H x 10^4	Source*
IC 1613	24.1	0.4	McAlary, et al. (1984)
	24.3±0.1		Freedman (1988a)
NGC 6822	23.4±0.2	1.7	McAlary, et al. (1983)
Sex A and B	25.5±0.1	0.2	Sandage & Carlson (1985b)
Sex A	26.0±0.4		Walker (1987a)
WLM	24.8±0.1		Sandage & Carlson (1985a)
NGC 3109	26.1±0.1		Demers, et al. (1985)

* These are the original references. The tabulated values are taken from Walker (1987b).

Returning to the composition dependence of the Population I distance scale, one notices in Table 1 the range of chemical composition exhibited by the dwarf galaxies in the Local Group with Cepheid distance moduli. These moduli have been summarized and put on a scale consistent with an LMC modulus of 18.47 by Walker (1987b). Table 1 indicates the oxygen abundance (Vigroux, et al. 1987) for the system and gives original reference for the Cepheid photometry. For most of these galaxies RR Lyrae stars are detectable from ground based telescopes. Given accurate RR Lyrae distances to these galaxies, the composition dependence of Cepheid distance moduli would then be apparent.

A Population II distance estimate for IC 1613 is already available. IC 1613 is a low metallicity dwarf with [O/H] = -1.2. Fitting the red giant branch tip to this galaxy yields $(m-M)_o = 24.2\pm0.2$ (Freedman 1988b) in agreement with the Cepheid modulus. It will be interesting to confirm this by observing RR Lyrae stars.

M 31 AND M 33

These galaxies provide consistency checks on our understanding of variable star P-L relations. Table 2 contains the most recent estimates of the distances of M 31 and M 33. For M 31 Walker (1987b) has updated earlier Cepheid data to a standard P-L relation and $A_V = 0.33$. This should not be considered inconsistent with $A_B = 0.31$ for halo RR Lyrae stars adopted by Pritchet and van den Bergh (1988), because of the intrinsic reddening of Cepheids (e.g., Freedman, 1985).

In M 33 $A_V = 0.3$ has been adopted for Cepheids in Table 2. The wavelength sequence of (m-M) suggests that a higher value might be appropriate. But the P-L relation for supergiant LPVs in M 33 (Mould 1987) is in opposition to this trend. Application of this form of the

Table 2. Distance Measurements for M 31 and M 33

Indicator	M 31	Source	M 33	Source
Cepheids B,V	24.2	Baade & Swope (1963)*	24.5±0.1	Christian & Schommer (1987)*
Cepheids R,I			24.5±0.1	Mould (1987)*
Cepheids H	24.3±0.1	Welch et al. (1986)*	24.2±0.1	Madore et al. (1985)
LPVs K			24.6±0.1	Mould (1987)
GB tip	24.4±0.3	Mould & Kristian (1986)	24.6±0.3	MK with $M_V = 0.8$
RR Lyr	24.2±0.2	Pritchet & van den Bergh (1987, 1988)	24.45±0.2	Pritchet (1988)
Novae	24.0±0.2	Cohen (1985)		

* Using P-L relation and $A_V = 0.33$ (Walker 1987b) for M 31 and $A_V = 0.3$ (Feast 1988) for M 33.

LPV distance indicator (Bessell and Wood 1984) without bias requires that supergiant and asymptotic giant branch LPVs should be separable. Supergiants and AGB stars are certainly in very different evolutionary phases (Wood, Bessell and Fox 1983), and the supergiant P-L relation has a different slope from that of AGB stars (cf. Feast, 1989). This distinction between extreme Population I LPVs and lower mass AGB Miras is an example of the symbiosis between the study of stellar evolution and the extragalactic distance scale.

The large number of Cepheids found by Kinman, Mould and Wood (1987) in the course of the LPV survey also deserve attention, particularly in the present context because six candidates for Population II Cepheids were identified in M 33. No follow-up work has been carried out on these stars as far as I know, but my estimate is that their bolometric magnitudes are in the range (-4.6, -3.8). The mechanism proposed by Gingold (1976) places these stars in the instability strip during thermal pulses of the helium burning shell of low mass AGB stars with very thin remnant envelopes of hydrogen fuel. These excursions are called "blueward noses" or "blue loops". Two of these stars would have the longest known periods of the W Vir class and certainly deserve further investigation. If this identification is correct, there may be a higher percentage of Population II Cepheids in earlier type galaxies.

To conclude the subject of the distances of these galaxies, from Table 2 the mean value of $(m-M)_o$ for M 31 is 24.2. The mean value for M 33 is 24.5. The consistency of these partially separate distance indicators is better than 10% in distance (r.m.s.). There are some obvious gaps in Table 2 to be filled and there is a need for a better study of the reddening problem.

H_o FROM CEPHEID DISTANCES TO GALAXIES

Because peculiar and infall velocities in the local supercluster are comparable to the Hubble flow, one cannot determine H_o by simply measuring the distances to a random assortment of nearby galaxies at redshifts of a few hundred $km \cdot sec^{-1}$. Consequently, one must use a high quality primary standard candle (Cepheids) in nearby galaxies to calibrate a reliable secondary distance indicator (such as the infrared Tully-Fisher method, or IRTF) which is luminous enough to reach galaxies with redshifts up to about 5000 $km \cdot sec^{-1}$, well beyond any substantial Hubble flow deviations. Thus, an important goal for HST is to derive accurate distances to a number of appropriately selected galaxies in the 3 to 20 Mpc range by using the Cepheid P-L relation. It will then be possible to use the distances to these galaxies to calibrate the IRTF and a number of other secondary distance indicators for extension into the smooth Hubble flow.

Figure 1 shows how an accurate Cepheid P-L relation lies at the heart of such a plan. Therefore it is essential to ensure an accurate calibration of the zero-point in this relation. This should be accomplished by a multi-pronged attack involving a) improved main sequence fitting to younger clusters in the LMC: b) horizontal branch distances to globulars in the LMC, M 31 and M 33; and c) application of the Baade-Wesselink method to Galactic Cepheids.

The main body of the work for HST, however, is the application of this P-L relation to measure the distances of a substantial sample of spiral galaxies. In the 1970s in their classical series of papers on the extragalactic distance scale, Sandage and Tammann talked about a "twilight zone" as one stepped out from the Milky Way towards the unperturbed Hubble flow, in which the distance scale became fuzzy. This twilight zone still exists; it extends from the Local group to the Virgo cluster; it is this region that HST can fill with accurately determined galaxy distances.

The direct route to H_o from a sample of spirals with Cepheid distances is to construct for the first time a reliable and fully populated calibration of the IRTF, and hence measure distances to a set of galaxy clusters located between 4000 and 10000 km·sec^{-1} away, and observed at 21 cm with the Arecibo radiotelescope. Although no perfect standard candle exists, one can nonetheless rank distance indicators by the degree to which they meet objective criteria, such as: 1) Is the candle luminous

Figure 1. In this program the Cepheid P-L relation is calibrated within 1 Mpc and applied to spiral galaxies at typically 10 Mpc. These measurements in turn calibrate secondary distance indicators which reach out to 100 Mpc.

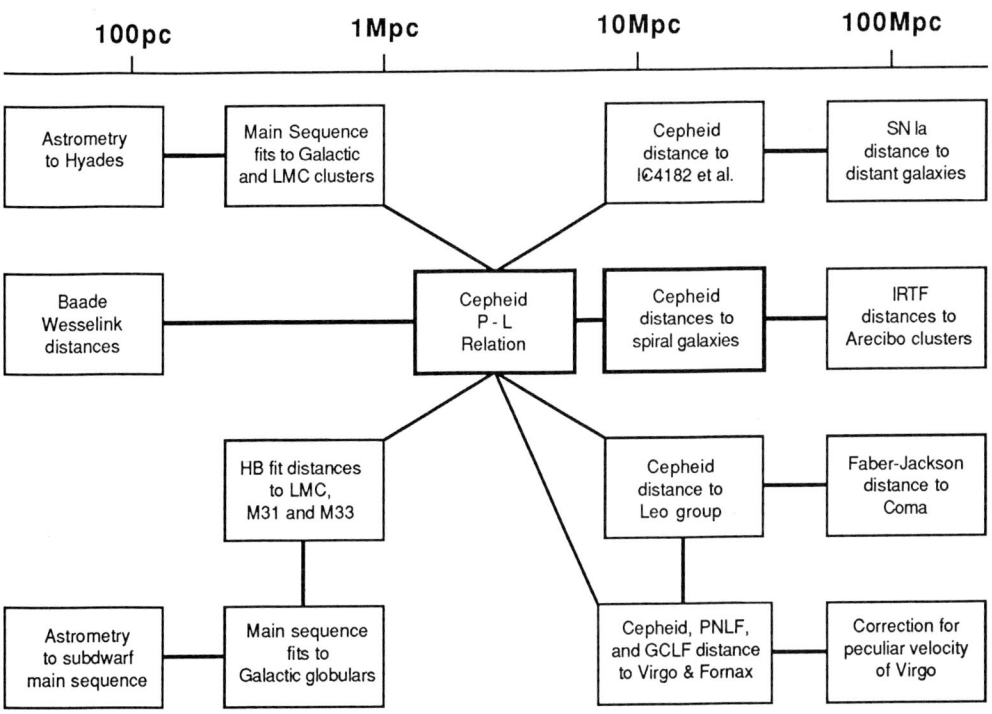

and easily identified? 2) Is a physical (rather than a statistical) basis for the luminosity criterion well understood? 3) Are the measurables objective, well-defined, and easy to determine? 4) Is there a demonstrably low dispersion based on large samples of data? Aaronson and Mould (1985) have made the case that Cepheids and the IRTF relation best meet these criteria.

The basis of the case for the IRTF as a secondary standard candle is Figure 2, which shows the velocity-distance relation for 11 nearby galaxy clusters from the work of Aaronson et al. (1986), derived from IRTF moduli to about 150 cluster spirals. To within the measurement errors there is no deviation from linearity. One simply needs to attach an absolute scale to the abscissa of Figure 2.

Of course, one distance indicator alone is not a basis for a conclusive measurement of H_0. Careful choice of the sample of spiral galaxies can support the calibration of a number of other secondary distance indicators. By including IC 4182 in the sample one can empirically calibrate the SN Ia standard candle with one of the few galaxies closer than Virgo which have measured light curves for such supernovae. This standard candle can then be applied to distant ($cz > 3000$ km·sec^{-1}) galaxies where several such supernovae have been recorded. The distances provided by such a program will make it possible to test the accuracy of the brightest resolved stars as standard candles. By measuring Cepheid distances to the Leo group, one can provide an additional calibration for three secondary indicators which are best applied to early type galaxies. The planetary nebula luminosity

Figure 2. The velocity-distance relation for eleven galaxy clusters, using moduli determined from the IR/HI relation. The velocity of Virgo (the lowest redshift cluster) has been corrected for infall and the remaining velocities have been corrected for the dipole anisotropy of the microwave background.

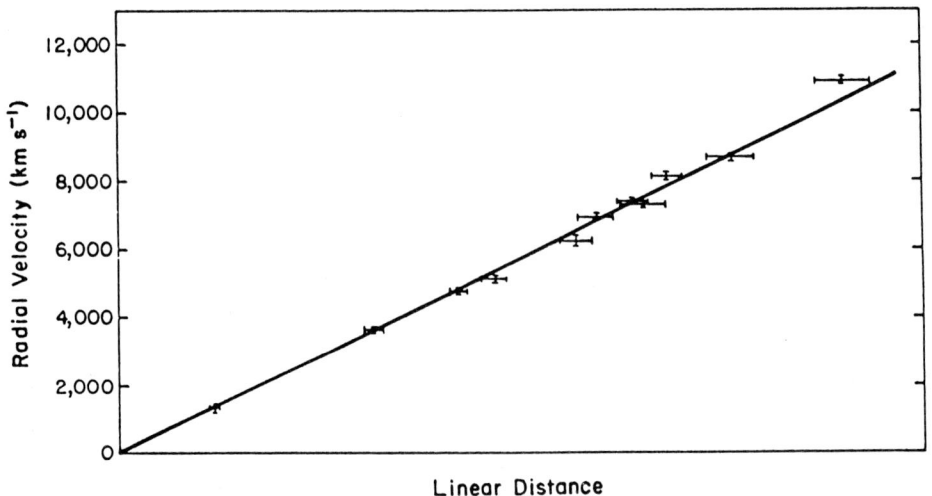

function (PNLF; see Ford et al. 1988) and the globular cluster luminosity function (GCLF; see Harris 1988) are two standard candles which can be easily and routinely observed in Virgo ellipticals. The Faber-Jackson relation, like the IRTF, can be applied to large distances, once appropriately calibrated (Dressler 1988).

The scope of this program is shown schematically in Figure 1. The other secondary indicators will provide important checks and interlocks on a determination of H_o through the IRTF. One can also attempt to measure direct Cepheid distances to spirals on the far side of the "twilight zone", namely in the Virgo and Fornax clusters. These observations will be difficult even with HST. However, secure distances to Virgo and Fornax may be almost sufficient to lay to rest the H_o controversy.

Partial support by NSF grants 85-02518 and 87-21705 is gratefully acknowledged.

REFERENCES

Aaronson, M., Bothun, G., Mould, J., Huchra, J., Schommer, R.A., and Cornell, M.E. (1986). Ap.J.,302, 536.
Aaronson, M., and Mould, J., (1985). Ap.J.,303, 1.
Baade, W. (1956). P.A.S.P.,68, 5.
Baade, W., and Swope, H. (1963). A.J.,68, 435.
Bessell, M. and Wood, P. (1984). P.A.S.P.,96, 247.
Böhm-Vitense, Garnavich, P., Lawler, M., Mena-Werth, J., Morgan, S., Peterson, E. and Temple, S. (1989). This volume.
Carney, B. (1988). Private communication.
Christian, C. and Schommer, R. (1987). A.J.,93, 557.
Clementini, G. and Cacchiari, C. (1989). This volume.
Cohen, J. (1985). Ap.J.,292, 90.
Demers, S., Kunkel, W., Irwin, M. (1985). A.J.,90, 1967.
Dressler, A. (1988). Ap.J.,317, 1.
Feast, M. (1988). Observatory, In press.
Feast, M. (1989). This volume.
Feast, M., and Walker, A. (1987). Ann. Rev. Astr. Ap.,25, 345.
Ford, H., Ciardullo, R., Jacoby, G., and Hui, X. (1988). In IAU Symposium 131. Dordrecht: Reidel, in press.
Freedman, W.L. (1985). In IAU Colloquium No. 82, Cepheids: Theory and Observation, ed. B. F. Madore, p. 225. Cambridge: Cambridge University Press.
Freedman, W. (1988a). Ap.J.,326, 691.
Freedman, W. (1988b). Preprint.
Freedman, W. (1988c). Private communication.
Gingold, R. (1976). Ap.J.,204, 116.
Harris, W.E. (1988). In The Extragalactic Distance Scale, eds. S. van den Bergh and C. Pritchet, A.S.P. Conference Series (in press).
Iben, I. and Renzini, A. (1984). Phys. Rep.,105, 329.
Jones, R., Carney, B., and Latham, D. (1988). Preprint.

Kinman, T., Mould, J., and Wood, P. (1987). A.J., **93**, 833.
Liu, T. and Janes, K. (1989). This volume.
Madore, B., McAlary, C., McLaren, R., Welch, D., Neugebauer, G. and Matthews, K. (1985). Ap.J., **294**, 560.
McAlary, C., Madore, B., Davis, L. (1984). Ap.J., **276**, 487.
McAlary, C., Madore, B., McGonegal, R., McLaren, R., and Welch, D. (1983). Ap.J., **273**, 539.
Mould, J. (1987). P.A.S.P., **99**, 1127.
Mould, J., and Kristian, J. (1986). Ap.J., **305**, 591.
Pritchet, C. (1988). In The Extragalactic Distance Scale, eds. S. van den Bergh and C. Pritchet, A.S.P. Conference Series (in press).
Pritchet, C., and van den Bergh, S. (1987). Ap.J., **316**, 517.
Pritchet, C., and van den Bergh, S. (1988). Ap.J., **331**, 135.
Sandage, A. (1989). This volume.
Sandage, A., and Carlson, G. (1985a). A.J., **90**, 1464.
Sandage, A., and Carlson, G. (1985b). A.J., **90**, 1019.
Stothers, R. (1988). Ap.J., **329**, 212.
Vigroux, L., Stasinska, G., and Comte, G. (1987). Astron. & Astrophys., **172**, 15.
Walker, A. (1987a). M.N.R.A.S., **224**, 935.
Walker, A. (1987b). SAAO Circulars, **11**, 125.
Welch, D., McAlary, C., McLaren, R., and Madore, B. (1986). Ap.J., **305**, 583.
Wood, P., Bessell, M. and Fox, M. (1983). Ap.J., **272** 99.

CEPHEIDS IN LOCAL GROUP GALAXIES

Edward G. Schmidt
Department of Physics and Astronomy
University of Nebraska
Lincoln, Nebraska, U.S.A.

Abstract. A program is underway to obtain accurate, well sampled light curves for Cepheids in several Local Group galaxies. We have succeeded in attaining accuracies adequate for light curve analysis for stars with periods as short as about 12 days. Photometry for six known Cepheids in NGC 6822 is discussed. Although better sampling in phase is needed to allow full light curve analysis, a Fourier decomposition was performed for one star and it appears to show some differences from galactic Cepheids. A preliminary comparison of some of the gross properties of the light curves with galactic Cepheids shows both similarities and interesting differences.

PAST STUDIES

The study of Local Group Cepheids has a long history going back at least three quarters of a century to the discovery of the period-luminosity relation in the Magellanic Clouds by Leavitt (1912). Since that time the relation has been basic both to the application of Cepheids to the determination of the cosmic distance scale and to studies of the properties of the stars themselves.

A decade and a half later, Hubble applied the period-luminosity relation to show that some objects which were then referred to as "nebulae" were in fact outside of the Milky Way and were galaxies in their own right. The first such "extragalactic nebulae" was NGC 6822 (Hubble 1925). We have obtained new photometric data for Cepheids in this dwarf irregular galaxy and will discuss them below.

Since the time of Hubble, the Cepheids have continued to play a central role in the determination of the distance scale within the Local Group. Some recent workers have recalibrated the older photographic photometry using modern techniques (for example, Sandage 1983 and Christian & Schommer 1987 for M 33; Freedman 1988 for IC 1613). Others have obtained new data in longer wavelength bands such as in R and I (for example, Freedman, Grieve & Madore 1985; Walker 1987; Freedman, 1988; Hodge, Lee & Mateo 1988) or in the near infrared (for example Madore et al. 1985). The emphasis has been on obtaining a few points which can be used in distance determination; of the cited authors, only Walker has attempted to obtain relatively complete light curves.

It is obvious from even this brief survey of the literature that most of the work on extragalactic Cepheids has been concerned with their use in distance determination while much less attention has been given to the study of the stars' own properties. No doubt this is due in part to the great importance of the distance scale in many areas of astronomy as well as the great difficulties it presents. The distance scale has both deserved and required all the attention it has received.

The faintness of extragalactic Cepheids has also been an impediment to their study. Except for the Magellanic Cloud Cepheids, only broad band light and sometimes color curves are available and they are not of good accuracy. This has limited most studies to the consideration of statistical properties derived from the light curves. For example Becker, Iben & Tuggle (1977) compared the period-frequency relations in several galaxies in an attempt to infer the star formation histories. However, selection effects can seriously distort period-frequency relations and other statistical properties of the Cepheid population.

A persual of photographic light curves for these stars (see for example Kayser 1961 for NGC 6822 and Sandage 1983 and Sandage & Carlson 1983 for M 33) shows that the scatter is several tenths of a magnitude at best and is often worse. This scatter differs from star to star and frequently light curves bottom out due to background effects. While this data is mostly adequate for the purpose for which it was collected, providing mean magnitudes and periods for distance determination, clearly the form of the light curves is lost in the noise.

A further difficulty in the study of light curves has been the lack of a firm theoretical basis for their interpretation. Hydrodynamic codes have not been able to reproduce observed light curves sufficiently well to restrict the parameters of the models (see Simon 1988). Additionally, comparisons of light curves have often been very subjective in nature.

CURRENT PROSPECTS

The situation described above is now changing. The theoretical interpretation of light curves is on firmer ground due to the development of Fourier decomposition while new detectors have greatly improved photometry at faint levels.

In the method of Fourier decomposition, the light curve of a variable star is fitted with a Fourier series. The amplitudes and phases of the various terms are combined to yield parameters which allow a quantitative comparison of light curves. This technique has been applied to a variety of stars including classical Cepheids, type II Cepheids, RR Lyrae stars in globular clusters and RR Lyrae stars in the field (see Simon 1988 for a review).

Simon & Moffett (1985) have presented Fourier decompositions for the classical Cepheids. In their plots of the Fourier parameters against period the changing asymmetry of the light curves and the presence of the period resonance near ten days are clearly seen. Overtone pulsators are also discriminated in those plots. Similar studies have been conducted for Cepheids in the Large and Small Magellanic Clouds by Andreasen & Peterson (1987) and Andreasen (1988a). Although the scatter in the diagrams is larger, it is possible to distinguish differences between those Cepheids and the galactic Cepheids. In particular the resonance at ten days appears at slightly longer periods among the Magellanic Cloud Cepheids.

The use of Fourier decomposition has partially although not totally alleviated the difficulties with light curve interpretation. The subjective nature of the comparisons have been removed while the appearance of the period resonance gives us information on the interiors of the stars. However, we are not yet able to reproduce the sequences of Fourier coefficients from hydrodynamic models. This should be an important goal for future theoretical studies.

Although, the Fourier technique has proved to be a great benefit to studies of light curves, it must be remembered that it is very demanding of the data. Good photometric accuracy is necessary to obtain reliable Fourier coefficients. Uniform phase coverage is required because gaps in the light curves result in unstable coefficients.

The introduction of solid state panoramic detectors, the CCD in particular, has opened the way to obtaining much better photometric data at faint levels than was possible with older techniques. These detectors offer better photometric properties than the photographic plate (for example, higher quantum efficiency, large dynamic range, linearity, good stability) while still being panoramic. This latter point is important in crowded fields where overlapping star images must be deconvolved.

Figure 1. Calculated accuracy of CCD photometry.

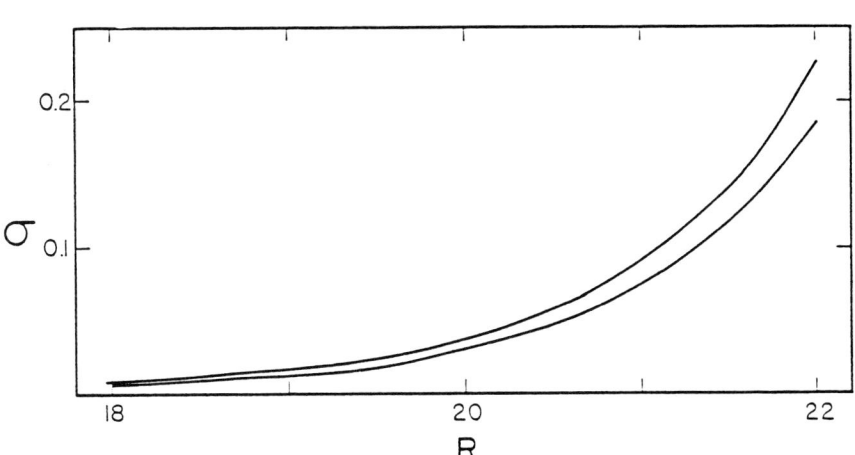

The good phase coverage required by the Fourier decomposition technique dictates the use of a small telescope. The quantum efficiency of the CCD makes it possible to obtain the needed accuracy in exposure times which are not prohibitive. In Figure 1 the photometric error in the Cousins R band due to photon statistics in the star signal and the background is plotted against the brightness of the star in the R band. This calculation assumes an 0.8-meter telescope and an exposure of 45 minutes. The background is an important contributor to the error and two values were used. The upper curve is with the measured background at Kitt Peak National Observatory while the lower is for the background brightness measured at McDonald Observatory. In each case, a contribution from the galaxy is included. This latter naturally varies from one galaxy to another and within a galaxy. We have used a brightness which is from our field #1 in NGC 6822. A seeing of two arc seconds (FWHM) has been assumed. It can be seen that with these assumptions, which are not particularly demanding for a good site, it is possible to obtain errors of a tenth of a magnitude at R = 21. This corresponds to minimum light for Cepheids with periods of about six days in NGC 6822 and nine days in M 33. Thus, a small telescope at a dark site is capable of obtaining data of sufficient quality to study light curves for extragalactic Cepheids.

OUR PROGRAM

Based on the considerations outlined above, we undertook a program aimed at obtaining accurate, well-sampled light curves for Cepheids in several Local Group galaxies. Observations have been made with CCD cameras on the 0.8-meter telescope at McDonald Observatory and the 0.9-meter and 2.1-meter telescopes at Kitt Peak National Observatory. We need to obtain data on enough stars in each system to allow us to learn the overall properties of the Cepheids but must minimize the number of fields observed in each to keep the program within feasible limits. The results of this program will be of more general use if they are reduced to a standard photometric system even though that is not necessary to realize the primary goals.

In order to carry out Fourier decomposition of a light curve, a minimum of about 30 well spaced data points are needed. In our fields there are a mixture of Cepheids of different periods. Thus, it is not possible to optimize the observing times to get even spacing in phase for more than a few of them. This, together with telescope scheduling, dictates that we must obtain enough data points to fill in the light curves in a random fashion. About 45 frames in each field will accomplish this for the bulk of the stars.

In Table 1 we list the galaxies for which we have obtained data. The two fields in each were selected to maximize the number of known Cepheids within the three by five arc minute area of the CCD. It can be seen from the table that we will be able to obtain data for between eight and forty Cepheids in each galaxy. It will also be noted that we have obtained nearly enough frames to meet our needs in three of the four galaxies on the list.

We have chosen to obtain our data through filters matched to the Cousins R and I bands. R will constitute the primary color for use in the Fourier decomposition and fewer I exposures will be obtained for the purpose of providing colors. The reliability of Fourier decompositions is dependent on the size of the photometric errors compared with the star's amplitude. Although the amplitude in the R band is 2/3 that in the V band, the greater sensitivity of the CCD at longer wavelengths allows us to obtain smaller relative errors in shorter times. In fact, with some of the CCD's we have used in this project, the exposures in V would have been prohibitive. The choice of I to provide a color was dictated by the fact that the amplitude in that band is sufficiently small to allow us to interpolate in the I light curve and form colors even if it is relatively poorly sampled.

Our experience with Fourier decomposition of light curves indicates that we need accuracy of several hundredths of a magnitude to carry out such an analysis. Based on the above discussion, we can expect to achieve this accuracy for Cepheids with periods longer than about twelve days in R with exposures of about an hour while the exposures in I will generally be less. Longer exposures will be used in some of the fields to extend our sample to shorter periods.

We will reduce the photometry to the R and I system of Cousins. With the CCD and suitable filters it has been possible to achieve accurate transformations to that system (Walker 1984; Schmidt 1988).

Although the number of known Cepheids in our fields are sufficient for our needs, clearly the value of this work would be enhanced if we could increase the number. There is reason to believe that, in fact, for M 33 and NGC 6822 our fields contain many undiscovered Cepheids.

In NGC 6822 there are thirteen known Cepheids. On the other hand, each of the Magellanic Clouds contains about 1100 known Cepheids. Scaling this by the galaxy masses or following Helfand's (1984) scaling of x-ray

Table 1
Fields in Local Group Galaxies

Galaxy	Field #	Number of frames R	Number of frames I	Number of Cepheids
NGC 6822	1	40	26	6
	2	31	16	4
M33	1	39	20	6
	2	35	14	2
M31	1	41	25	20
	2	27	17	25
IC 1613	1	10	8	10
	2	4	2	8

binary rates or supernova rates, we find that there should be between 30 and 200 Cepheids in NGC 6822. Our sample could thus be increased by a factor of at least two with a complete search.

An examination of the period-frequency relations of Becker, Iben & Tuggle (1977) also indicates that there are short period stars missing in the samples for M 33 and NGC 6822. Extrapolations using the relations for M 31 and the Milky Way show that there should be between eight and thirteen additional Cepheids with periods between six and twenty days in our field 1 in NGC 6822 and thirteen to sixteen with periods between nine and twenty days in our field 1 in M 33. Again an increase of more than a factor of two is indicated over the present sample.

Finally we note that a study by Freedman (1988) supports these conclusions. She finds that stars within the instability region of the HR diagram in IC 1613 outnumber the known Cepheids by a factor of about two. Additionally, preliminary searches of some of our data for new Cepheids have revealed many new suspected variables (Schmidt & Spear 1988a).

Based on these considerations, it appears that the number of Cepheids in our sample will be at least twice that listed in Table 1 for M 33 and NGC 6822.

FIRST RESULTS

We have used the DAOPHOT image reduction package (described by Stetson 1987) to extract magnitudes from part of our data set. This package is useful in crowded fields.

DOAPHOT forms groups of stars with overlapping images which are then fit jointly to a point spread function. In the fields we are studying the groups generally grow to include most of the stars on the frame and are too large to be handled by DAOPHOT. It is obvious that any given program star is not greatly affected by stars some distance away even if there is an overlapping chain of stars between. We have thus limited the groups to several image diameters and fitted each program star in a group with between a half dozen and several dozen stars.

A serious source of error in photometry at faint levels is the presence of background stars. In the case of variable star observations we have a distinct advantage because we are analyzing many exposures of the same field. Additionally, we obtained some exposures with the 2.1-meter telescope at Kitt Peak which offered a larger image scale and thus a better definition of the background around the program stars. Following a first pass in which we allowed DAOPHOT to identify neighboring stars and obtain preliminary magnitudes, we formed a collated list of stars around each program object. We then used this as input to a second pass in which DAOPHOT fit the background with the same set of stars for each frame. The scatter from frame to frame was reduced by about half in the second pass.

A third difficulty was the unavailability of standard stars in our field. We have overcome this by selecting a number of local comparison stars and have defined instrumental magnitudes for them from a selection of the best frames. In NGC 6822 we have made a preliminary calibration of these stars using the photometry of Hoessel & Anderson (1986) but will obtain a new calibration using the CCD at Behlen Observatory. In the meantime, the data discussed here are reduced to the r magnitude in the Thuann and Gunn system.

To determine how close we have come to the expected accuracy discussed above and to determine whether the final accuracy of the extracted magnitudes is adequate to our purposes we have compared the photometry from 32 frames. Since all were accurately reduced to the same system through the use of the local comparison stars, the internal scatter among the various frames will give us an indication of the random errors. The results are plotted against the r magnitude in Figure 2. The solid circles indicate the standard deviations estimated from the local comparison star magnitudes while the open circles and x's are the errors from other stars which happened to be in the groups with program stars. The + signs indicate the rms scatter in Cepheid light curves during intervals when the star appeared to have little variation. The solid curve is from Figure 1 shifted to correspond to the Thuann and Gunn r magnitude. It can be seen that the actual errors are only about 50% higher than those estimated from photon statistics alone. This is not surprising in view of the need to fit the backgounds and star images. Methods are being investigated to further reduce the internal errors.

In our field 1 in NGC 6822 we have extracted data for nine known variables of which six are Cepheids. In plotting the new data we found that the periods given by Kayser (1961) for these stars generally produce excess scatter. We have therefore adjusted the periods for all except one of the Cepheids. The older periods and the revised values

Figure 2. Errors of CCD photometry. Various symbols show internal errors as discussed in the text and the solid line is from Figure 1.

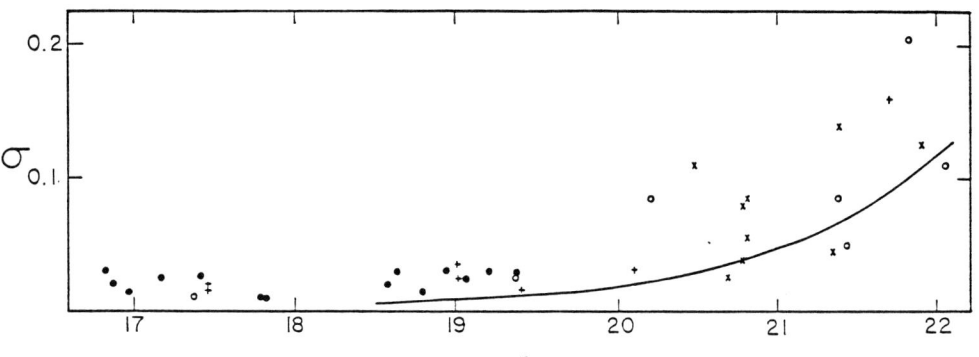

are listed in Table 2. Since the data used by Kayser were obtained over nearly fifty years and there are often gaps of the order of a year in the series, it might be speculated that the erroneous periods result from aliasing with one year. In the last column of Table 2 we list the periods obtained under this assumption; surprisingly only one of the periods, that for V 6, can be explained in this way. Unfortunately, our data is not ideally spaced in time to determine good periods and our revised values should not be considered definitive.

The light curves for the six Cepheids using data from two observing seasons, 1985 and 1986, are shown in Figure 3. A comparison of these light curves with the previous photographic light curves (Kayser 1961) shows that we have achieved a scatter which is smaller by two to six times. The shapes of the light curves are thus well defined.

It will be noticed that in several of the light curves there are points which scatter by an abnormally large amount compared with the rest. For example in the light curve of V 4 there are three points just after maximum light which scatter by between a tenth and two tenths of a magnitude from the remainder while in V 21 there are two points which fall below the minimum defined by other points by a similar amount. Problems of this type are common in CCD data and reflect undiagnosed cosmic ray hits or defective pixels. Unfortunately, the statistical tests applied by DAOPHOT to the fitting do not seem to effectively discriminate against these occurences. Other means of eliminating or correcting such data points are being investigated but in the mean time we can ignore the few obviously discrepant points.

Although the phase coverage in our light curves is not yet sufficient to carry out a definitive Fourier decomposition, it appeared worthwhile to attempt a very preliminary analysis for two stars, V 4 (with three discrepant points omitted) and V 6. The parameters derived from the fits are displayed in Table 3. For each star we give the Fourier parameters for fits of order one through five. It will be noted that the fit for V 4 is stable in the sense that the Fourier parameters change very little from one order to the next. On the other hand the

Table 2
Periods of the NGC 6822 Variables

Star	Period (Kayser)	Period (revised)	Alias with one year
V 5	13.3550	13.412	12.94, 13.94
V 4	17.3471	17.3471	16.56, 18.21
V 21	17.457	17.51	16.71, 18.39
V 6	21.1450*	19.986	18.95, 21.14
V 28	34.6672	34.625	31.63, 38.25
V 7	65.45	66.13	55.99, 80.75

* Alias with one year

parameters for V 6 can be seen to change appreciably and the fitting is clearly unstable. This is no doubt due to the gap in phase coverage near and after maximum light. We can not conclude anything from the Fourier fit to this star.

In Figure 4 the fitted curve for V 4 is plotted with the original data. It can be seen that the fit represents the data well; the scatter about the line is ±0.033 magnitudes. Some of the parameters formed from the Fourier coefficients are plotted in Figure 5 on the diagrams for classical Cepheids (from Simon & Moffett 1985). For R_{21} and Φ_{21}, V 4 matches the classical Cepheids well. However, the discrepancy for Φ_{31} is too large to attribute to uncertainties in the fit and may indicate differences between the classical Cepheids and the Cepheids in NGC 6822.

Figure 3. Light curves for Cepheids in NGC 6822.

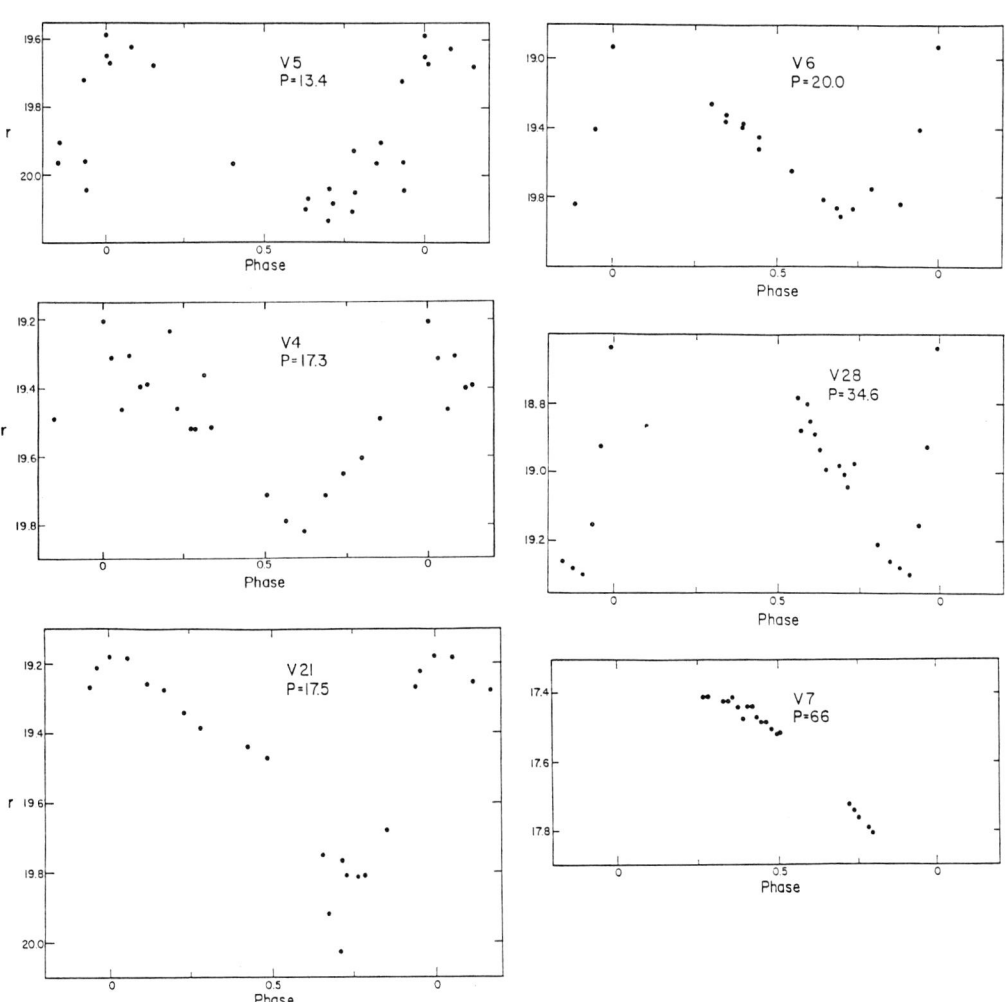

Fourier decompositions of more stars are needed to determine the significance of this tantalizing result.

Although there is not yet enough data to carry out Fourier decompositions, we can make some comparisons of gross light curve

Table 3
Fourier Fits for Variable V 4

Order	Std.Dev. (0.01 mag)	R_{21} (mag)	Φ_{21} (rad)	R_{31} (mag)	Φ_{31} (rad)
1	5.3				
2	4.1	0.26	4.76		
3	3.6	0.28	4.79	0.13	0.76
4	3.0	0.34	4.84	0.12	0.75
5	3.3	0.32	4.76	0.11	1.15

Fourier Fits for Variable V 6

Order	Std.Dev. (0.01 mag)	R_{21} (mag)	Φ_{21} (rad)	R_{31} (mag)	Φ_{31} (rad)
1	13.0				
2	9.6	0.33	4.27		
3	5.5	0.42	3.83	0.21	1.03
4	4.2	0.44	4.12	0.27	1.44
5	3.0	0.38	5.21	0.37	2.98

Figure 4. Light curve of V4 with the Fourier fitted curve.

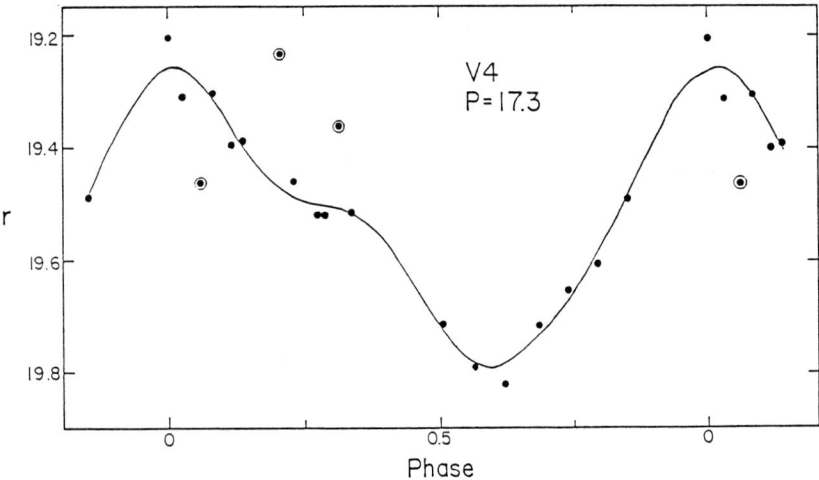

properties between the Milky Way and NGC 6822. Figure 6 shows several of these. There are both similarities and differences between the galactic and the NGC 6822 Cepheids. For example V 21 has a light curve which has a greater skewness and smaller acuteness than any galactic Cepheid of similar period while the rest match the galactic stars quite well. This is obvious from the form of its light curve in Figure 3. There are also some larger amplitudes and faster decline rates seen in NGC 6822 than for galactic Cepheids.

Figure 5. Fourier diagrams for galactic Cepheids (filled circles) and V4 in NGC 6822 (+ signs).

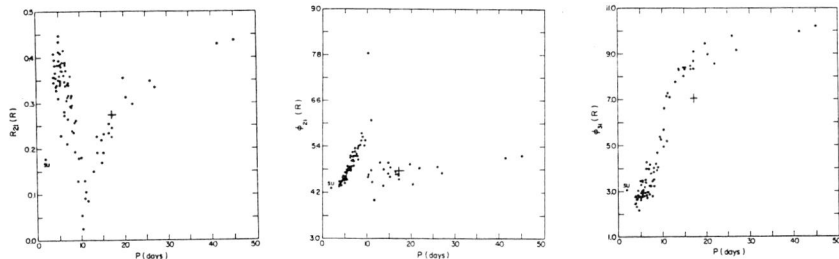

Figure 6. The decline rate, amplitude, skewness and acuteness of galactic Cepheids (filled circles) and Cepheids in NGC 6822 (open circles). For the galactic Cepheids the decline rates were derived from the data of Moffett & Barnes (1984) while the amplitudes are from Moffett & Barnes (1985). The skewness and acuteness data for the galactic Cepheids are from Andreasen (1988b).

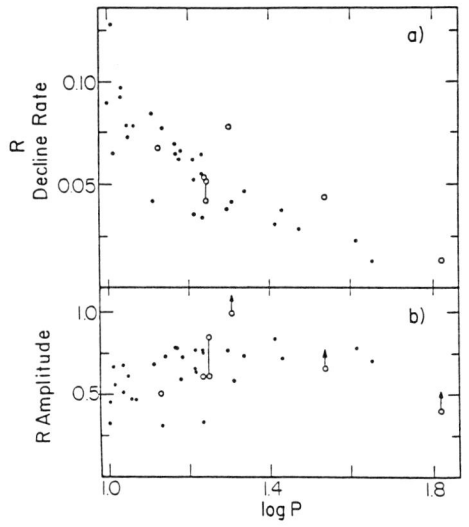
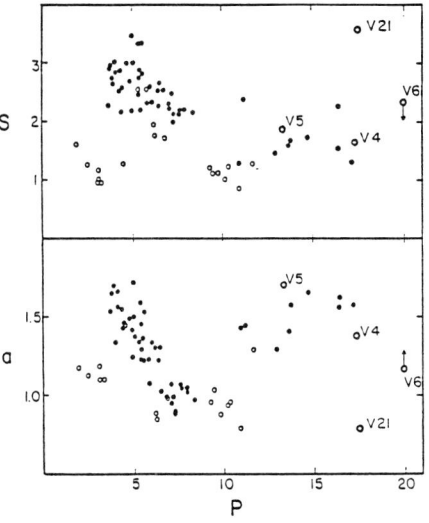

The results so far obtained are not sufficient to attack astrophysical problems. However, it is clear that we are now in a position to utilize the Cepheids in Local Group galaxies as a probe of the stellar content of those galaxies due to the improvements in both interpretation of light curves and in the observational techniques available to obtain them.

ACKNOWLEDGEMENTS

A number of people have contributed to the program described here and deserve to be thanked. N.R. Simon and T.J. Teays suggested this investigation and provided the impetus to undertake it. G.G. Spear carried out much of the analysis of the data. B.J. Anthony-Twarog developed the image processing facility at the MidAmerica Image Processing Laboratory at the University of Kansas which was used for the reductions. She also provided considerable aid in its use and instructed me in the art of image processing. The observations were carried out at McDonald Observatory of the University of Texas and at Kitt Peak National Observatory. The project was supported by NSF grant AST-8312649.

REFERENCES

Andreasen, G. K. (1988a). Astron. & Astrophys.,191, 71.
Andreasen, G.K. (1988b). Astron. & Astrophys.,196, 159.
Andraesen, G.K. & Petersen, J.O. (1987). Astron. & Astrophys.,180, 129.
Becker, S.A., Iben, I. & Tuggle, R.S. (1977). Ap.J.,218, 633.
Christian, C.A. & Schommer, R.A. (1987). A.J.,93, 557.
Freedman, W.L. (1988). Ap.J.,326, 691.
Freedman, W.L., Grieve, G.R. & Madore B.F. (1985). Ap.J.Suppl.,59, 311.
Helfand, D.J. (1984). P.A.S.P.,96, 913.
Hodge, P., Lee, M.G. & Mateo, M. (1988). Ap.J.,324, 172.
Hoessel, J.G., & Anderson, N. (1986). Ap.J.Suppl.,60, 507.
Hubble, E. (1925). Ap.J.,62, 409.
Kayser, S.E. (1961). Astron.J.,72, 134.
Leavitt, H.S. (1912). Harvard Circ. no. 173.
Madore, B.F., McAlary, C.W., McLaren, R.A., Welch, D.L., Neugebauer, G., & Matthews, K. (1985). Ap.J.,294, 560.
Moffett, T.J., Barnes, T.G. (1984). Ap.J.Suppl.,55, 389.
Moffett, T.J., Barnes, T.G. (1985). Ap.J.Suppl.,58, 843.
Sandage, A. (1983). A.J.,88, 1108.
Sandage, A. & Carlson, G. (1983). Ap.J.,267, L25.
Schmidt, E.G. (1988). In Proceedings of the Ninth Annual Fairborne-Smithsonian-IAPPP Symposium, ed. D. Hayes. Phoenix: Fairborn Observatory Press.
Schmidt, E.G. & Spear, G.G. (1988a). B.A.A.S.,19,1036.
Schmidt, E.G. & Spear, G.G. (1988b). in press M.N.R.A.S.
Simon, N.R. (1988). In Stellar Pulsation and Mass Loss, eds. R. Stalio & L.A. Wilson, p. 27. Dordrecht: D. Reidel.
Simon, N.R. & Moffett, T.J. (1985). P.A.S.P.,97, 1078.

Stetson, P.B. (1987). P.A.S.P., 99, 191.
Walker, A.R. (1984). M.N.R.A.S., 209, 83.
Walker, A.R. (1987). M.N.R.A.S., 225, 627.

THE BAADE-WESSELINK TECHNIQUE

Thomas J. Moffett
Department of Physics, Purdue University
West Lafayette, Indiana, 47907

INTRODUCTION

The historical foundations of the Baade-Wesselink (BW) method were established through the work of Baade (1926), Becker (1940), and Wesselink (1947). All modern versions of the BW-method are constructed upon these foundations. Since the radii of pulsating stars are fundamental to many areas of astronomical research (theories of stellar structure and evolution, galactic structure, and the cosmic distance scale), it is paramount that we strive to improve their observational determination. At the present time, the BW-technique is the only "direct" method for determining the radii of pulsating stars. Until a new technique is devised, better determinations of this important physical parameter will only occur through improvements in the BW-method.

By necessity, the scope of this review of the BW-technique will be limited in scope. The historical development along with the spectroscopic problems associated with the BW-technique are more fully discussed in the excellent review by Gautschy (1987).

THE BASIC IDEA

Using the definitions of effective temperature and apparent bolometric magnitude, the ratio of the radii at different phases in the pulsation cycle can be expressed as:

$$\log \left[\frac{R_2}{R_1} \right] = -0.2 \left[(V_2 + B.C._2) - (V_1 + B.C._1) + 10 \log \frac{T_{eff_2}}{T_{eff_1}} \right] \quad (1)$$

where the subscripts refer to the two phases, V, B.C., and T_{eff} refer to the apparent visual magnitude, the bolometric correction, and the effective temperature. This equation describes the behavior of the "photometric radii" over the pulsation cycle.

Integration of the observed radial velocity variation of the star throughout its cycle can be used to describe the behavior of the "spectroscopic radii":

$$R_2 - R_1 = -p \, \tau \int_{\Phi_1}^{\Phi_2} (V_r - \gamma) \, d\Phi \quad (2)$$

where p is the conversion factor from radial velocity to pulsation velocity, τ is the pulsation period in seconds, V_r is the observed radial velocity, and γ is the velocity of the star's center of mass. The center of mass velocity can be determined from the observed radial velocities by demanding that the following equation be satisfied:

$$\int_{t}^{t+\tau} (V_r - \gamma) \, dt = 0. \qquad (3)$$

Equations (1) and (2) can now be solved, in principle, for the radius of the pulsating star.

$$R_1 = \frac{-p\,\tau \int_{\Phi_1}^{\Phi_2} (V_r - \gamma) \, d\Phi}{10^{-0.2\,[(V_2 + B.C._2) - (V_1 + B.C._1) + 10 \log(T_{eff2}/T_{eff1})]} - 1} \qquad (4)$$

Equation (4) represents the formal solution of the BW-method provided one knows the values of all the quantities on the right side. In practice, the values of the bolometric correction and effective temperature as functions of pulsational phase are difficult to determine from the observational data. Various schemes have been devised, which Gautschy (1987) refers to as "realizations" of the BW-method, to solve this equation. The fundamental problem faced in all realizations of the BW-method is that the flux emitted by the star is the result of both temperature (more correctly surface brightness) and radius changes during the pulsation cycle. One must devise means to separate these two effects.

Consider the idealized case of a pulsating star which maintains constant temperature and atmospheric structure throughout its cycle. This is the simplest realization since equation (4) would reduce to:

$$R_1 = \frac{-p\,\tau \int_{\Phi_1}^{\Phi_2} (V_r - \gamma) \, d\Phi}{10^{-0.2(V_2 - V_1)} - 1} \qquad (5)$$

Since the surface brightness is constant, all of the quantities on the right side of the above equation are now easily determined from the observations. In the real world, this idealized case is approximated by working in the infrared region. The flux emitted by the star has a weaker coupling to the temperature in the infrared ($L_{IR} \propto T^{1.5}$) as compared to the optical region ($L_V \propto T^4$) of the spectrum. If the star is allowed to vary its temperature during the pulsation cycle but still required to be a blackbody, then the simple realization of equation (5) can still be used. By selecting pairs of phases with equal colors, one has selected pairs of equal surface brightness since the temperature and bolometric correction is uniquely determined by the color for a blackbody. In practice, this case is approximated by working in regions of the spectrum where the influence of non-blackbody effects (i.e. shock waves, etc.) is minimized.

Most realizations of the BW-technique are formulated in terms of surface brightness defined by Wesselink (1969) as:

$$S_V \equiv m_V + 5 \log \theta \tag{6}$$

or,

$$S_V \equiv M_V + 5 \log R + \text{const.} \tag{7}$$

where θ is the angular diameter, R the linear radius, and the constant is defined by the adopted zero points of S_V and M_V, plus the units used for the linear radius. The surface brightness can be expressed in a variety of forms. For example, since the surface brightness is the flux expressed in magnitudes per unit surface area of the star, one can write,

$$S_{bol} = -2.5 \log \left[\frac{4 \pi R^2 \sigma T_{eff}^4}{4 \pi R^2} \right] = -10 \log T_{eff} + \text{const.} \tag{8}$$

or, in terms of the visual magnitude of the star,

$$S_V = -10 \log T_{eff} - \text{B.C.} + \text{const.} \tag{9}$$

Barnes, Evans and Moffett (1978) express the surface brightness in terms of the visual surface brightness parameter, F_V, defined as,

$$F_V \equiv \log T_{eff} + 0.1 \text{ B.C.} \tag{10}$$

Comparing eqs. (9) and (10), we see that the relationship between the surface brightness and the visual surface brightness parameter is:

$$F_V = -0.1 (S_V + \text{const.}) \tag{11}$$

Similar to eq. (6), the visual surface brightness parameter can also be expressed in terms of the angular diameter of the star:

$$F_V = 4.2207 - 0.1 V_0 - 0.5 \log \theta \tag{12}$$

where the constant was determined by adopting solar values, V_0 is the un-reddened apparent visual magnitude, and θ the angular diameter expressed in arc milliseconds.

Using the surface brightness approach is the most natural way to formulate a realization of the BW-technique since it separates the temperature dependent effects from the radius variations. In the classical BW-realization, pairs of phases with equal surface brightnesses are used to reduce eq.(4) to the simple form of eq.(5). Realizations which rely on pairs of points suffer from two problems. First, not all of the observed points are used in the final radius determination. Observations near the turning points in the light curve are not utilized, for practical reasons, in the pair selection process. Secondly, the selected pairs of phases are composed of inhomogeneous members. In general, one member of the pair will represent a point during the expansion portion of the pulsation cycle while the other member corresponds to a contraction portion (see Gautschy 1987).

On the other hand, if S_V or F_V is explicitly known, then data over the entire cycle can be used in the determination of the mean radius. The most popular approach is to parameterize the surface brightness in terms of a linear relation using some color index,

$$F_V = a + b \ (C.I.) \tag{13}$$

If the zero-point and the slope in the above equation are known, then measurements of the apparent magnitude and appropriate color index over the entire pulsation cycle allows one to determine the variation of the star's angular diameter over the cycle by means of eqs.(13) and (12). Integration of the radial velocity curve provides the changes in linear diameter over the cycle. The distance to a pulsating star, r (in parsecs), is related to its linear and angular diameter, θ, through the equation,

$$\Delta D + D_m = 10^{-3} r \theta \tag{14}$$

where ΔD is the instantaneous linear displacement from the mean diameter, D_m, both expressed in Astronomical Units. A regression analysis of θ against ΔD yields both the star's distance and mean radius. In this realization, the radius depends only on the value of the slope, b, but the distance depends on both the slope and the zero-point, a, in eq.(13).

MODERN REALIZATIONS

Since a major objective of radius determination via the BW-technique is to serve as a check on theory, one would like to maintain observational "purity". It is obvious that a radius determined without recourse to theory serves as the best observational check. In all realizations of the BW-method, theory enters through the calculation of the velocity conversion factor p, so absolute purity is never realized. The question which must be addressed is how much additional theoretical input is acceptable before observational purity is severely compromised?

Two philosophies have evolved in formulating modern realizations. In the empirical approach, one tries to find a region of the spectrum where the effects of surface gravity, abundance, shocks, temperature etc. are minimized. An example of this approach is the use of the infrared region to minimize the effect of temperature changes to approximate the simple realization of eq.(5). The advantages of the empirical approach are simplicity and observational purity. The major disadvantage is that one intentionally avoids some of the interesting physics (e.g. shock waves, surface gravity effects, etc.). A concern is the compromising of physical reality in favor of observational purity.

In the modeling approach, one tries to handle some of the non-blackbody effects by invoking theory (model atmospheres) to some degree. An example of the modeling approach is the CORS method of Caccin et al. (1981) in which two color indices, calibrated by the model atmosphere grids of Kurucz (1979), are used to track variations in T_{eff}, B.C., and g_{eff} over the pulsation cycle. The advantages are a more realistic realization and the inclusion of interesting physics. The disadvantages are the realization is more complex and observational purity is compromised to some degree. This last point is often overlooked or not fully appreciated. Simon (1988) and Fernley et al. (1988) both stress that model atmospheres used to track surface gravity effects etc. may contain inaccurate or missing physics which could lead to systematic errors in the radii determined in these realizations.

THE EMPIRICAL APPROACH

The realizations of the BW-technique used by Gieren (1986) and Moffett and Barnes (1987) are identical in that they both use the visual surface brightness method. The method used by Balona (1977) and Balona and Martin (1978a,b) is essentially the same realization except that the method of linear least squares is not used in the solution.

In the realizations of Moffett and Barnes (1987) and Gieren (1986), the Thompson (1975) method is used to determined the slope of the visual surface brightness relation, eq.(13), for each Cepheid. These individual values are then used to determine a mean slope relation which is then used in a linear least squares solution for the mean radius. In Simon's (1988) nomenclature, this is a "consecutive" realization since the mean radius and slope of the visual surface brightness are not found in a "concurrent" manner.

Balona (1977) made a major contribution by emphasizing that the method of linear least squares does not correctly take into account the effects of errors, in all the observational quantities, on the final mean radius determination. Balona's formulation of the surface brightness method uses the following working equation:

$$V_o = a + b \ (B-V)_o - 5 \ \log(R_o + \Delta r) \tag{15}$$

This gives the star's apparent magnitude as a function of the surface brightness variation (first two terms) and the surface area variation (third term) at each phase during the pulsation cycle. The constant a contains a distance term which is different for each star. The last term is linearized under the assumption that Δr is small compared to the mean radius R_0. Coulson, Caldwell and Gieren (1986) later improved this realization by developing an iterative procedure which no longer required that Δr be small. Since Balona's realization is a concurrent solution, the slope of the surface brightness relation, b, is allowed to vary star by star.

Balona states that the Principle of Maximum Likelihood is used to solve eq. (15). Strictly speaking, this is incorrect since the Principle of Maximum Likelihood requires that both the errors in all observables and the functional form of their distribution be known. If one assumes, as Balona does, that the errors follow a Gaussian distribution, then the Principle of Maximum Likelihood reduces to the method of nonlinear least squares.

The above realizations, and in fact all realizations both empirical and modeling, have a common unresolved problem. The spectroscopic and photometric radii curves do not agree in shape and/or phase for most stars. The correct phasing between the photometric and spectroscopic curves is very difficult to disentangle because there are two different sources of phase mismatch. The first is a "real" phase shift due to the fact that the photometric and spectroscopic data were acquired at difference epochs. Inaccuracies in the period or real changes in the period of the star produce a real phase shift. The second source is an "artificial" phase shift introduced by an incorrect slope in the surface brightness relation, eq.(13), as pointed out by Barnes et al. (1977) and Balona and Martin (1978a). The real phase shifts can only be eliminated by having simultaneous photometric and spectroscopic data which is usually not the case.

Balona and Martin (1978a,b) clearly demonstrate that introducing an arbitrary phase shift between the velocity and light curves is equivalent to changing the value of the slope in eq.(13). The converse is also true. If the adopted value of the slope of the surface brightness relation is slightly off, then an artificial phase shift is required to bring the spectroscopic and photometric curves into phase alignment. In a consecutive approach using a mean value for the slope, artificial phase shifts are inevitable since the individual slopes will scatter about the mean value. Leaving the slope as a free parameter, which forces phase alignment, is claimed to be one of the virtues of concurrent solutions, but this virtue is more aesthetic than substantive. As Balona and Martin (1978b) show, these two approaches are equivalent ways of obtaining phase alignment. Since real phase shifts can also be present, it could be argued that a final phase adjustment between the spectroscopic and photometric curves might be a better approach.

The real problem lies in the parameterization of the surface brightness by a linear relation of a single color index, eq.(13), which is an oversimplification of physical reality. This problem is present in all realizations, both empirical and modeling. The breakdown of this simple assumption is clearly seen in stars with very dynamic atmospheres, e.g. RR Lyrae stars. In these stars, the photometric curves are in very poor agreement with the spectroscopic curves near maximum light. In most applications, these problem phases are excluded in the final solution for the mean radius. The same shortcoming is present, but not as pronounced, in the Classical Cepheids. If the surface brightness relation is correct, then the photometric and spectroscopic curves should be in both phase and shape agreement. A plot of photometric versus spectroscopic displacement should show a one-to-one correspondence i.e. a straight line. Figure 1 shows such a plot where the equivalence is approximately, but not strictly, true.

Since the photometric radii, or angular diameters, are very sensitive to errors in the photometry, it is difficult to decide if the scatter from a straight line in Figure 1 is real, or merely due to observational uncertainty. In order to differentiate between the two possibilities, we solved eq.(15), using the $(V - R)_o$ index, for a, b, and R_o for each of the 63 Cepheids in our sample (Moffett & Barnes 1987). The residuals of the observed value minus the calculated visual magnitude, as given by eq. (15), were determined for each Cepheid. The residuals for all the Cepheids were then combined into small phases bins (Barnes et al. 1987). Combining the residuals in this manner reduces the influence of random

Figure 1. The angular diameter as determined from the surface brightness relation versus the linear displacement from integration of the velocity curve.

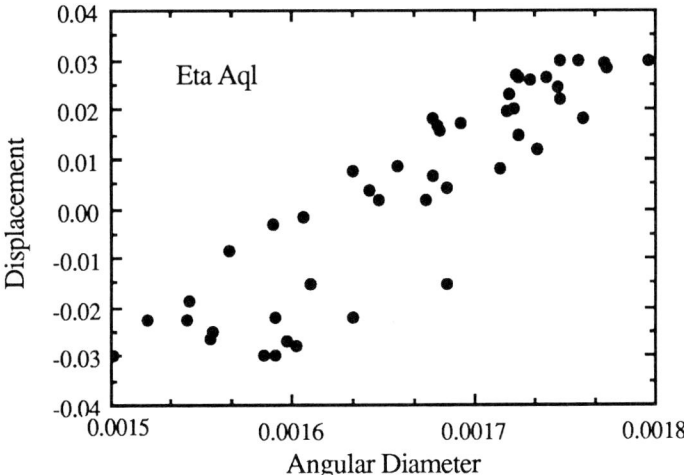

observational errors. The results of this comparison are shown in Figure 2 which clearly indicates that there are systematic effects present.

Despite the trend shown in Figure 2, the residuals produced by the simple linear version of the surface brightness relation are not bad. This simple empirical approach produces residuals with a peak-to-peak spread of less than ± 0.02 magnitudes. As a point of comparison, Hindsley and Bell (1987) used static model atmospheres to produce synthetic infrared magnitudes of Cepheids which were systematically different from the observed values by 0.15 magnitudes. The empirical approach using the simple linear surface brightness relation appears to be a good, but not perfect, approximation for Cepheids.

The empirical calibration of the surface brightness, F_V, in eq.(13) using $(V-R)_o$ is very good for non-variable stars (see Barnes et al. 1978). This is confirmed by the theoretical work of Bell and Gustafsson (1980) in which static model atmospheres were used to show that the combined effects of metal abundance and gravity are smallest in the (V-R) index. This empirical calibration starts to breakdown for variable stars because of their dynamic nature. Several researchers have shown that the influence of dynamic effects can be reduced by the judicious selection of both the magnitude and color index employed. Jones et al. (1987a), working with RR Lyrae stars, found that the size of the required artificial phase shift is smallest for the (V-K) index when used in combination with the V-mag. They found even better

Figure 2. The combined residuals for all stars as a function of pulsation phase.

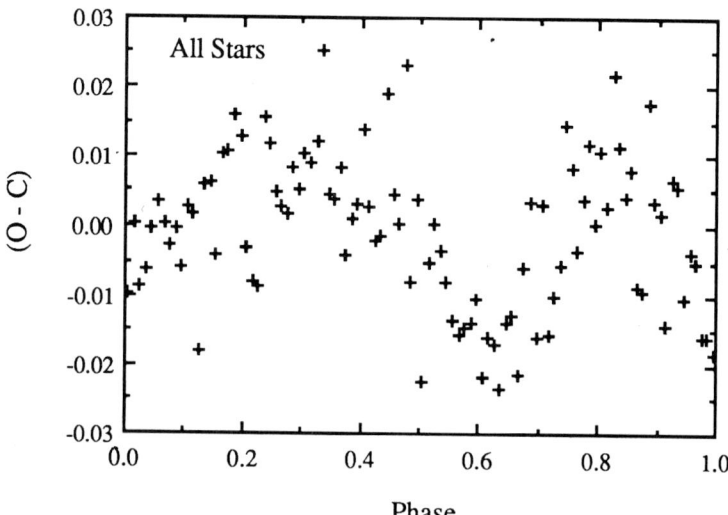

agreement between the photometric and spectroscopic curves when the
K-mag. replaced the V-mag. Fernley et al. (1987) pointed out the
advantages of the infrared over the optical region for radius
determination. For RR Lyrae stars, the optical region is near the peak
of their flux distribution so $L_V \propto T^4$ but in the infrared one is on the
Rayleigh-Jeans tail of the Planck function, where $L_K \propto T^{1.5}$. In other
words, the luminosity variation in the infrared is dominated by the
changes in surface area rather than temperature, but the opposite is
true in the optical region. Errors in the temperature description have
a lesser effect in the infrared than in the optical region. In
addition, the effects of metallicity and gravity are an order of
magnitude less sensitive in the infrared than in the optical.

Coulson, Caldwell, and Gieren (1986) used the method of Balona (1977) to
find the mean radius of the Cepheid TT Aql employing a variety of
magnitude and color index combinations. Their study gave a range of 68
to 107 $R_\odot$ for the mean radius depending on the particular combination
used in the solution. Their mean period-radius relation for Cepheids,
based on a nonlinear least squares solution using the V-mag. and
Cousins (V-I) index, is in excellent agreement with the mean relation of
Moffett and Barnes (1987), based on a linear least squares solution
using the V-mag. and Johnson (V-R) index. For example, the mean
period-radius relation of Coulson et al. (1986) predicts a 68.2 $R_\odot$
radius for a 10 day Cepheid and the corresponding relation of Moffett
and Barnes (1987) predicts 70.8 $R_\odot$. The particular color index used has
a much greater effect on the radius determination than does the
particular mathematical technique employed in the realization.

All realizations of the BW-technique will benefit from the high
precision radial velocity measurements being made of pulsating stars.
The CORAVEL velocities, for example, have uncertainties of ± 1 km·sec^{-1}
or less for a single velocity measurement. Integrations of radial
velocity curves of this quality yield linear displacements of very high
precision.

Simon (1987) has proposed a realization which takes full advantage of
these improved velocities by inverting the classical BW-method. Instead
of trying to pick pairs of phases at which the surface brightnesses are
equal, Simon picks pairs of phases at which the linear displacements are
equal, since these can now be determined with high precision. Once
phase pairs of equal radii are known, one can determine the values of
the coefficients of the adopted surface brightness relation by noting
that any difference in the flux levels is due to the surface brightness.
The formulation of this realization is very general, since the adopted
parameterization of the surface brightness can be of any form: linear
or nonlinear, single color index, multiple color indices, dynamic
parameters (kinetic energy, acceleration), etc. In his method, both the
radial velocity and photometric curves are represented by a Fourier
series. This realization requires photometric data of equivalent
precision (± 0.002 mag.) as the radial velocities. In addition, rather
complete phase coverage over the pulsation cycle is needed to determine

an accurate Fourier series representation of the velocity and
photometric curves. The use of pairs of points has the same
disadvantage as mentioned earlier, inhomogeneous pair members. However,
Simon (1988) described a consecutive solution version of his method
which removes the phase pair problem. Since this is a new realization,
it has not been fully tested, but it does show great promise for
improvements in the empirical approach.

THE MODELING APPROACH

All of these realizations use model atmospheres in some
fashion to determine the photometric displacement curve. Realizations
of this class suffer from the same problems encountered in the empirical
approaches: poor phase and shape agreement between the photometric and
spectroscopic displacement curves, strong sensitivity to the spectral
region used, and difficulty in assigning realistic error estimates to
the final mean radius.

Fernley et al. (1988) have used a very direct method which matches the
observed flux to the predicted model atmosphere flux in order to
determine the mean radius of the RR Lyrae star X Arietis. The total
flux of the star at any phase, $L(\phi)$, can be determined from the observed
de-reddened flux at the top of the earth's atmosphere, ℓ_λ, from:

$$L(\phi) = \int \ell_\lambda(\phi) \, d\lambda \qquad (16)$$

The effective temperature can be determined from the standard relation:

$$L(\phi) = \theta^2(\phi) \, \sigma \, T_{eff}^4(\phi) \qquad (17)$$

if an initial guess of the angular diameter, θ, is made. Using this
estimate of the effective temperature, an improved determination of the
angular diameter can be obtained by using the predicted flux from a
model atmosphere via the equation:

$$L_{IR}(\phi) = \theta^2(\phi) \, F_{IR} \{T_{eff}(\phi)\} \qquad (18)$$

where L_{IR} is the observed value and F_{IR} is the model atmosphere
prediction. Iteration between eqs. (17) and (18) yields the final
values of $T_{eff}(\phi)$, and $\theta(\phi)$. In evaluating the integral in eq. (16),
observations over the range of 2,000 to 22,000 Å were used. This gives
temperature determination based on a wide spectral range encompassing
over 95% of the total flux rather than a temperature estimate based on
the rather narrow spectral region defined by a color index. Equation
(18) makes use of the fact that most of the variation in the infrared
magnitudes, H and K bands, is due to radius (surface area) changes with
reduced sensitivity to temperature, metallicity, and gravity effects.

Cohen and Gordon (1987) used a similar realization to study RR Lyrae
stars in the globular cluster M5. The major difference is that they use
two narrow regions in the spectrum, one in the optical and the other in

the infrared. The observed fluxes in these two regions are fitted to
Kurucz (1979) model fluxes to find the angular diameter as a function of
phase.

Burki and Benz (1982) and Burki and Meylan (1986) adopt a quadratic
surface brightness relation calibrated by means of model atmospheres.
The shape mismatch between the two displacement curves is still present,
and the final mean radius excludes the phase interval of poor fit.

The most complex modeling realization is the CORS method developed by
Caccin et al. (1981), Sollazzo et al. (1981), and Onnembo et al. (1985).
They assume that the atmosphere of the pulsating star can be described
by the QSA approximation (quasi-static approximation), the
plane-parallel approximation for the radiative transfer and radiative
and hydrostatic equilibrium. The simple assumption that the surface
brightness can be described by a linear relation of a single color index
is abandoned. The effective temperatures and gravities are described by
two color indices calibrated via model atmospheres and the physical
calibration of the Walraven photometric system. The CORS realization is
currently the only one which uses all observations over the entire cycle
to calculate the radius at one particular phase. In principle, the CORS
method treats the BW-problem in a realistic rather than oversimplified
fashion, but in practice, it is only as reliable as the model
atmospheres and calibration of the photometric system used in its
application. The fundamental problem remains; a dynamic situation is
described by a series of static models.

The most popular version of modeling realizations uses synthetic colors
produced from model atmospheres to effectively determine a surface
brightness relation. Examples of this approach are the papers by:
McNamara and Feltz (1977), Manduca et al. (1981), Carney and Latham
(1984), Liu and Janes (1988), Jones (1988), and Jones et al. (1987a,b,
1988).

Most researchers using the modeling approach have pointed out the major
limitation of this technique. The final mean radii are only as good as
the theoretical model input parameters. The only available model
atmospheres are static models which do not satisfactorily reproduce the
dynamic effects present in real stars. Faced with describing a dynamic
atmosphere with a static model, one is forced to either pick a spectral
region where the dynamic effects (shock waves, rapid changes in
effective gravity, etc.) are minimized or exclude certain phase regions,
near minimum radius in the case of the RR Lyrae stars, or do both. The
fundamental requirement for better mean radii via the modeling approach
is better model atmospheres which include dynamic effects.

QUALITY OF RESULTS

Assigning realistic uncertainties to the final mean radius
has been a problem for all of the realizations. Balona (1977) correctly
points out that the method of linear least squares is not strictly valid

when all of the variables are subject to observational errors. He recommends that the Principle of Maximum Likelihood be used in place of linear least squares since the observational errors are treated in the correct manner. However, the full advantage of the Maximum Likelihood method is only realized when both the standard errors and the form of their distributions are known for all observational quantities. This latter requirement has not been included in any of the so called Maximum Likelihood solutions. Hawley et al. (1987) and Fernley et al. (1987) recommend an iterative nonlinear least squares with outlier rejection as a better mathematical technique in BW-solutions. It is clear that linear least squares is not the best mathematical method to use. Nonlinear least squares is better and the full application of the Principle of Maximum Likelihood may be the best.

Simon (1988) has the best procedure for estimating the uncertainty due to the methodology of the particular realization. One would like to have a good estimate of the true uncertainty of the mean radius due to all sources (observational, methodological, physical aptness of the adopted surface brightness relation), but these are difficult to estimate. Error estimates for the modeling approaches are even more complex. In addition to the uncertainties referred to above, one has to include errors due to incorrect or missing physics in the model atmospheres, errors in constructing synthetic colors and magnitudes, and errors due to interpolation between grids in the model atmospheres.

Using current BW-realizations, the true uncertainty of individual mean radii are probably not better than ±10% (Gautschy 1987). If a large sample of stars is used to determine a mean period-radius relation, the result may be better than ±10%. However, as Gautschy (1987) points out, it is very difficult to compare the various period-radius relations since the distribution of stars, both in terms of the number of stars and their period distribution, makes direct comparisons hard to interpret.

CONCLUSIONS

We must concur with Gautschy's statement that, **THE** BW-method does not yet exist. Of the two general approaches, empirical and modeling, there are no compelling reasons to pick one as superior to the other. Since the two approaches actually complement each other, we strongly recommend that improvements to both be vigorously pursued. Knowledge of mean radii at the 10% level is simply not good enough to address current problems in pulsating stars. For example, the current precision in the radii implies an uncertainty of 30% in the mass determination.

On the observational side, several things are needed to improve all types of BW-realizations. More high quality radial velocities with good phase coverage and, ideally, with simultaneous photometry to eliminate the phasing problem would be a great help. Photometric data, again with good phase coverage and with a precision to match the radial velocities, are needed. Additional, high precision infrared observations are needed to realize the advantages offered in the infrared region of the spectrum.

On the theoretical side, better model atmospheres which include dynamic effects represent the most urgent need. More work on the projection factor for converting observed radial velocities to pulsational velocities would benefit all realizations.

REFERENCES

Baade, W. (1926). Astron. Nachr., 228, 359.
Balona, L.A. (1977). M.N.R.A.S., 178, 231.
Balona, L.A. & Martin, W.L. (1978a). M.N.R.A.S., 184, 1.
Balona, L.A. & Martin, W.L. (1978b). M.N.R.A.S., 184, 11.
Barnes, T.G., Dominy, J.F., Evans, D.S., Kelton, P.W., Parsons, S.B. & Stover, R.J. (1977). M.N.R.A.S., 178, 661.
Barnes, T.G., Evans, D.S. & Moffett, T.J. (1978). M.N.R.A.S., 183, 285.
Barnes, T.G., Moffett, T.J., Jefferys, W.H. & Hawley, S.L. (1987). Bull. Amer. Astron. Soc., 19, 754.
Becker, W. (1940). Zeitschr. f. Astrophys, 19, 269.
Bell, R.A. & Gustafsson, B. (1980). M.N.R.A.S., 191, 435.
Burki, G. & Benz, W. (1982). Astron. & Astrophys., 115, 30.
Burki, G. & Meylan, G. (1986). Astron. Astrophys., 156, 131.
Caccin, B., Onnembo, A., Russo, G. & Sollazzo, C. (1981). Astron. Astrophys., 97, 104.
Carney, B.W. & Latham, D.W. (1984). Ap. J., 278, 241.
Cohen, J.G. & Gordon, G.A. (1987). Ap. J., 318, 215.
Coulson, I.M., Caldwell, J.A.R. & Gieren, W.P. (1986). Ap. J., 303, 273.
Fernley, J.A., Longmore, A.J. & Jameson, R.F. (1987). In Stellar Pulsation: A Memorial to John P. Cox, eds. A.N. Cox, W.M. Sparks, and S.G. Starrfield. p. 239. Berlin: Springer-Verlag.
Fernley, J.A., Lynas-Gray, A.E., Skillen, I., Jameson, R.F., Marang, F., Kilkenny, D. & Longmore, A.J. (1988). M.N.R.A.S., (in press).
Gautschy, A. (1987). Vistas in Astronomy, 30, 197.
Gieren, W. P. (1986). M.N.R.A.S., 222, 251.
Hawley, S.L., Barnes, T.G. & Moffett, T.J. (1987). In Stellar Pulsation: A Memorial to John P. Cox, eds. A.N. Cox, W.M. Sparks, and S.G. Starrfield, p. 235. Berlin: Springer-Verlag.
Hindsley, R. & Bell, R.A. (1987). In Stellar Pulsation: A Memorial to John P. Cox, eds. A.N. Cox, W.M. Sparks, and S.G. Starrfield, p. 212. Berlin: Springer-Verlag.
Jones, R.V. (1988). Ap. J., 326, 305.
Jones, R.V., Carney, B.W. & Latham, D.W. (1988). Ap. J., 326, 312.
Jones, R.V., Carney, B.W., Latham, D.W. & Kurucz, R.L. (1987a). Ap. J., 312, 254.
Jones, R.V., Carney, B.W., Latham, D.W. & Kurucz, R.L. (1987b). Ap. J., 314, 605.

Kurucz, R.L. (1979). Ap. J. Suppl.,40, 1.
Liu, T. & Janes, K.A. (1988). In Calibrating Stellar Ages, Contributions of the Van Vleck Observatory No. 6, ed. A.G.D. Philips, (in press). Schenectady, N.Y.: L. Davis Press.
Manduca, A., Bell, R.A., Barnes, T.G., Moffett, T.J. & Evans, D.S. (1981). Ap. J.,250, 312.
McNamara, D.H. & Feltz, K.A. (1977). P.A.S.P.,89, 699.
Moffett, T.J. & Barnes, T.G. (1987). Ap.J.,323, 280.
Onnembo, A., Buonaura, B., Caccin, B., Russo, G. & Sollazzo, C. (1985). Astron. & Astrophys.,152, 349.
Simon, N.R. (1987). P.A.S.P.,99, 868.
Simon, N.R. (1988). M.N.R.A.S., (in press).
Sollazzo, C., Russo, G., Onnembo, A. & Caccin, B. (1981). Astron. & Astrophys.,99, 66.
Thompson, R.J. (1975). M.N.R.A.S.,172, 455.
Wesselink, A.J. (1947). B.A.N.,10, 256.
Wesselink, A.J. (1969). M.N.R.A.S.,144, 297.

MIRA VARIABLES, STELLAR EVOLUTION AND GALACTIC STRUCTURE

M.W. Feast
South African Astronomical Observatory
P O Box 9 Observatory 7935
SOUTH AFRICA

1. INTRODUCTION

The Mira variables make important contributions to four of the main problems under discussion at this meeting, (1) stellar pulsation, (2) stellar evolution, (3) the morphology and history of the Galaxy, (4) the comparative study of different galaxies. The Miras also show how these rather different fields of study overlap, so that it is no longer possible to deal with any one field in isolation.

2. THE MIRAS AND STELLAR POPULATIONS

It has long been known that the Miras (defined as red giant variables of large amplitude, conventionally $\Delta V \geqslant 2.5$ mag) show, in the solar neighbourhood, both a period-asymmetrical drift and a period-velocity dispersion relation (see Table 1, data from Feast et al. 1972). The velocity dispersion perpendicular to the galactic plane (σ_W) varies from ~60 kms^{-1} at 200 day period to ~26 kms^{-1} at 400 day period, corresponding to scale heights of about 1000pc and 300pc respectively. Table 1 shows that if we adopt a velocity of the sun in the direction

Table 1

Asymmetrical drift with respect to sun (V) and total velocity dispersion (σ_T) for Miras and local M giants

(a) Miras

Range of Periods (days)	Mean Period (days)	V kms^{-1}	σ_T kms^{-1}	No. of stars
<140	131	-33 ± 13	81	22
145-200	179	-111 ± 22	180	46
200-250	225	- 61 ± 9	101	71
250-300	270	- 33 ± 10	88	77
300-350	324	- 32 ± 6	69	83
350-410	382	- 23 ± 8	58	54
>410	454	- 15 ± 8	50	35

(b) local M giants

-	-	- 18.3	42	266

of galactic rotation of -13 kms^{-1} relative to the circular velocity then Miras with periods between 145 and 200 days ($\bar{P}$ = 179 days) have an asymmetrical drift of 98 ± 22 kms^{-1}, distinctly less than that of halo objects (e.g. 220 ± 23 kms^{-1} derived by Oort (1965) for RR Lyrae variables with $\Delta\delta$ > 5) but greater than that of Gilmore's thick disk (~30 kms^{-1}, Sandage 1987, Gilmore this volume). Presumably indicating the existence of objects of age intermediate between that of the halo and that of the thick disk. Miras with periods of ~250 days have kinematics similar to that of the thick disk.

A number of Zinn's (1985) disk globular clusters contain Mira variables. Their periods range from 191 days (NGC 6712) to 265 days (NGC 6553), and possibly to 310 days (NGC 5927) (cf. Feast 1981). Over this period range the kinematics of the local Miras change rapidly with period (cf. Table 1 and Feast et al. 1972 Fig.2) and this suggests that there is a range of kinematic subgroups amongst the disk globular clusters. The recent analysis of Armandroff (1988) would also lead to this same conclusion. He finds an asymmetrical drift of ~30 kms^{-1} for the disk globular clusters as a group (a considerable revision from earlier estimates). However the results just discussed strongly suggest that clusters containing Miras of ~200 day period (47 Tuc, NGC 6712, 6637, 6356) must belong to a population with a distinctly larger drift (~70 kms^{-1}).

3. MIRAS, SEMIREGULARS, STELLAR EVOLUTION AND STELLAR PULSATION

In any globular cluster with more than one Mira variable, the periods are all close to one another and the mean period tends to increase with the cluster metallicity (cf. Feast 1981). The Miras are also bolometrically the most luminous stars in such a cluster and evidently mark the end of AGB evolution. The kinematic results of the last section indicate that the period of a Mira is a function of its age and hence of its initial mass and/or metallicity. From these results it is possible to infer initial masses ranging from about $0.9M_\odot$ at a period of 200 days to about $1.1M_\odot$ at a period of 400 days with some dependence of the masses on the adopted metallicity (cf. Feast and Whitelock 1987).

The Mira variables in the Large Magellanic Cloud have a well defined Period-Luminosity (PL) relation. The relation at K(2.2μm) for M and S type stars has a remarkably small scatter (σ = 0.13 mag). This is an upper limit since there are still not good infrared light curves for all the stars. Work is in progress to see whether the true width of the relation can be established. Bolometric luminosities (from integrated JHKL fluxes) are somewhat less well determined but there is still an excellent PL relation (σ = 0.16 mag) (Glass et al. 1987). The dependence of kinematics on period for Miras in the solar neighbourhood shows that the Mira PL relation is not an evolutionary sequence but the locus of the end points of AGB evolution of stars of different masses and metallicities. This is further demonstrated by the data on bolometric luminosities of Miras and of smaller amplitude semiregular red variables in globular clusters, assembled and discussed

by Whitelock (1986). The data for metal rich clusters show that this
subset of semiregular variables define a line in the PL diagram (Fig.1)
of shallower slope than the Mira PL relation. This shallower line is
evidently an evolutionary track along which a star either evolves from
short to long period or perhaps oscillates back and forwards over all
or part of its length as it undergoes thermal pulsing. We might
expect semiregular variables of different mass/metallicities to lie on
tracks roughly parallel to that defined by the metal rich clusters.

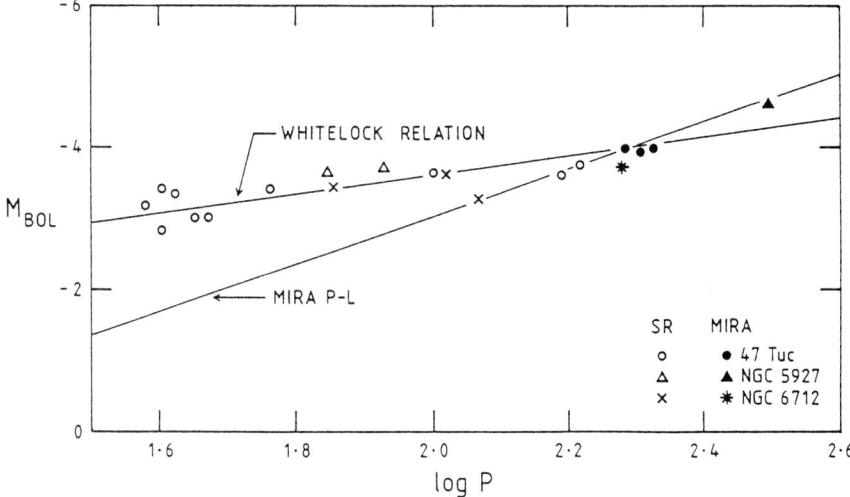

Fig.1. The low amplitude semiregular variables and Miras in metal rich
globular clusters (from Whitelock 1986). The Mira PL relation is also
shown.

Whitelock's work has revealed another interesting result. If one
plots semiregular variables from metal poor and metal rich clusters on
the same PL diagram (e.g. Whitelock 1986 Fig.2) one finds that they
define roughly the same line. If the cluster distances are derived
using a constant horizontal branch luminosity at all metallicities then
the variables in metal poor clusters are slightly less luminous at a
given period than those in metal rich clusters. If the absolute magnitudes of the horizontal branches are taken as, constant +0.2 [Fe/H],
(a relation for which there is some support, see summaries in Feast
1988, Sandage this volume) then the metal rich and metal poor variables
lie rather close to the same line. These possible changes are not
important for the present purpose since they leave any difference in
luminosity between metal poor and metal rich variables small at a given
period. Whitelock also finds that in a plot of log P against log T
(where T is the temperature derived from infrared photometry) there is
a rather clear separation between metal rich and metal poor variables,
amounting to Δ log T ~ 0.1 at a given period (see Whitelock 1986
Fig.3). It is very unlikely that errors in the calibration of log T
can be sufficient to account for this difference. In fact, as White-

lock points out, an error in log T can hardly explain this result for
the following reason. The variables fall on or close to the giant
branches of the clusters in which they lie. Since the luminosities of
the two groups of variables are the same, or nearly the same, at a
given period, making the log T values also the same would force together the giant branches of different metallicities in the HR diagram.
However a metallicity dependent separation of these branches is expected theoretically and observationally confirmed (e.g. Frogel et al.
1983). It must therefore be concluded that there is a real separation
of the two groups in the period - log T plot. The pulsation equation,
$\log Q = 0.5 \log M + \log P + 0.3 M_{bol} + 3 \log T - 12.71$,
then shows that this result implies a difference in mass (M) or pulsation constant (Q) (or both) between the two groups. To explain the
result in terms of a mass difference would require the masses of the
metal rich variables to be a factor four greater than the masses of the
metal poor ones. So large a difference seems quite unlikely. Whitelock points out that a change of Q by a factor of two will produce the
desired result. This implies that the small amplitude metal poor
variables pulsate in the fundamental mode and the metal rich ones in
the first overtone.

4. MIRAS IN DIFFERENT SYSTEMS

It is useful to compare the Miras in the LMC, SMC, globular
clusters and the galactic bulge. The LMC Miras show a narrow PL relation at K even when carbon Miras are included. The LMC data give the
best determined slope and the Mira data in the other systems can be
compared with it. The relatively small amount of data in the SMC
(mainly from Lloyd Evans et al. 1988) yields, within the errors, a PL
relation of the same slope as the LMC and, equally important, the same
zero point (Feast 1988). The range in periods of Miras in globular
clusters is not large enough to get a particularly accurate value for
the slope, but within the uncertainties, the LMC slope fits the cluster
data. This is true whether one uses a cluster distance scale based on
a fixed horizontal branch luminosity or on one depending on [Fe/H]
(cf. Menzies & Whitelock 1985, Feast, 1984, Feast 1987). In several
of these discussions the 149 day variable in ω Cen is found to lie
below the mean relation by ~0.3 mag. This variable will no longer be
so discrepant if one adopts the revised (increased) distance modulus
for this cluster discussed by Dickens (this volume). The Miras in the
galactic bulge (the NGC 6522 window) observed by Glass & Feast (1982)
and Wood et al. (1985) also show a PL relation of the same slope as the
LMC variables (Feast 1986). The bulge stars show a considerable scatter about this relation due to the spread of the stars along the line
of sight. A comparison of the Mira PL zero points in the LMC, the
globular clusters and the galactic bulge depends to some extent on
whether RR Lyrae variables (or horizontal branches) have luminosities
which depend on [Fe/H] or not. Within the uncertainties introduced by
this situation there are no evident zero point discrepancies (cf. Feast
1987, 1988).

5. MIRAS, IRAS SOURCES AND THE AGE OF THE GALACTIC BULGE

This topic has been reviewed several times recently (Feast 1986, Feast & Whitelock 1987, Feast 1987) and only a brief summary is given here. Optical Miras in the bulge (the Baade Windows) have periods ranging up to 400-500 days (Lloyd Evans 1976). These periods correspond (Feast & Whitelock 1987) to initial masses of about $1.1 M_\odot$ and ages of about 5 Gyr or greater (depending on the metallicity). Are there evolved stars in the bulge which are younger, or have higher initial masses, than this?

In the Galactic disk generally we find objects which apparently extend the optical Mira sequence to longer periods. These are the OH/IR stars with periods of 800 to 2000 days. It has been suggested that these OH/IR stars extend the Mira PL relation to longer periods (Feast 1985). Independent of this, any OH/IR star with M_{Bol} in the range -5.4 to -6.3 (the expected range on the PL relation, for 800 to 2000 day variables) will have higher initial mass (about 1.5 to 2.5 $M_\odot$) and younger ages than the optical Miras (cf. Iben & Renzini 1983 Fig.7, Feast & Whitelock 1987). Since such objects are dust enshrouded they are likely to have been missed in optical searches of the bulge. The IRAS survey provides an excellent opportunity to discover whether or not bright, long period OH/IR stars are present in the bulge. It was quickly realized that most IRAS sources in the bulge were Miras or at least Mira-like (Habing 1986, Feast 1986). In the Baade windows most of the IRAS sources are known optical Miras (Feast 1986, Glass 1986). This latter result suggests that the relative number of IRAS Mira-type objects in the bulge with period greater than 500 days must be small. This conclusion is strengthened by a ground-based JHKL survey of IRAS sources in a strip at $|b|$ between $7°$ and $8°$ (b = galactic latitude) being carried out by Whitelock, Catchpole and Feast (see Whitelock, Feast & Catchpole 1986, Feast & Whitelock 1987). These results suggest that there are relatively few Mira-like objects with M_{Bol} brighter than -4.7 which, on the PL relation, corresponds to a period of about 400 days. Harmon & Gilmore (1987) have carried out a statistical analysis of data from the IRAS survey. Making certain simplifying assumptions they suggest that there may be a significant fraction of the bulge (IR) Miras with periods greater than 600 days. If this is so it may put the upper limit to the initial masses of bulge stars slightly higher than the value derived above ($1.1 M_\odot$). Work is in progress to determine the periods of a significant sample of IRAS bulge sources so as to place firmer upper limits on the initial masses.

The above discussion refers to the bulge region in general. Some years ago an OH survey (Winnberg et al. 1985) revealed a population of OH/IR stars within $0.3°$ of the galactic centre. These objects seem distinctly more luminous than the Mira-like objects found further out in the bulge. The data of Jones & Hyland (1986) (cf. Feast 1987) yield a mean M_{Bol} of about -5.4 for these objects corresponding to a mean initial mass of about $1.5 M_\odot$. Further observation are required to investigate the possibility of a range in masses for these objects. More recent OH surveys (Winnberg 1988) over a wider area reveal many more objects which seem likely to be of the same type. They are found to be strongly concentrated to the galactic plane. Possibly these objects are best regarded as belonging to the inner part of the galactic disk rather than the bulge proper.

6. POST-MIRA EVOLUTION

If Miras lie at the tip of the AGB, they are expected to evolve rapidly into planetary nebulae (PN). Thus to a first approximation we expect the mass of the PN envelope to be equal to the difference between the (pulsation) mass of the Mira and the final (white dwarf) mass. Observations and theory (Weidemann 1984, Mazzitelli and D'Antona 1986) indicate that the final white dwarf mass varies rather little with initial mass. It is perhaps as low as $0.55 M_\odot$ for an initial mass of $0.9 M_\odot$ but generally lies close to $0.65 M_\odot$ out to initial masses of about $5 M_\odot$. Thus the mass of the PN envelope will increase quite rapidly with increasing initial mass. Data compiled from various sources by Pottasch (1988a) show that the mass of ionized gas in a PN envelope has an upper limit (in the available data) of $\sim 1 M_\odot$ for PN in the solar neighbourhood and in the Magellanic Clouds. This ionized mass is obviously a minimum value for the upper limit to the total envelope mass and implied stellar masses at the Mira stage of at least $1.7 M_\odot$.

PN in the galactic bulge provide an interesting confirmation of this general picture of post-Mira evolution. Both radio (Gauthier et al. 1983) and optical (Kinman et al. 1988) surveys of bulge PN show that there is an upper limit of close to $0.3 M_\odot$ for the ionized mass in the envelopes of these objects. In the case of the optical sample of Kinman et al., Pottasch (1988b) has shown that if one assumes the envelopes are optically thick then the central stars of these objects must lie in a region of the HR diagram well away from the expected evolutionary tracks for these objects. It would seem best to interpret this result as indicating that these objects are not optically thick. In that case the ionized mass is equal to the total mass in the envelope and we deduce an upper limit to the total mass of $0.3 M_\odot$. As indicated above the masses of some PN envelopes in other regions can be higher than this. This shows that the upper range of PN progenitor masses is missing in the galactic bulge. This is of course what was found in section 5 (an upper limit of $\sim 1.1 M_\odot$ for the initial masses of bulge objects). There is in fact good quantitative agreement with predictions. A combination of pulsation masses of Miras and evolutionary theory (Feast & Whitelock 1987) predicts that an object of initial mass $1.1 M_\odot$ will yield a PN envelope of $\sim 0.3 M_\odot$.

Recently, Webster (1988) has published an important study of line intensities and abundances for PN in the bulge. She suggests that a few of these objects may qualify for classification as Type I PN (PN with high helium and nitrogen abundances). It seems likely (cf. Peimbert & Torres-Peimbert 1983) that at least some Type I PN evolve from massive objects (initial masses of $2-5 M_\odot$). In the light of the previous discussion, and also the work of Terndrup (1988) on the main sequence turn-off in the bulge, it seems unlikely that there can be objects with initial masses as large as 2-5$M_\odot$ in the bulge. Further work is desirable to establish whether Webster's candidates are indeed of type I and also to place the mass determination of Type I PN on a firmer basis. It would obviously be of interest to derive the envelope masses for the bulge Type I PN candidates.

7. POSSIBLE MIRA PROGENITORS

Since short period (~200 day) Miras occur in metal rich globular clusters, the immediate progenitors of such Miras must be stars (small amplitude variables and constant stars) like those on the upper part of the AGB in these clusters. It has not however been possible to identify so easily the immediate progenitors of the longer period Miras. The galactic bulge with its large population of Miras of all periods might well be expected to yield important clues to these progenitors. The large number of M giants present in the bulge (e.g. Blanco et al. 1984) suggests these as candidate progenitors. In the past it was not possible to identify these stars as Mira progenitors because their surface distribution across the bulge (Blanco & Blanco 1986) seemed quite different from that of the Miras (cf. Feast 1987 esp. Table 7). However Blanco (1987) has considerably revised this derived surface distribution and the results are now quite similar to that of the Miras and of some other bulge objects (cf. Table 2). It would now seem reasonable to believe that these bulge M giants are progenitors of at least the longer period bulge Miras.

If this is the case we are left searching for the progenitors (presumably M giants) of the longer period Miras in the solar neighbourhood. Frogel & Whitford (1987) have emphasised that the Blanco et al. M giants in the bulge are different (e.g. in infrared colour-spectral type relations) from a sample of bright M giants in the solar neighbourhood. However it seems unlikely that these latter stars can, as a group, be considered the progenitors of the Miras in the solar neighbourhood. The kinematics of solar neighbourhood M giants is different from that of Miras. This is illustrated in Table 1 where it will be seen that the total velocity dispersion (σ_T) of the local M giants (cf. Delhaye 1965, Parenago 1951) is less than that of any of the Mira groups quoted in the table and that the group motion of the M stars with respect to the sun in the direction of galactic rotation is distinctly different from all but the two longest period groups of Miras. It seems likely that the bright local M giants are younger, more massive stars than is required for the Mira progenitors.

TABLE 2

Ratio of the number of Objects per unit area in the Bulge Window at
$\ell = 1°$ b = -3.9 (NGC 6522 field) to that at $\ell = 0°$ b = -8.5 (Plaut field 3)

    Optical Miras   = 10
    IRAS Sources   = 13
    RR Lyrae variables = 17
    Late M type stars   = 12

In these circumstances the most promising candidates for the progenitors of local long period Miras are the late M stars identified by Stephenson (1986) in a survey of a region at galactic latitudes greater than $10°$ (or perhaps a subset of these late M stars). Investigations are in progress by a group at SAAO to see how these stars are related to the local Mira population and to the M giants in the galactic bulge.

8. ACKNOWLEDGEMENTS

I am grateful to Dr P.A. Whitelock for helpful discussions and to Professor S.R. Pottasch for important preprints.

REFERENCES

Armandroff, T.E. (1988). Verbal report at 20th I.A.U. General Assembly, Baltimore, Md.
Blanco, V.M. (1988). Astron. J., 95, 1400.
Blanco, V.M. & Blanco, B.M. (1986). Astrophys. Space Sci. 118, 365.
Blanco, V.M., McCarthy, M.F. & Blanco, B.M. (1984). Astron. J. 89, 636.
Delhaye, J. (1965). In Galactic Structure, ed. A. Blaauw & M. Schmidt, p.61. (Stars and Stellar Systems, v.5). Chicago: University of Chicago Press.
Feast, M.W. (1981). In Physical Processes in Red Giants, ed. I. Iben & A. Renzini, p.193. Dordrecht, Reidel.
Feast, M.W. (1984). Mon. Not. R. astr. Soc., 211, 51P.
Feast, M.W. (1985). Observatory, 105, 85.
Feast, M.W. (1986). In Light on Dark Matter, ed. F.P. Israel, p.339.
Feast, M.W. (1987). In The Galaxy, ed. G. Gilmore & B. Carswell, p.1. Dordrecht: Reidel.
Feast, M.W. (1988). In The Extragalactic Distance Scale (Victoria meeting), ed. S. van den Bergh. In press.
Feast, M.W. & Whitelock, P.A. (1987). In Late Stages of Stellar Evolution, ed. S. Kwok & S.R. Pottasch, p.33. Dordrecht: Reidel.
Feast, M.W., Woolley, R. & Yilmaz, N. (1972). Mon. Not. R. astr. Soc., 158, 23.
Frogel, J.A., Persson, S.E. & Cohen, J.G. (1983). Astrophys. J. Suppl., 53, 713.
Frogel, J.A. & Whitford, A.E. (1987). Astrophys. J., 320, 199.
Gathier, R., Pottasch, S.R., Goss, W.M. & van Gorkom, J.H. (1983). Astron. Astrophys., 128, 325.

Glass, I.S. (1986). Mon. Not. R. astr. Soc., 221, 879.
Glass, I.S., Catchpole, R.M., Feast, M.W., Whitelock, P.A. & Reid, I.N. (1987). In Late Stages of Stellar Evolution, ed. S. Kwok & S.R. Pottasch, p.51. Dordrecht: Reidel.
Glass, I.S. & Feast, M.W. (1982). Mon. Not. R. astr. Soc., 199, 245.
Habing, H.J. (1986). In Light on Dark Matter, ed. F.P. Israel, p.329. Dordrecht: Reidel.
Harmon, R. & Gilmore, G. (1987). In Comets to Cosmology, ed. A. Lawrence. Berlin: Springer.
Iben, I. & Renzini, A. (1983). Ann. Rev. Astr. Astrophys., 21, 271.
Jones, T.J. & Hyland, A.R. (1986). Astron. J. 92, 805.
Kinman, T.D., Feast, M.W. & Lasker, B.M. (1988). Astron. J. 95, 804.
Lloyd Evans, T. (1976). Mon. Not. R. astr. Soc., 174, 169.
Lloyd Evans, T., Glass, I.S. & Catchpole, R.M. (1988). Mon. Not. R. astr. Soc., 231, 773.
Mazzitelli, I. & D'Antona, F. (1986). Astrophys, J., 311, 762.
Menzies, J.W. & Whitelock, P.A. (1985). Mon. Not. R. astr. Soc., 212, 783.
Oort, J.H. (1965). In Galactic Structure, ed. A. Blaauw & M. Schmidt, p.455. (Stars and Stellar Systems, v.5). Chicago: University of Chicago Press.
Parenago, P.P. (1951). Pub. Sternberg Inst. 20, 26.
Peimbert, M. & Torres-Peimbert, S. (1983). In Planetary Nebulae, ed. D.R. Flower, p.233. (IAU Symp. 103). Dordrecht: Reidel.
Pottasch, S.R. (1988a). Preprint.
Pottasch, S.R. (1988b). In Planetary Nebulae (IAU Symp. 131). In press.
Sandage, A. (1987). In The Galaxy, ed. G. Gilmore & B. Carswell, p.321.Dordrecht: Reidel.
Stephenson, C.B. (1986). Astrophys. J., 301, 927.
Terndrup, D.M. (1986). Ph.D. Thesis, University of California, Santa Cruz.
Webster, B.L. (1988). Mon. Not. R. astr. Soc., 230, 377.
Weidemann, V. (1984). Astron. Astrophys. 134, L1.
Whitelock, P.A. (1986). Mon. Not. R. astr. Soc., 219, 525.
Whitelock, P.A., Feast, M.W. & Catchpole, R.M. (1986). Mon. Not. R. astr. Soc., 222, 1.
Winnberg, A. (1988). Verbal report at 20th I.A.U. General Assembly, Baltimore, Md.
Winnberg, A., Baud, B., Matthews, H.E., Habing, H.J. & Olnon, F.M. (1985). Astrophys. J., 291, L45.
Wood, P.R., Bessell, M.S. & Paltoglou, G. (1985). Astrophys. J., 290, 477.
Zinn, R. (1985). Astrophys. J., 293, 424.

ANOMALOUS CEPHEIDS AND POPULATION II BLUE STRAGGLERS

James M. Nemec
Department of Geophysics & Astronomy
University of British Columbia
Vancouver, B.C. V6T 1W5, Canada

Abstract. Recent studies of anomalous Cepheids (ACs) and Pop II blue stragglers (BSs), including photometrically variable BSs (VBSs), are reviewed. The VBSs represent about 25% of the BSs, the majority of which are SX Phe short-period variables in the Cepheid instability strip. Mass estimates derived using various techniques suggest that both ACs and BSs are relatively massive (about 1.0-1.6 $M_\odot$). The recent discovery that two BSs in the globular cluster NGC 5466 are contact binaries, and the earlier discovery that one of the BSs in ω Cen is an eclipsing binary, provide direct evidence that at least some BSs are binary systems. If all BSs are binaries, and the time scale for coalescence is a few Gyr, then the majority are likely to be coalesced. Because ACs and BSs are found in the same stellar systems, and are probably related through their evolution, it is highly likely that most ACs are also coalesced binary systems. The fact that ACs and BSs are found only in low density environments, suggests that they were primordial binaries.

1. INTRODUCTION

Anomalous Cepheids and Population II blue stragglers have several features in common, most notably, proliferation in low density stellar environments (such as low central concentration globular clusters and Local Group dwarf galaxies), and relatively large masses, about 1.3 $M_\odot$ (Christy 1970; Zinn & Searle 1976). Since their discovery over 30 years ago, various single and binary star models have been proposed to explain the structure and evolution of both types of stars. One of the first and most obvious models is that they are massive single stars of intermediate age (i.e. a few Gyr). Such a model seems plausible for the BSs in open clusters (see Eggen & Iben 1988), and for the BSs and ACs in those luminous dwarf spheroidal galaxies that contain a substantial population of young stars (Mould 1983; DaCosta 1988), but is less convincing when it comes to explaining the large number of BSs that are found in about 15 Gyr old globular clusters, such as NGC 5466 (Nemec & Harris, 1987) and NGC 5053 (Nemec & Cohen 1988), that show little evidence of recent star formation. On the other hand, it is not necessary to assume that BSs and ACs are much younger than the systems in which they are found if they are relatively massive stars whose lifetimes were extended through mixing (McCrea, 1964; Wheeler 1979; Saio & Wheeler 1980; Da Costa & Demarque 1982; VandenBerg & Smith 1988).

McCrea (1964), and later van den Heuvel (1968) and Collier & Jenkins (1984), considered the possibility that BSs are close binary systems in which mass transfer has occurred between the evolving component stars. Under the assumption that the stars form with initially unequal masses, these studies found that a blue straggler sequence can be explained, and that the sequence could extend about 2.5 magnitudes above the main sequence turnoff. The influence of mass exchange on the orbital period of the binary system was also investigated. Renzini, Mengel & Sweigart (1977; hereafter RMS) were the first to demonstrate that a mass transfer model can explain both BSs and ACs, provided that the initial separation of the component stars is appropriate, i.e. during the red giant phase of the original secondary, mass is not transferred back to the the original primary (Norris & Zinn 1975). RMS predicted that because the frequency of stellar collisions in globular clusters increases with stellar density, ACs and BSs ought to be found preferentially in low density environments. They also predicted that the ratio of the number of BSs to ACs in the same stellar system should be greater than the ratio of the respective lifetimes, which they estimated to be in the range 1-10. This prediction provides a useful test of the binary hypothesis (see Nemec, Wehlau & Oliveira 1988; hereafter NWO). In the limiting case, it has been suggested that BSs are coalesced stars, which were originally detached binary systems (Zinn & Searle 1976; Webbink 1976a,b, 1979; Wallerstein & Cox 1984; Campbell 1986).

Further information on the theoretical models for ACs and BSs can be found in the reviews by Zinn (1980,1985), Hirshfeld (1980), Da Costa, Norris & Villumsen (1986), Nemec & Harris (1987), Harris (1987) and Da Costa (1988b). The emphasis of the present review is on recent observational developments concerning ACs (§2), BSs (§3), and variable BSs (§4).

2. ANOMALOUS CEPHEIDS
"The cause is hidden, but the result is well known."
  Ovid (Metamorphoses IV, 287)

Baade & Swope (1961), van Agt (1967, 1968), and Swope (1968) recognized from their pioneering studies of the variable stars in the Draco, Ursa Minor and Leo II dwarf galaxies that there exists in these galaxies a population of Cepheid-like variables that does not follow the period-luminosity (P-L) relationships of Pop I and Pop II Cepheids. These stars have become known as "anomalous Cepheids" (Norris & Zinn 1975; Zinn & Searle 1976). Unlike globular cluster Cepheids, which typically have masses appropriate for single stars in post-horizontal-branch evolutionary phases, i.e. M approximately 0.6 $M_\odot$ (Böhm-Vitense et al. 1974), the pulsational periods of ACs require that their masses be about 1.2 to 1.8 $M_\odot$ (Christy 1970; Norris & Zinn 1975; Demarque & Hirshfeld 1975; Zinn & Searle 1976; Zinn & Dahn 1976).

Over 50 ACs are known, almost all of which are located in the nearest Local Group dwarf galaxies. The only AC to have been identified in a galactic globular cluster is V19 in NGC 5466 (Zinn & Dahn 1976, Zinn & King 1982). A list of ACs, and their properties, is given in Table IX of NWO. The only nearby dwarf spheroidal galaxy whose variable stars have been completely surveyed for periods is Ursa Minor.

Much work remains to be done identifying ACs in other nearby dwarf galaxies (such as the Fornax dwarf galaxy; Demers & Irwin 1987, 1989), as well as in more distant Local Group dwarf galaxies (such as the three dwarfs spheroidals in the direction of M31; van den Bergh 1972a,b). The ACs in other environments, such as the Small and Large Magellanic Clouds (see Smith & Stryker 1986) have been studied to some degree; however, further study would be valuable.

2.1 Periods, Luminosities and Pulsation Modes

The shortest and longest period ACs have periods of 0.26 day (V69 in Carina) and 2.37 day (V8 in Leo I), with a uniform distribution of periods throughout this range. Absolute B magnitudes of ACs, assuming that the mean M_B level of the RR Lyrae stars in each system is 0.80 mag, range from a high of $M_B = -1.7$ (again, V8 in Leo I) to a low of $M_B \sim +0.5$ (several stars).

Since the first anomalous Cepheids were discovered, the determination of their pulsation modes has been a major source of uncertainty in deriving their masses (see Wallerstein & Cox 1984; Cox & Proffitt 1988). By assuming that the degree of symmetry of the light curve depends on the pulsation mode and on the T_{eff} in the same manner for ACs as for RR Lyrae stars, Zinn & Searle (1976) suggested that among the ACs in the Draco dwarf galaxy, V141 and V157 pulsate in the fundamental mode, and V134 and V204 pulsate in the first-overtone mode. Support for this suggestion was provided by RMS, who showed that the location of the first-overtone blue edges (for models with helium abundance Y = 0.2 and 0.3, and an assumed mass of 0.65 $M_\odot$) were consistent with the above pulsation mode assignments.

It is often assumed that ACs define a single P-L relationship. Figure 1 shows the P-L diagram constructed by NWO (with minor changes, as noted in the figure caption), illustrating that the best studied ACs obey one of two distinct P-L relationships: $M_B = -2.294 \log P - 0.374$, (with a standard deviation of the individual points about the adopted regression line of only $\sigma = 0.11$ mag), or $M_B = -4.213 \log P - 1.498$ ($\sigma = 0.09$ mag). The small dispersions about the two lines suggest that the assumed M_B values (hence the distance estimates to the various dwarf galaxies) must be approximately correct. NWO associate the first of the two relationships with fundamental mode pulsation, and the second with first-overtone pulsation. The pulsation modes for the Draco variables mentioned above, and for V19 in NGC 5466, can be read from Figure 1. These agree with the modes determined by Zinn & Searle (1976) and Zinn & King (1982), respectively. Thus, it follows that period and M_B information alone may be sufficient for determining pulsation mode. If this is the case, the pulsation modes for the Sculptor, Fornax and Ursa Minor variables, which were previously unknown, can be read from the graph. Of course, at the lowest luminosities where the two lines merge the pulsation modes are uncertain.

It is obviously of interest to compare the observed P-L relationships (Figure 1) with the P-L relations predicted by the fundamental equations of stellar pulsation appropriate for ACs. The best theoretical estimates of the coefficients in the pulsation equations are those given by Cox (1987), which for the King Ia composition (Y = 0.299, Z = 0.001) including the effects of convection, are:

$$\log P = 11.331 - 0.655 \log M/M_\odot + 0.829 \log L/L_\odot - 3.430 \log T_{eff}$$

for the fundamental mode, and

$$\log P_1/P_0 = -0.4227 + 0.0438 \log M/M_\odot - 0.0240 \log L/L_\odot + 0.0806 \log T_{eff}$$

for the ratio of the first-overtone to fundamental-mode periods. In the absence of bolometric corrections for the anomalous Cepheids, the

Figure 1. Log P vs. M_B for 21 well-studied ACs. Also shown in the figure are the RR Lyrae stars in Ursa Minor, and schematic outlines of where Pop I and Pop II Cepheids would lie. (Note: Several minor errors were made in Table IX and Figure 13 of NWO. For the three Sculptor variables, V25, V26 and V119, the $\langle m_B \rangle$ should have been 19.10, 18.96 and 19.03, and the M_B should have been -0.48, -0.62 and -0.55:, respectively. For the 9 Carina ACs the M_B should have been -0.05, -0.13, -0.68, -1.03, -0.06, +0.10, -1.21, -0.93, +0.15. For V149 in Carina, $\langle m_B \rangle$ should have been 20.35; and, for NGC5466-V19 M should have been -1.07. The equations of the P-L relations given above are the corrected equations. The main conclusions reached by NWO are unchanged).

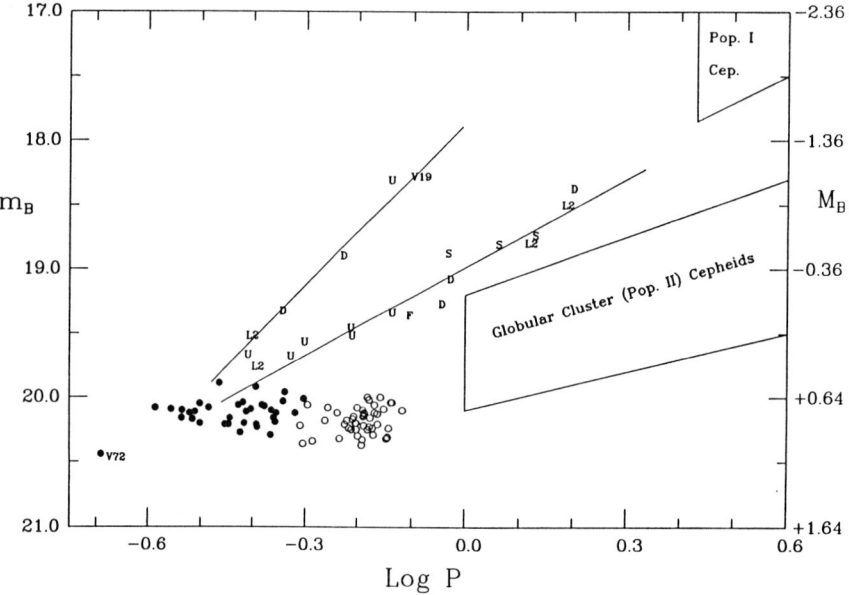

observed (log P)-M_B relations cannot be directly compared with the model results. However, luminosities are known from spectroscopic observations for NGC 5466-V19, and for the ACs in Draco (Searle & Zinn 1976). Assuming the pulsation modes given above, the following P-L relations were derived:

$$\log(L/L_\odot) = 2.226 + 1.412 \log P,$$

with $\sigma = 0.003$, for the three fundamental-mode ACs in Draco; and

$$\log(L/L_\odot) = 2.537 + 1.436 \log P,$$

with $\sigma = 0.007$, for NGC5466-V19 and the two first-overtone ACs in Draco. Note that the two lines are approximately parallel, and that the observed slope is in approximate agreement with the slopes 1.24 and 1.21 predicted for the first-overtone and fundamental modes, respectively.

2.2 Light Curves of Anomalous Cepheids

The pulsation of ACs in the fundamental and first-overtone modes is expected to lead to a dichotomy in the morphology of their light curves. To investigate this, the Fourier decomposition technique (see Simon & Lee 1981) is presently being used to evaluate the shapes of the light curves of ACs (Simon & Nemec, in preparation). Some preliminary results for the ACs in Ursa Minor are given in Table 1. Column (7) gives R_{21}, which is the ratio of the second largest to the largest amplitude of the fitted sinusoids. In other types of stars, such as dwarf Cepheids (see Fernley et al. 1987), it is found that the R_{21} ratio is large for fundamental mode pulsators, and small for first-overtone pulsators. The same appears to be true for ACs. The five stars in Table 1 with large R_{21} ratios do, in fact, lie along the fundamental mode relationship in the P-L diagram, and the star V6 lies along the first-overtone P-L line. The two stars V59 and V80 stand out in that they have intermediate R_{21} values, which makes it difficult to assign a mode. In Figure 1 it is also difficult to assign a mode since they have relatively low luminosities. The light curve of V59 is

Table 1.
Fourier coefficients for anomalous Cepheids in Ursa Minor

Star	Pulsation Mode	Period (days)	A_B	$<B>$	R_{21}	Φ_{21}	R_{31}	Φ_{31}	Φ_{41}
(1)	(2)	(3)	(4)	(5)	(6)	(7)	(8)	(9)	(10)
59	F	0.389981	1.59	19.70	0.37	3.91	0.08	1.87	0.64
1	F	0.470	1.13	19.64	0.57	3.93	--	--	--
80	?	0.498746	0.52	19.50	0.28	4.68	0.10	4.34	--
11	F	0.675	1.53	19.43	0.41	3.68	0.22	1.77	--
56	F	0.612494	1.65	19.34	0.50	3.67	0.35	1.58	--
6	H	0.725586	0.93	18.26	0.16	4.34	0.10	2.23	--
62	F	0.729	1.68	19.33	0.46	3.87	--	--	--

asymmetric and appears to be that of a fundamental mode pulsator. V80 has been problematic for some time; Kholopov suggested that it might be an eclipsing binary, but this remains to be seen. Further observations of V59 and V80 may help resolve these problems.

2.3 Masses of Anomalous Cepheids

The derivation of masses for ACs, using the fundamental equation of stellar pulsation, and T_{eff} and log g information, depends on knowledge of the pulsation modes of the stars. Zinn & Searle (1976) derived masses for the ACs in Draco, assuming pulsation modes based on T_{eff}, L, and an examination of the associated light curves, and derived masses ranging from about 1.0 to 1.8 $M_\odot$. Hirshfeld (1980), Wallerstein & Cox (1984) and Cox & Proffitt (1988) analyzed the same data and confirmed the masses derived by Zinn & Searle. However, the mass estimates would have been considerably different if the assumed pulsation modes were incorrect. For example, if the pulsation is in the first-overtone or the second-overtone mode, and not the fundamental mode, the derived mass drops by a factor of about 1.6 or about 2.5, respectively, from the fundamental mode value. Thus, uncertainy in the pulsation mode translates into a large uncertainty in the corresponding mass.

With the recent observation that fundamental and first-overtone pulsators separate in the P-L plane, and the development of Fourier methods for discriminating between the two groups, the pulsation modes, and hence the masses, of ACs can now be derived with confidence. Using the T_{eff} and log g values given by Searle & Zinn, and pulsation modes based on the P-L diagram, NWO determined that the minimum, maximum and mean masses for the five ACs in the Draco dwarf galaxy are 1.04, 1.77 and 1.38 $M_\odot$, respectively. These agree well with the earlier mass estimates, but the uncertainty is now much smaller (assuming that the T_{eff} and log g are accurate).

Despite the fact that the uncertainty due to error in the assignment of the pulsation mode has essentially been eliminated, there remain two controversies concerning the masses of ACs. First, the fact that the derived maximum mass, 1.77 $M_\odot$, is larger than twice the mass of the main sequence turnoff stars in Draco argues against the binary hypothesis. However, as noted by Cox & Proffitt, the discrepancy can be explained if there is an error in T_{eff}. They calculate that an increase in the measured T_{eff} of only 100 K is sufficient to reduce the derived mass to 1.6 $M_\odot$, which is consistent with the binary hypothesis. On the other hand, Wallerstein & Cox (1984) see no inconsistency if ACs were formed by the coalescence of two or more stars. The second controversy surrounds the minimum mass quoted above. Hirshfeld (1980) found that computer models predict that only those core helium-burning stars with masses between 1.3 and 1.6 $M_\odot$ can enter the instability strip. In this case, the estimated minimum mass for ACs is lower than expected. Again, it is necessary to appeal to errors in the T_{eff} or log g data, or errors in the assumptions of the models (e.g., Hirshfeld's models also show that helium abundance variations make a big difference), or a combination of both, to explain the apparent discrepany. The identification of double-mode ACs (if they exist), and the derivation of independent mass estimates using the Petersen diagram, might help resolve these issues.

2.4 Anomalous Cepheids as Distance Indicators

Figure 2 (adapted from NWO) shows the log P-M_B diagrams for the ACs in the Leo I dwarf galaxy (Hodge & Wright 1983), in the Carina dwarf galaxy (Saha, Monet & Seitzer 1986), in the Small Magellanic Cloud (Graham 1975), and in the field of the Galaxy (Teays & Simon 1985). Assuming that the fundamental and first-overtone P-L relationships given in §2.1 hold for all ACs (in all systems), and the scatter seen in Figure 2 can be reduced with improved accuracy of the apparent B magnitudes, it should be possible to estimate distances to the respective systems accurate to better than ±0.10 magnitudes. If ACs are found in the And I, II and III dwarf galaxies, the P-L method can be used to determine their distances, which are, at present, poorly known.

Figure 2. P-L diagrams for the ACs in Leo I, Carina, the Small Magellanic Cloud, and in the field of our Galaxy. The lines are the fundamental and first-overtone P-L relationships defined by well-studied ACs.

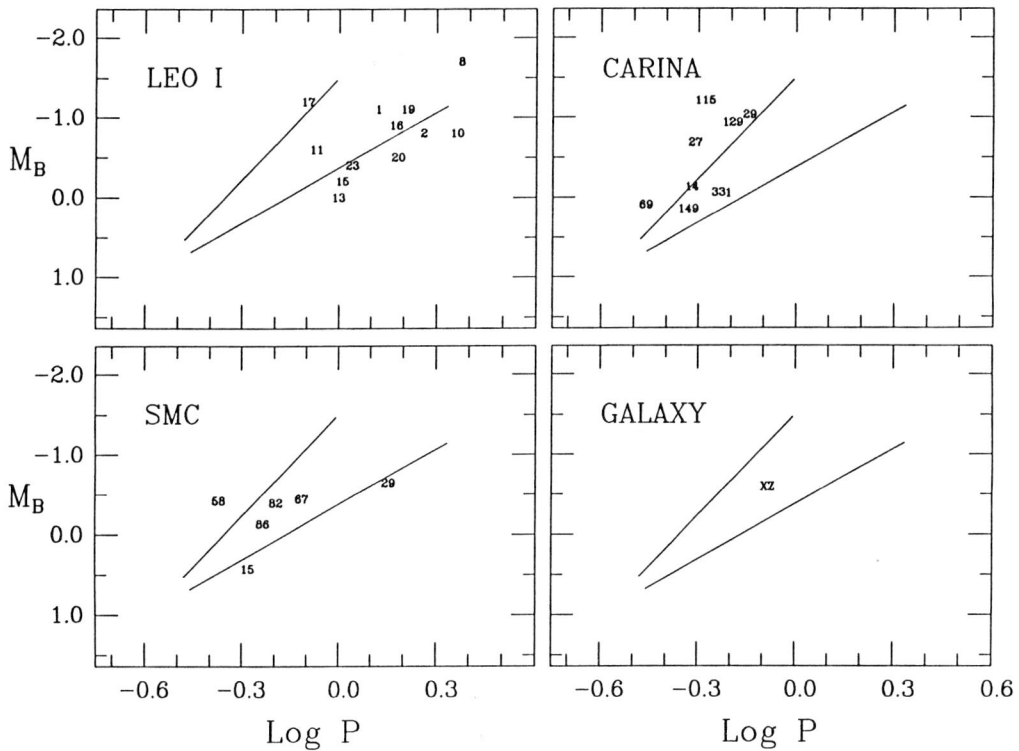

3. POPULATION II BLUE STRAGGLERS

Thirty-five years ago the first halo population BSs were discovered, in the globular cluster M3 (Sandage 1953). Over 300 BSs are now known in at least 17 globular clusters; several have been identified in the halo field of the Galaxy; and large populations are known in nearby dwarf spheroidal galaxies. In general, dedicated efforts have not been made to identify complete samples of BSs in individual clusters. Deep photometric investigations have identified relatively complete samples within a given surface area, but the areas surveyed have tended to be small and to not represent a large range of projected radial distances from the cluster centers.

Table 2 lists the number of BSs that have been identified in 17 globular clusters, and in the four nearest dwarf galaxies. The globular clusters are ordered according to increasing central concentration, $c=\log(r_t/r_c)$ (Peterson & King 1975). Column (3) contains the number of BSs that have been observed in the cluster, column (4) notes whether or not the

Table 2.
Stragglers in Globular Clusters and Dwarf Galaxies

Cluster (1)	c (2)	N(BSs) (3)	Center? (4)	Reference (5)
Pal 15	0.60	15	Yes	Seitzer & Carney (1988, unpubl.)
Pal 5	0.7	9	Yes	Smith et al. (1986)
Pal 14	0.75	5	Yes	Da Costa, Ortolani & Mould (1982)
E3	0.75	16	Yes	van den Bergh et al. (1980)
		16	Yes	McClure et al. (1985)
		17	Yes	Hesser et al. (1984)
NGC 5053	0.75	24	Yes	Nemec & Cohen (1988)
Pal 4	0.76	18	Yes	Christian & Heasley (1986)
NGC 5897	0.82	24	Yes	Nemec, Richer & Fahlman (1989)
NGC 288	0.89	9	(Yes)	Buonanno et al. (1984a)
Pal 12	0.90	12	Yes	Harris & Canterna (1980)
Pal 3	0.97	10:	Yes	Gratton & Ortolani (1984)
NGC 7492	0.98	5	Yes	Buonanno et al. (1987)
Pal 13	1.00	6	Yes	Carney & Inman (unpublished)
NGC 5466	1.25	48	Yes	Nemec & Harris (1987)
NGC 2419	1.38	few	No	Christian & Heasley (1988)
		few	No	Nemec & Rich (1985)
M71	1:50:	6	No	Arp & Hartwick (1971)
		50:	Yes	Richer & Fahlman (1987)
N2808	1.76	20	No	Buonanno et al. (1984b)
M3	1.90	30:	No	Sandage (1953)
		53	No	Buonanno et al. (1988)
Ursa Minor	-	28	-	Olszewski & Aaronson (1985)
Draco	-	100:	-	Carney & Seitzer (1986)
Sculptor	-	35:	-	Da Costa (1984)
Carina	-	100:	-	Mould & Aaronson (1983)

central region of the cluster was observed, and column (5) contains the reference to the published c-m diagram. In each cluster (or galaxy) the number of BSs is a lower bound for the true number of BSs, since only NGC 5466, NGC 5053, and possibly E3, have been systematically searched for BSs over their entire surface areas, and even in these three systems the incompleteness due to crowding at faint magnitude levels is not insignificant.

The binary hypothesis, as formulated by RMS, predicts that BSs (and ACs) ought to be found preferentially in low density environments, where the expected frequency of two- and three-body stellar encounters is low. The RMS argument is based, in part, on the calculations of Hills and Day (1976), which showed that less than one collision is expected to have occurred over the entire lifetime of a globular cluster with a very-low central concentration, and ≥ 2000 collisions may have occurred in the centers of the densest globular clusters (such as M80 and M15). It follows, therefore, that one test of the RMS hypothesis is to determine whether or not BSs (and ACs) are more likely to be found in low density environments. For the test to be unbiased, a large number of representative globular clusters would have to be surveyed, and the absence and presence of BSs at all distances from the cluster centers would have to be investigated. At present, such information exists for only a few of the most open globular clusters in Table 2 (such as NGC 5053, NGC 5466 and E3), and is not available for most clusters. High resolution observations will be required to identify BSs in the most dense regions of globular clusters.

Although the available data on the statistics of BSs is incomplete and biased, Nemec & Harris (1987), using a subset of the data in Table 2, computed that the mean central concentration of globular clusters with BSs is $<c> = 1.15$, with $\sigma = 0.12$. This was compared with the mean concentration for 117 globular clusters in Webbink's (1985) catalog ($<c> = 1.53$, $\sigma = 0.17$), and with the mean central concentration for the globular clusters that do not apparently contain BSs ($<c> = 1.91$, $\sigma = 0.18$). Recognizing that these mean central concentrations for clusters with and without BSs are highly uncertain, Nemec & Harris concluded that "the available numbers appear to support the RMS hypothesis that there is a correlation between central concentration and the presence of blue stragglers". The newer information contained in Table 2 gives no reason for altering this conclusion. In fact, factoring it in reduces the mean central concentration for the globular clusters that contain BSs to $<c> = 1.04$, with $\sigma = 0.09$.

The discussions that follow focus on three important issues concerning BSs: the morphology of blue straggler sequences (§3.1), the mean dynamical mass of a sample of BSs (§3.2), and field Pop II BSs (§3.3).

3.1 Morphology of Blue Straggler Sequences

Figure 3 shows a deep c-m diagram for NGC 5053 (Nemec & Cohen 1988), fitted with theoretical isochrones appropriate for the Thuan-Gunn photometric system (Bell & VandenBerg 1987). The isochrones correspond to the evolutionary tracks of single stars having ages of 6, 8, 10, 12, 14 and 16 Gyr, and an assumed helium abundance $Y = 0.30$. Most of the 24 BSs clearly lie along a sequence that is an extension of the main sequence. The five most luminous BSs appear to define a redward turnoff. If the BSs in NGC 5053 are coeval, and one extrapolates the isochrones shown in Figure 3 to younger ages, a fit of the isochrones to the observed blue straggler sequence gives an age 2 to 4 Gyr. This is in sharp contrast to the mean age of the majority of the cluster stars, which is about 18 Gyr. The theoretical models (appropriate for single, unmixed stars) also indicate that at the "blue straggler turnoff" the BSs have a mass about 1.5 $M_\odot$, compared with 0.8

Figure 3. A c-m diagram for NGC 5053 (from Nemec & Cohen 1988).

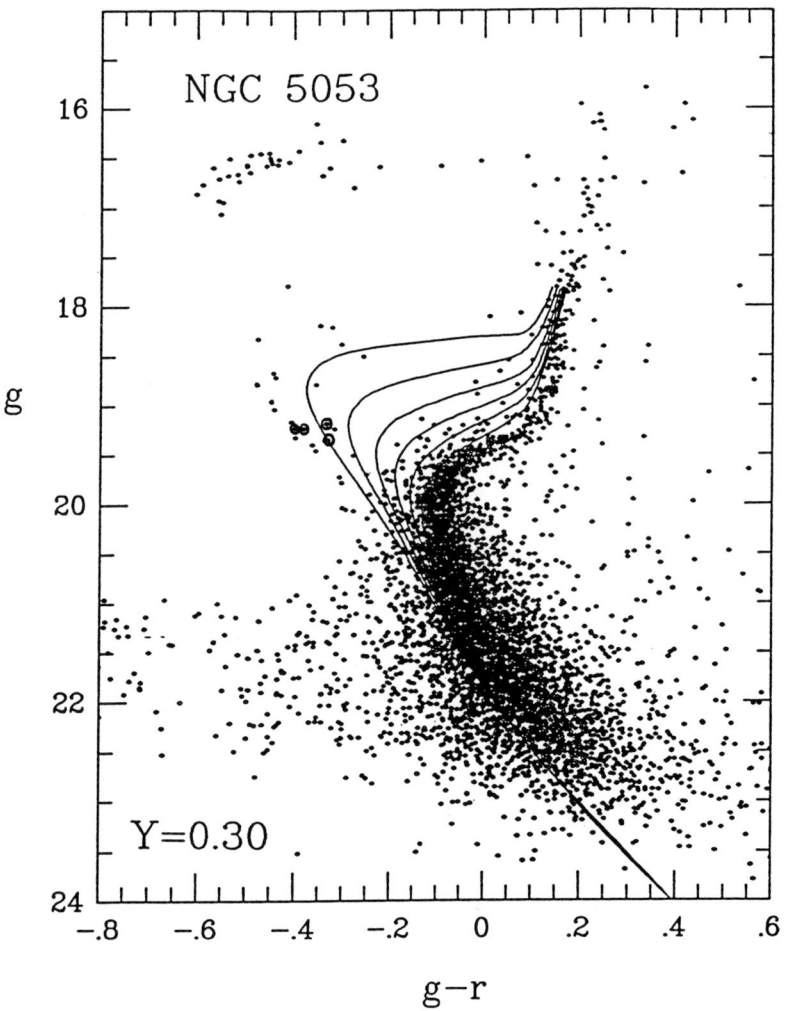

$M_\odot$ for the main sequence turnoff stars.

The apparent blue straggler turnoff seen in NGC 5053 is not unique. In M3, redward turnoffs are seen in both the Sandage (1953) and the Buonanno et al. (1988) c-m diagrams. Blue straggler turnoffs are also seen in the c-m diagrams for NGC 5466 (Nemec & Harris 1987), the Draco dwarf galaxy (Carney & Seitzer 1986), and the old open cluster NGC 188 (McClure & Twarog 1977). Presently, not enough data exists to investigate trends; however, as the number of known BSs increases, and photometry improves, the morphology of the sequence will become better revealed. Eventually, measurements of the structure along the blue straggler sequence should provide important information for modelling the evolution of BSs.

Another aspect of the morphology question concerns the globular cluster E3 (van den Bergh, Demers & Kunkel 1983; Frogel & Twarog 1983), and its population of about 10 unexplained "yellow stragglers" (Hesser et al. 1984). In the c-m diagram, the yellow stragglers (YSs) are situated above the flat portion of the main sequence turnoff, and are redder than the BSs. The additional detection in E3 of a sequence of main-sequence binaries about 0.75 mag above the standard main sequence (McClure et al. 1985), lends support to the binary hypothesis explanation for both YSs and BSs. A search for short-period variability among the blue and yellow stragglers, and among the candidate main sequence binaries in E3, might be a most profitable endeavor. Yellow stragglers appear also to be present in NGC 2808 (Buonanno et al. 1984b) and in NGC 7492 (Buonanno et al. 1987).

3.2 Mass Segregation and Dynamical Masses of Blue Stragglers

The masses of BSs, like the masses of ACs, can be calculated in a number of ways. One of the most direct methods is based on elementary kinetic theory and the principle that in a stellar system that is in thermal equilibrium there will be equipartition of kinetic energy among the stars, and mass segregation (i.e., the massive stars will be more centrally concentrated than the less massive stars). In practice, by comparing the observed radial distributions of any two samples of stars, one can determine if one sample is, on average, more massive than another. To compute the amount of the mass difference, it is necessary to compare the observed distributions with projected radial distributions (for stars of different masses) calculated with theoretical models of the stellar system. In globular clusters, where relaxation times range from about 0.1 Gyr in the centers of the most dense globular clusters, to about 5.0 Gyr in the centers of the least dense clusters, mass segregation is expected to occur. And, if BSs are more massive than the main sequence turnoff stars, then one expects them to be more centrally concentrated than the turnoff stars.

Dynamical masses for the relatively complete samples of BSs in the globular clusters NGC 5466 and NGC 5053 have been computed in just this manner. In NGC 5466, Nemec & Harris (1987) found the 48 BSs to be more centrally concentrated than the subgiant stars in the same magnitude interval. (The subgiants were chosen for comparison purposes because they have approximately the same mass as the main sequence turnoff stars, and they were believed to be as incomplete a sample of stars as the BSs). By fitting multi-mass King models to the observed cumulative radial distribution function, the mean mass of the NGC 5466 BSs was determined to be 1.3 ± 0.3 $M_\odot$. In NGC 5053 (Nemec & Cohen 1988) a similar result was found for the 12 most luminous BSs, suggesting a mean dynamical mass for them of 1.3 ± 0.3 $M_\odot$. That the 12 lowest luminosity BSs showed no evidence for central concentration is attributed to a range of masses of the BSs, with the lower luminosity BSs having lower masses than the higher luminosity BSs.

In retrospect, one can see that the brightest BSs in Palomar 12 also appear to be located in the central region of that cluster (Harris & Canterna 1980). An increased central concentration of BSs has also been seen in the old open cluster M67 (Mathieu & Latham 1986). In M67 the BSs follow the radial distribution of the spectroscopic binaries. The mean dynamical mass is about 2.0 $M_\odot$ for the BSs in M67, compared with about 1.2 $M_\odot$ for the main sequence turnoff stars. Of course, with all of these mass estimates a critical factor is that the models must adequately represent the real clusters, and this means correctly estimating, among other things, the amount of dark matter in the clusters. The fact that the M3 BSs discovered by Sandage (1953) are not centrally concentrated (Sandage & Katem 1968) can readily be explained by the lack of dynamical equilibrium in the outer regions of M3 (see Chaffee & Ables 1983; Peterson, Carney & Latham 1984), and the relatively high frequency of stellar collisions in the central regions of M3 (Hills & Day 1976).

3.3 Field Pop II Blue Stragglers

The identification of nearby (and therefore bright) field Pop II BSs is obviously important, and several candidate stars have been identified (Bond & MacConnell 1971; Eggen 1970). However, the list includes mostly variable stars, such as VW Ari, SU Crt and BS Tuc (these latter two stars both have broad lined spectra and may be detached binary systems). Carney & Peterson (1981) concluded that the only known non-variable BSs brighter than $V = 12$ mag, with metal abundances less than 0.1 solar, are BD$-12°$2669 and BD$+25°$1981.

4. VARIABLE BLUE STRAGGLERS

Potentially the most valuable development in the study of Pop II BSs has been the discovery of short period (44 min to 5 hr) photometric variables among known BSs. These include: several SX Phe variables (or Pop II dwarf Cepheids) in the halo of the Galaxy (§4.1); at least 14 SX Phe stars in the globular clusters ω Cen (§4.2), NGC 5466 (§4.3), NGC 5053 (§4.4), M3 (§4.5), and in the Draco dwarf galaxy (§4.6); two W UMa-like contact binaries in NGC 5466 (§4.7); and, an eclipsing binary in ω Cen (§4.8).

Table 3 summarizes the pulsational and photometric properties of the known variable BSs. Column (1) contains the star name; column (2) gives either the pulsational or the orbital period; column (3) contains the logarithm of the period (days); column (4) contains the amplitude of the photometric variations (V filter); columns (5), (6) and (7) contain the mean V mag, the mean B mag, and the mean B minus the mean V color of the star; and column (8) lists the absolute V magnitude.

Table 3.
A Summary of Properties of Known Variable Blue Stragglers

Star	Period (days)	log P	A_V	$\langle V \rangle$	$\langle B \rangle$	$\langle B \rangle - \langle V \rangle$	M_V
(1)	(2)	(3)	(4)	(5)	(6)	(7)	(8)
	ω Cen (Jorgensen and Hansen 1984)						
E39	0.057	-1.252	0.25	17.03	17.34	0.31	3.11
NJL 220	0.047	-1.328	0.50	17.04	17.33	0.29	3.12
NJL 79	0.063	-1.201	0.38	16.79	17.12	0.33	2.86
NJL 5	1.38*	0.139	0.50	15.96	16.22	0.26**	2.9
	NGC 5466 (Mateo et al. 1988)						
NH38	0.054	-1.268	0.44	18.79	19.07	0.28	2.79
NH29	0.040	-1.398	0.12	18.93	19.15	0.23	2.93
NH27	0.051	-1.292	0.12	18.84	19.08	0.24	2.84
NH35	0.0504	-1.298	0.51	18.96	19.27	0.31	2.96
Anon.1	0.0455	-1.342	0.30	19.32	19.49	0.18	3.32
NH19	0.42*	-0.356	0.4	18.67	18.79	0.12	2.67
NH30	0.33*	-0.481	0.38	19.42	19.64	0.22	3.42
	NGC 5053 (Nemec et al. 1988)						
NC14	0.0379	-1.421	0.08	19.42			3.53
NC11	0.0357	-1.447	0.25	19.55			3.63
NC7	0.0347	-1.460	0.12	19.20			3.37
NC13	0.0369	-1.433	0.13	19.41	19.73	0.32	3.59
	M3 (DaCosta 1987)						
Anon	0.031	-1.51	0.08	18.0:	18.2:	0.2:	3.34
	Field SX Phe Stars						
BL Cam	0.039	-1.409	0.50	13.10		0.21	3.2
DY Peg	0.073	-1.137	0.54	10.25		0.24	2.2
SX Phe	0.055	-1.260	0.50	7.0	7.24	0.24	2.7
CY Aqr	0.061	-1.215	0.70	10.78		0.22	2.5
KZ Hya	0.060	-1.225	0.80	9.97		0.22	2.6

* Orbital period,   ** At maximum light (phase 0.25)

4.1 Field SX Phe Variables

SX Phoenicis, discovered by Eggen (1952a,b), is a seventh magnitude, high proper motion (0.8"/yr) star with an ultra-short period and a variable amplitude ($A_v \sim 0.3$-0.6 magnitudes). Walraven (1955) showed that it pulsates simultaneously in the fundamental mode ($P_0 = 0.05496$ day) and in the first-overtone mode ($P_1 = 0.04277$ day), with a ratio of its two pulsation periods ($P_1/P_0 = 0.7782$) that is not unlike that for Pop I double-mode δ Sct variables (Fitch 1970; Fitch & Szeidl 1976; and Cox, King & Hodson 1979). SX Phe has long been recognized as the brightest example of the class of low-metal abundance, high-amplitude Pop II dwarf Cepheids, other members of which include BL Cam, CY Aqr, KZ Hya and DY Peg. (The existence of low-amplitude field Pop II dwarf Cepheids remains questionable - see McMillan et al. 1976; Carney & Peterson 1981; and Halprin & Moon 1983). A list of 178 dwarf Cepheids in the field of the Galaxy has been compiled by Halpren & Moon (1983), while Frolov & Irkaeav (1984) have attempted to identify which of these are Pop II stars. Properties of selected field Pop II dwarf Cepheid candidates are summarized in Table 3.

The evolutionary status of SX Phe variables, like that of ACs and BSs, is controversial. Bessell (1969) proposed that they might be low-mass (0.5 $M_\odot$), post-helium-flash Pop II stars. Another hypothesis is that they are short period δ Scuti (or AI Velorum) variables with relatively large masses (1.5 to 2.5 $M_\odot$) similar to those of δ Scuti stars. The most likely explanation (Eggen 1970, 1971, 1979) is that they are variable BSs (VBSs), analogous to Pop I δ Scuti variables. These hypotheses, and others, are discussed by Bessell (1969), McNamara & Feltz (1978), Breger (1980), Fernley et al. (1987), and Eggen & Iben (1988).

The pulsation periods of SX Phe stars allow several different mass estimates to be made, depending on the available information. One can compute the theoretical mass (using P and T_{eff}), the pulsation mass (using P, L and T_{eff}), the mass relative to the RR Lyrae stars in the same system (using P, and Teff and m_{bol} relative to the RR Lyraes), and, if the star is also double-periodic, it may be possible to compute its mass using the Petersen method. The reader is referred to Cox, King & Hodson (1979) and Jorgensen & Hansen (1984) for detailed discussions of these mass estimation techniques. If SX Phe stars are VBSs, then they are particularly valuable for improving our understanding of BSs. The masses determined for individual SX Phe stars provide an important check on the dynamical mass estimate discussed above (which provides only a mean mass for an appropriately selected sample of BSs), and the evolutionary mass estimate derived by a comparison of theoretical isochrones with position in in the c-m diagram. By establishing the physical characteristics of SX Phe stars (and other faint variable stars in globular clusters), many key questions regarding BSs will almost certainly be answered.

4.2 Variable Blue Stragglers in ω Centauri

The first VBSs to be identified in a globular cluster were the three Pop II dwarf Cepheids found in ω Cen. Niss, Jorgensen & Laustsen (1978) surveyed approximately 25% of ω Cen's surface area and identified 29 candidate short-period variables. Two of the stars, NJL 79 and NJL 220, and a third star not in the NJL catalog, E39, were subsequently found to be "dwarf Cepheids" with periods between 0.047 and 0.063 day (Jorgensen 1982; Jorgensen & Hansen 1984; Da Costa & Norris 1988). A fourth star, NJL 5, is an eclipsing binary (see §4.9). All four stars lie in the blue straggler region of the ω Cen c-m diagram (Da Costa, Norris and Villumsen 1986), which also includes five BSs that appear to be photometrically constant to 0.03 mag (DaCosta & Norris 1988; Da Costa 1988, personal communication).

Figure 4. A c-m diagram for NGC 5466 (from Mateo et al. 1988) based on one pair of 300 sec B and V Palomar 60-inch CCD exposures. The blue straggler sequence is obvious above and to the left of the main sequence turnoff. The open circles denote the mean V magnitude and (B-V) colors of the seven variables found in NGC 5466. Note that there is little doubt that these variables are each also blue stragglers.

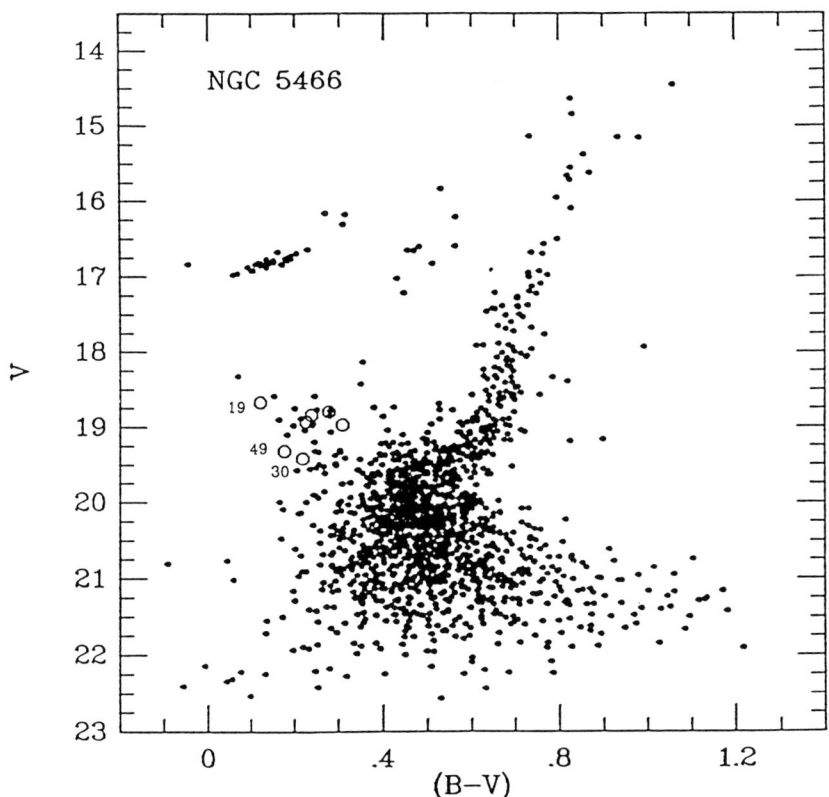

Depending on the assumed pulsation modes of the three dwarf Cepheids (see §4.7), estimates of their pulsation masses range from 2.24 $M_\odot$ for NJL 220, to 0.57 $M_\odot$ for NJL 79 (Jorgensen 1982), and 0.72 $M_\odot$ for E39 and 0.74 $M_\odot$ for NJL 220 (Jorgensen & Hansen 1984). The mean mass, computed relative to the RR Lyrae stars, is 1.2 ± 0.2 $M_\odot$. Since the pulsation modes of these stars are not known (the fundamental mode was assumed), the uncertainties are large. Properties of the four VBSs in ωCen are summarized in Table 3. Clearly, there is little to distinguish the three dwarf Cepheids from the field SX Phe stars, and NJL 5 stands out as an exceptional stellar system.

4.3 Variable Blue Stragglers in NGC 5466

With four out of nine BSs in ωCen known to be either dwarf Cepheids or eclipsing binaries, it was obviously of interest to establish whether or not a comparable fraction of the recently discovered BSs in NGC 5466 (Nemec & Harris 1987) and NGC 5053 (Nemec & Cohen 1988) are variable. CCD programs to monitor the variability of the BSs in these two systems, and to search for short-period variables, are underway (Nemec et al. 1988a; Mateo et al. 1988a, b). To date, hundreds of short exposure frames have been taken by a team of observers using the Palomar Observatory 60-inch telescope, the Canada-France-Hawaii 3.6-meter telescope, and the Steward Observatory 90-inch telescope. Over 30 hours of observing time has been devoted to this project over a two year period. A preliminary report on the results of both surveys is given here.

Mateo et al. (1988a) have recently discovered seven short-period variable stars among the BSs in NGC 5466 (Nemec & Harris 1987). The stars have periods ranging from 0.04 day to 0.21 day (0.42 day if the variations are interpreted as orbital motion), and amplitudes between 0.12 mag and 0.51 mag. Figure 4 identifies the seven variables in one of the Mateo et al. c-m diagrams, and light curves for the seven stars are plotted in Figure 5. (One of the variable BSs was not in the NH-catalog of BSs in NGC 5466, and was labelled "Anon. 1" by Mateo et al. In Figures 4 and 7 it has been labelled "49" to indicate that it is the 49th known blue straggler). With the exception of Nos. 19 and 30, which deserve special mention and are discussed in §4.8, the periods, amplitudes and light curves of the five shorter-period variables resemble those of SX Phe variables, i.e. the light curves are Cepheid-like, showing some flatness at minimum light and some asymmetry. Basic information on the seven VBSs is summarized in Table 3. The B-V color range of the five SX Phe variables is from 0.18 to 0.31 (and is only from 0.23 to 0.31 if No. 49 is excluded). The variables tend to be among the brighter BSs in NGC 5466. Photometrically constant BSs appear to sit side-by-side in the c-m diagram with the variables. Of course, these "non-variable" stars may turn out to vary, with amplitudes ≤0.10 mag. (Eggen 1970 has shown that non-variable and variable old disk BSs share a common location in the c-m diagram).

4.4 SX Phe Variables in NGC 5053

The search for photometric variables among the 24 BSs in NGC 5053 (Nemec & Cohen 1988) has resulted in the recent discovery of four SX Phe stars, with periods ranging from 0.0347 to 0.0379 day, and V amplitudes in the range 0.19 to 0.35 mag (Nemec et al. 1988). All four stars are (in projection) in the innermost region of the cluster. Preliminary details are given in Table 3, and light curves for the four stars are plotted in Figure 6. In the NGC 5053 c-m diagram shown in Figure 3, the four stars occupy a small region about half way down the blue straggler sequence. It is curious that none of the highest luminosity BSs (i.e. those with g ≈ 18.5, g-r ≈ -0.30) appears to be variable, since these appear to lie between the RR Lyrae stars and the SX Phe stars in the instability strip. As was also the case in NGC 5466, variable and non-variable BSs appear to be mingled together at a similar position in the c-m diagram. The range in the <V>-magnitudes of the four SX Phe stars is relatively small, from 19.27 to 19.53 mag, and the range of mean colors is also small, from g-r about -0.40 to -0.33. This localization of the four variables in the c-m diagram offers the hope that perhaps with identification of more SX Phe stars in NGC 5053, and in other systems, and determination of their colors and luminosities, the edges of the instability strip will become well established.

Figure 5. Light curves for the VBSs in NGC 5466 (from Mateo et al. 1988).

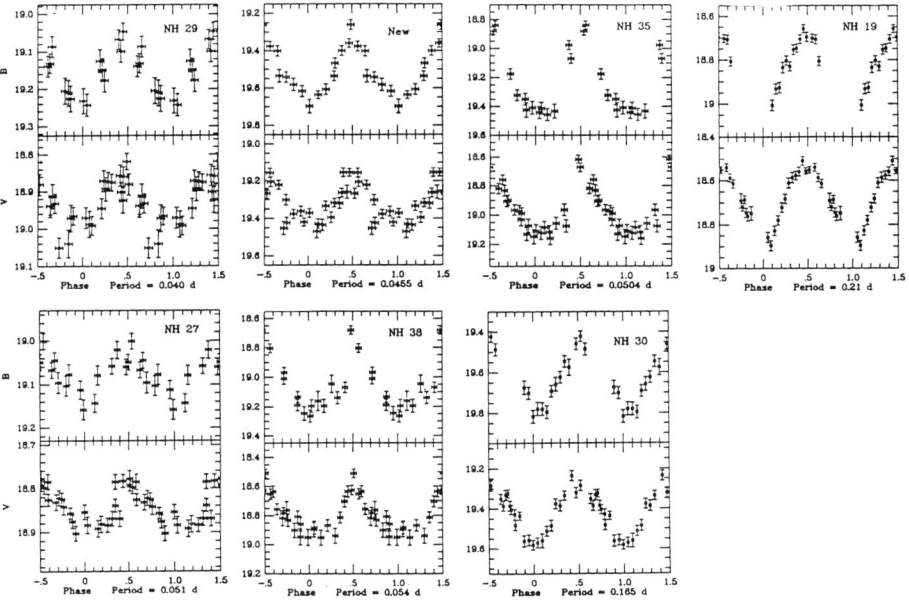

4.5 SX Phe Variables in M3

As discussed above, about 30 BSs can be identified in Sandage's (1953) c-m diagram for M3. More recently, Buonnanno et al. (1988) note that they have identified 53 BSs in their 10,000 star c-m diagram for M3. Da Costa (1987) reports that photometry has been obtained for 11 BSs in M3, and that one of the monitored stars appears to be variable, with a pulsation period about 44 min (P = 0.031 day), and a B-amplitude about 0.07 magnitudes. The published light curve for Da Costa's variable is sinusoidal in shape, and there is every reason to believe that it is an SX Phe star. Adopting $<B>$ = 18.56, and assuming $<B-V>$ = 0.30 and $(m-M)_o$ = 14.83, the absolute V magnitude of this star is M_V = 3.3. If the 44 min period is confirmed, this variable would be the shortest period SX Phe star known. Its very low amplitude suggests that precise photometric measurements will be necessary if other such VBSs are to be identified in M3, and in other stellar systems. A complete search for BSs over the entire surface area of M3, and further study of the BSs that have already been identified, would be of great value.

4.6 Variable Blue Stragglers in Nearby Dwarf Galaxies

The Ursa Minor, Draco, Sculptor and Carina dwarf galaxies, in addition to their populations of ACs, contain substantial populations of BSs. The (total) number of BSs in Ursa Minor, Draco and Sculptor is estimated to be about 300, 700 and 1400, respectively (Olszewski & Aaronson 1985; Carney & Seitzer 1986; Da Costa 1984). Counting only those BSs brighter than M_V = +2.5, Da Costa (1988) has determined that

Figure 6. Light curves (in V) for the four SX Phe variables in NGC 5053 (from Nemec et al. 1988).

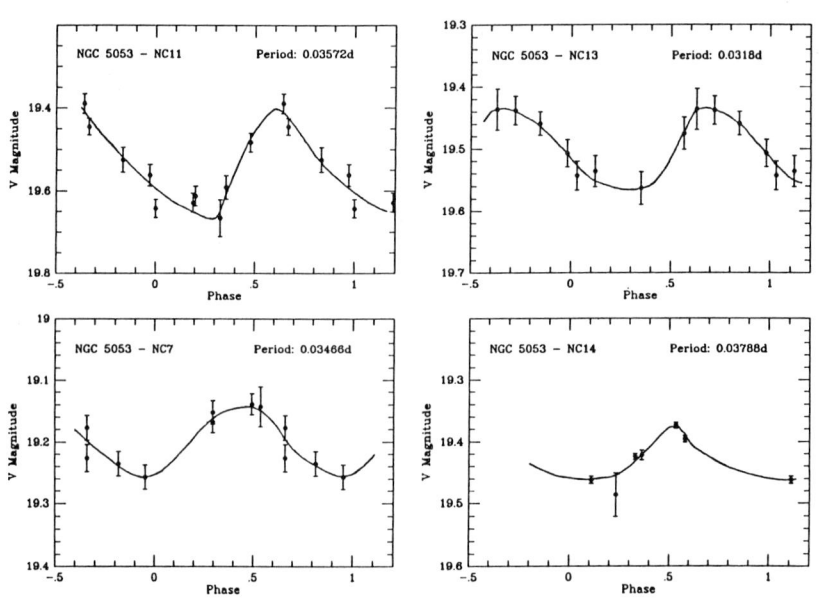

these three galaxies contain, respectively, 45, 70 and 200 BSs bright
enough to produce ACs. Mould & Aaronson (1983) argue that the stars in
the blue straggler region of their Carina c-m diagram may be of
intermediate age (about 7.5 Gyr).

It is noteworthy that Carney and Seitzer (1986) have found that one of
the luminous Draco BSs appears to be either a dwarf Cepheid or an
eclipsing binary. Given that Draco, Ursa Minor, Sculptor and Carina
have such large populations of BSs, a search for more such variables
would be of great value for determining whether the physical properties
of the VBSs in these galaxies are similar to those of the VBSs in
globular clusters, and for establishing the connection with the many
other kinds of chemically peculiar stars, including CH and CN stars, in
dwarf galaxies (see Zinn 1981, Smith & Dopita 1983, Stetson 1984).
Given that the V mags of the BSs in these three systems range from 21 to
23 mag, such a search is currently within reach of large ground-based
telescopes with CCD cameras.

4.7 Pulsational Properties of SX Phe Stars

With the identification of relatively large numbers of SX
Phe stars it should be possible to establish P-L and P-A relationships,
and determine pulsation modes for the stars. The lower panel of Figure
7 shows all the variable BSs in NGC 5466, NGC 5053, M3, and ω Cen,
plotted in a (log P)-(M_V) diagram; the upper panel shows the
corresponding (log P)-(A_V) diagram. Also plotted in the diagrams are
the field SX Phe variables DY Peg, CY Aqr, KZ Hya, SX Phe and BL Cam,
the known W UMa contact binaries in the old open cluster NGC 188 (age 5
to 11 Gyr), and two variable BSs in M67 (age about 3 Gyr). Several
large amplitude δ Scuti variables also are represented in the upper
panel (for comparison with the SX Phe stars). A number of interesting
points are raised by these diagrams:

(1) The VBSs in the globular clusters do not constitute a homogeneous
sample, but divide into three types of variables: SX Phe stars, contact
binaries, and eclipsing binaries. The SX Phe variables have M_V between
+2 and +4, and periods in the range 44 to 110 min; the contact binaries
appear to be found over a wide range of luminosities, with orbital
periods in the interval 0.28 to 0.41 day; the eclipsing binaries have
relatively long orbital periods. The contact and detached binaries are
discussed in §4.8 and §4.9, respectively.

(2) All the known SX Phe stars define an approximately linear
relationship between log P and M_V. Upon closer inspection, it appears
that the scatter might be systematic. Excluding NJL 79 and E39 in
ω Cen, which appear to have longer periods at a given luminosity than
the bulk of the SX Phe stars, No. 29 in NGC 5466, BL Cam, No. 7 in NGC
5053, and Da Costa's star in M3, tend to have shorter periods at a given
luminosity than the bulk of the variables. Since the periods are short
and well determined, the greatest uncertainties are in the absolute V
magnitudes. However, with the exception of BL Cam, these four outlier
stars are in well-studied clusters, and hence the M_V values should be

Figure 7. (Lower panel) Period-luminosity diagram for variable blue stragglers. Nos. 19 and 30 in NGC 5466 are plotted with the assumed orbital periods. Both the fundamental and the first-overtone periods are plotted for SX Phe (connected with a line). Contact and detached binary systems are labelled. (Upper panel) Period-amplitude diagram for the SX Phe variables, the binary systems, and for selected δ Sct stars. The SX Phe stars and the δ Scuti variables are labelled.

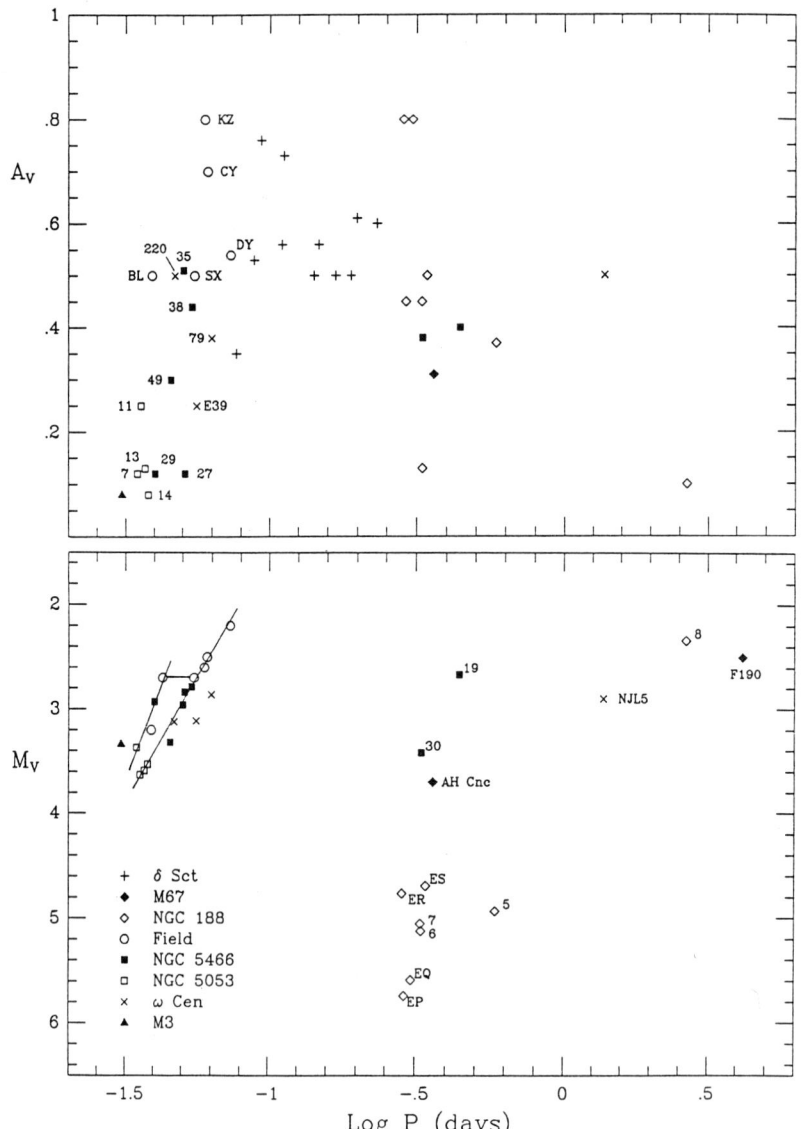

secure. Thus, the scatter appears to be real. For the stars with periods
short for their luminosities, including SU Crt, one explanation of the
large scatter at a given luminosity might be that they are oscillating
in a higher pulsation mode. Given that we know that SX Phe pulsates
simultaneously in the fundamental and first-overtone modes, and not in
the first- and second-overtone modes (in which case the period ratio
would be expected to be much larger), it is reasonable to postulate that
the four high-amplitude SX Phe variables are first-overtone pulsators.
Further evidence for this is that the shapes of the light curves for all
four short-period stars are sinusoidal, which is to be expected for
first-overtone pulsators. Support for the notion that some SX Phe stars
pulsate in the first-overtone mode comes from Fourier decomposition
studies of field dwarf Cepheid light curves (Antonello et al. 1986;
Fernley et al. 1987), which show that dwarf Cepheids with R_{21} < 0.30 are
first-overtone pulsators, and those with R_{21} > 0.30 are fundamental mode
pulsators. When more photometry becomes available, it will be of
considerable value to subject the data on the globular cluster SX Phe
stars to a Fourier decomposition analysis. Assuming that the SX Phe
stars can be sorted into fundamental and first-overtone pulsation modes,
the possible relationships are as plotted in Figure 7.

(3) Pulsation in different modes does not explain NJL 79 and E39 in
ω Cen. For these two stars, a number of possibilities come to mind (none
of which is appealing): (a) the two stars may have incorrect
luminosities. However, if this is the case, and their periods are
correct (which they seem to be), the adopted M_V values are too faint by
about 0.5 magnitudes; (b) if the two stars are fundamental mode
pulsators, then the bulk of the SX Phe stars would have to be
first-overtone pulsators, and the stars with periods short for their
luminosities would have to be second-overtone pulstors. Although there
seems to be no theoretical reason why SX Phe stars cannot pulsate in
modes from the fundamental up to the fifth-overtone (see Figure 1 of
Stellingwerf 1979), a problem with this interpretation is that the
period ratio for SX Phe would then be anomalous; (c) the two stars are
binary systems. Here however, the light curves are those of dwarf
Cepheids, and not contact binary systems (see next section). Given that
ω Cen is famous for its chemical inhomogeneity and other anomalies, more
work is necessary to understand these stars.

(4) The P-A diagram (upper panel of Figure 7) shows that all of the
shortest period SX Phe stars have amplitudes less than about 0.3 mag,
and as the periods become longer the amplitudes increase (up to about
0.5 mag). The largest amplitude SX Phe variables are KZ Hya, CY Aqr,
GP And and DY Peg, all of which are intermediate metal-abundance
variables (some of which are old disk stars). Traditionally, SX Phe
variables have been identified by their high space motions, low
luminosities, high galactic latitudes, and low metal abundances. Because
most of the field SX Phe variables that are known have amplitudes larger
than about 0.40 mags, it is commonly thought that SX Phe variables, in
general, have large amplitudes. Figure 7 also shows several
large-amplitude δ Scuti variables. Recognizing that this sample

of Pop I dwarf Cepheids excludes most δ Scuti stars, which have relatively small amplitudes, and is not representative of δ Scuti stars in general, Figure 7 nevertheless suggests that most (and maybe all?) large amplitude dwarf Cepheids with periods longer than about about 0.08 day are Pop I dwarf Cepheids. The reader is referred to Figure 1 of Breger (1980) for the over-all period distribution of δ Scuti stars.

(5) In analyses involving the periods and amplitudes of SX Phe stars there are a number of potential problems. Many SX Phe and δ Sct dwarf Cepheids exhibit cycle-to-cycle amplitude variations, reminiscent of Blazhko variability in RR Lyrae stars. These variations are often caused by multi-mode radial pulsations, but many variations are complex and not understood. The brightest double-mode Pop II dwarf Cepheid is SX Phe, itself (which was discussed earlier). Some SX Phe stars, such as Nos. 27 and 35 in NGC 5466, have similar periods and luminosities, but there is a large difference in their amplitudes. One possibility is that these two stars have variable amplitudes and are doubly-periodic, and the observations have been made at high and low amplitude states. Other problems involve the question of non-radial pulsations, and the effects of rotation and magnetic fields. Certainly much more work is needed to identify and deal properly with the question of amplitude variations.

4.8 Contact Binary Systems Among Blue Stragglers

The long period VBSs in NGC 5466, Nos. 19 and 30, represent a type of variable star not previously seen in a globular cluster. They have periods considerably longer than those of the other VBSs in NGC 5466, light curves with shapes unlike those of the shorter-period variables in NGC 5466, and B-V colors (0.1 < B-V < 0.2) bluer than most of the SX Phe stars discussed above (see Figure 4). Mateo et al. (1988) conclude that both stars are contact binary systems, based on the relatively long periods, and the morphology of the light curves. The light curve of No. 19 is asymmetric, and at minimum appears to come to a sharp point. The light curve for No. 30 is almost symmetrical. The light curves of both stars resemble those of W UMa-type binary systems (see Eggen 1976; Kaluzny & Shara 1987). Under the W UMa contact binary hypothesis, the pointed minimum for No. 19 suggests that a partial eclipse may be occurring, possibly like that seen in Figure 17 of Binnendijk (1970). The flat bottom of the light curve for 30 is more suggestive of a much more complete eclipse in an edge-on system. This discovery provides direct evidence that at least some BSs are contact binaries. In Figure 7, the similar orbital periods of these two stars, and of the certain contact binary (and blue straggler) AH Cnc in M67 (Eggen 1976; Whelan et al. 1979), and of the lower-luminosity W UMa-type contact binaries in NGC 188 (Baliunus & Guinan 1985; Kaluzny & Shara 1987), lends considerable support to this interpretation. It is of added interest that the well-defined light curve of AH Cnc resembles those of 19 and 30. The fact that the two NGC 5466 stars are very luminous relative to these other binaries is not sufficient evidence to rule out the possibility that there are lower luminosity contact binaries in NGC 5466. A systematic survey of M67 to identify W UMa binaries similar to those found in NGC 188 would be helpful in determining the properties of these stars. Obviously, much more work needs to be done to exploit a comparison between the binary systems in old open clusters, and the newly discovered candidates in NGC 5466.

4.9 Detached Binary Systems Among Blue Stragglers

The eclipsing binary NJL 5 in ωCen, with an orbital period of 1.376 days and a well defined light curve (Liller 1978, Jensen & Jorgensen 1985), is a blue straggler in the ωCen c-m diagram (DaCosta, Norris & Villumsen 1986). Since NJL 5 is a radial velocity member of the cluster (Jensen & Jorgensen 1985), it is unique among BSs in globular clusters. The fact that the light curve indicates that the component stars are detached provides direct evidence that some BSs in globular clusters are detached binaries. The existence of only one such star suggests that detached (and eclipsing) binaries are relatively rare. Since ωCen also contains many spectroscopically peculiar stars, including six CH stars (see Cowley & Crampton 1985) which are probably spectroscopic binaries (see McClure 1985), and since CH stars and other chemically peculiar stars are also present in many dwarf spheroidal galaxies (see Mould 1983), it is not implausible that those BSs that are detached binaries (such as NJL 5) are the progenitors of such chemically peculiar giant stars. Circumstantial evidence for this is provided by a parallel situation among old disk stars: the Ba II stars, which are believed to be the old disk counterparts of the CH stars, are known to be detached binary systems (McClure, Fletcher & Nemec 1980; Böhm-Vitense, Nemec & Proffitt 1984; McClure 1985) and to probably have as their progenitors old disk BSs.

Detached binary stars with orbital periods similar to that of NJL 5 are also known in NGC 188 (Baliunas & Guinan 1985) and M67. It is enlightening to review these stars, since they are possible counterparts of NJL 5. Two candidate detached binaries are known in NGC 188: V8, which is a yellow straggler and V5, which sits below the subgiant turnoff in the c-m diagram of McClure & Twarog (1977). Kaluzny & Shara (1987), while recognizing that the 2.67 day period for V8 is somewhat uncertain, conclude that "the time scale and amplitude of the light variations suggest that V8 is either an RS CVn-type star or an FK Com-type star. (FK Com-type stars are rapidly rotating single giants exhibiting light variations of the order of 0.1 mag. Their variability is most probably due to the presence of large starspots on their surfaces)". Since coalesced W UMa binaries are believed to be the progenitors of FK Com stars (Bopp & Stencel 1981; Guinan & Bradstreet), it follows that a search for low-amplitude variability in yellow stragglers in different environments might be quite productive. The period of V5, 0.586 day, is "extraordinarily long for a contact binary of spectral type K2" (Kaluzny & Shara, 1987), and needs to be checked.

M67 is well-known for its large populations of BSs and spectroscopic binaries. The fact that the BSs follow the same radial distribution as the spectroscopic binaries (Mathieu & Latham 1986) implies that the mean mass of the BSs is similar to that of the spectroscopic binaries.

Periods for seven of the long period binaries in M67 (see Mathieu & Mazeh 1988) range from 4.2 day to 60.5 days. The shortest period variable, F190 (see Eggen & Iben 1988), has recently had its 4.2 day period confirmed (Latham et al. 1988, private communication). The location of F190 in the c-m diagram is similar to those of V8 in NGC 188, and to the high-luminosity BSs in NGC 5053 (Nos. 2, 3, 4, and 24 in Nemec & Cohen 1988). None of the NGC 5053 stars appears to be photometrically variable; however, the photometric reductions are not complete. Since F190 is a blue straggler, it is tempting to speculate that some of these long period binaries may have evolved from the BSs with the largest separations. Of course, the BSs that have coalesced could not be the progenitors.

5. CONCLUSION

The derived masses for ACs leave little doubt that they are stars with relatively large masses, 1.0 to 1.7 $M_\odot$. The recent discovery that there is a dichotomy in the P-L relationship for ACs, and the application of the Fourier decomposition method, help considerably in identifying the pulsation modes of ACs. This has all but eliminated what was previously a major source of uncertainty in the computation of the pulsation masses of ACs. In addition to facilitating mode identification, the fundamental and first-overtone P-L relationships for ACs can be used as distance indicators, provided that they are universal in nature.

Large numbers of BSs are now known in about 20 globular clusters, and in all the nearest dwarf galaxies. Although an unbiased statistical test has yet to be carried out, the RMS prediction that BSs should occur preferentially in globular clusters with low central concentrations appears to be supported by the observations. Observed blue straggler sequences tend to show evidence for a redward turnoff at high luminosities. In both NGC 5466 and NGC 5053, the mass of the BSs from dynamical consideratons is substantially larger than the mass of the main sequence stars. In NGC 5053, the most luminous BSs appear to have larger masses than the least luminous BSs.

Surveys of photometrically variable BSs indicate that about 25% of all BSs are variable, with periods ranging from about 45 min to 10 hours. At least 14 SX Phe stars, with periods between 45 min and 120 min and amplitudes between 0.1 and 0.6 mag, have been identified in four globular clusters. (There is some evidence that a similar star exists in the Draco dwarf galaxy). The shapes of their light curves and their P-L relationships, suggest that the pulsation modes of the SX Phe stars are probably the fundamental and the first-overtone modes. However, this conclusion is not certain since there are still some questions concerning the pulsation modes of NJL 79 and E39 in ω Cen. The derived pulsation mass, and mass relative to the RR Lyrae stars, of the SX Phe stars in ω Cen, NGC 5466 and NGC 5053, is consistently large, about 1.3 $M_\odot$. Thus, there is good agreement between the masses of non-variable BSs and the SX Phe stars.

Two contact binaries, with orbital periods of about 8 and about 10 hours, have been discovered in NGC 5466. These stars are similar to the W UMa contact binaries in the old open clusters M67 and NGC 188, and provide direct evidence that some BSs are binary systems. Further evidence for this conclusion is provided by the eclipsing binary NJL 5 in ωCen, which has an orbital period of 1.38 day. Almost certainly there are more contact binaries, eclipsing binaries and SX Phe stars to be discovered among BSs, and these will yield important information.

Finally, the observational evidence concerning the origin and evolution of Pop II BSs and ACs seems to favor the binary hypothesis. It is highly likely that both types of stars have as their progenitors primordial binary (or possibly multiple) systems. Direct observation of both types of stars in very low-density stellar systems, where collisions are extremely rare, argues strongly in favor of this theory. Because of angular momentum losses, and the tendancy for close binaries to coalesce on times scales of a few Gyr (Mochnachi, 1985), most BSs are likely to be coalesced systems. However, the observed diversity of variable star types among BSs indicates that at present there is a range of separations of the binary components.

ACKNOWLEDGEMENTS

The following people are gratefully acknowledged for having contributed to this review: E. Böhm-Vitense, M. Breger, G. Brent, M. Burke, B. Carney, J. Cohen, G. Da Costa, G. Fahlman, H. Harris, J. Hesser, M. Mateo, R.D. McClure, C. Oliveira, H. Richer, A. Sandage, N. Simon, P. Stetson, and A. Wehlau. This research was supported by an operating grant from the Natural Sciences and Engineering Research Council of Canada.

REFERENCES

Antonello, E., Broglia, P., Conconi, P. & Mantegazza, L. (1986). Fourier decomposition of the light curves of high amplitude Delta Scuti and SX Phe stars. Astron. & Astrophys.,169, 122-132.

Arp, H.C. & Hartwick, F.D.A. (1971). A photometric study of the metal-rich globular cluster M71. Ap.J.,167, 499-509.

Baade, W. & Swope, H.H. (1961). The Draco system, a dwarf galaxy. A.J.,66, 300-347.

Baliunas, S.L. & Guinan, E.F. (1985). The old galactic cluster NGC 188 and the origin of the W Ursae Majoris-type contact binaries. Ap.J.,294, 207-215.

Bell, R.A. & VandenBerg, D.A. (1987). Theoretical isochrones for globular clusters with predicted Thuan-Gunn photometry. Ap.J.Suppl.,63, 335-364.

Bessell, M. (1969). An investigation of short-period variable stars. Ap.J.Suppl.,18, 195-220.

Binnendijk, L. (1970). The orbital elements of W Ursae Majoris systems. Vistas in Astronomy,12, 217-256.

Böhm-Vitense, E., Nemec, J.M., & Proffitt, C. (1984). The problem of the Barium stars. Ap.J.,278, 726-738.
Böhm-Vitense, E., Szkody, P., Wallerstein, G., & Iben, I. (1974). Masses and Luminosities of Population II Cepheids. Ap.J.,194, 125.
Bond, H.E. & MacConnell, D.J. (1971). The nature of the field blue straggler stars. Ap.J.,165, 51-55.
Bopp, B.W. & Stencel, R.E. (1981). The FK Comae Stars. Ap.J.,247, L131-L134.
Breger, M. (1980). The nature of dwarf Cepheids. V. Analysis and Conclusions. Ap.J.,235, 153-162.
Buonanno, R., Buzzoni, A., Corsi, C.E., Fusi Pecci, F. & Sandage, A.R. (1988). High Precision Photometry of 10,000 Stars in M3. In The Harlow-Shapley Symposium on Globular Cluster Systems in Galaxies, IAU Symp.126, eds. J.E. Grindlay & A.G.D. Philip, pp.621-622. Dordrecht: Reidel.
Buonanno, R., Corsi, C.E., Ferraro, I. & Fusi Pecci, F. (1987). CCD Photometry in globular clusters. II. NGC 7492. Astron. & Astrophys.Suppl.,67, 327-340.
Buonanno, R., Corsi, C.E., Fusi Pecci, F., Alcaino, G. & Liller, W. (1984a). The color-magnitude diagram of NGC 288. Astron. & Astrophys.Suppl.,57, 75-90.
Buonanno, R., Corsi, C.E., Fusi Pecci, F. & Harris, W.E. (1984b). Main-sequence photometry in NGC 2808. A.J.,89, 365.
Campbell, B. (1986). Strong cyanogen stars: the result of binary coalescence? Ap.J.,307, 750-759.
Carney, B.W. & Peterson, R.C. (1981). Field Population II blue stragglers. Ap.J.,251, 190-200.
Carney, B.W. & Seitzer, P. (1986). Deep photometry of the Draco dwarf spheroidal galaxy. A.J.,92, 23-42.
Chaffee, F.H. Jr. & Ables, H.D. (1983). Radial velocities of blue stragglers in M3. P.A.S.P.,95, 835-838.
Christian, C.A. & Heasley, J.N. (1986). Color-magnitude diagram of Palomar 4: CCD photometry. Ap.J.,303, 216-225.
Christian, C.A. & Heasley, J.N. (1988). CCD photometry of NGC 2419. A.J.,95, 1422-1452.
Christy, R.F. (1970). Lectures on variable stars. III. J.R.A.S.C.,64, 8-31.
Collier, A.C. & Jenkins, C.R. (1984). Close binary stars and old stellar populations: the blue straggler problem revisited. M.N.R.A.S.,211, 391-419.
Cowley, A.P. & Crampton, D. (1985). A new CH star in Omega Centauri. P.A.S.P.,97, 835-837.
Cox, A.N. (1987). Modes, masses, metallicities, and magnitudes of RR Lyrae variables. In Second Conference on Faint Blue Stars, Proc.I.A.U.Colloq. 95, in press. Schenectady, N.Y: L. Davis Press.
Cox, A.N., King, D.S. & Hodson, S.W. (1979). Theoretical periods and masses of double-mode dwarf Cepheids. Ap.J.,228, 870-874.
Cox, A.N. & Proffitt, C. (1988). Theoretical interpretations of anomalous Cepheid pulsations. Ap.J.,324, 1042-1047.

Da Costa, G.S. (1984) The age(s?) of the Sculptor dwarf galaxy. Ap.J.,285, 483-494.
Da Costa, G.S. (1987). CCD photometry of blue stragglers in the globular cluster M3. In The Second Conference on Faint Blue Stars, IAU Colloquium 95, eds. A.G.D. Philip, D.S. Hayes & J.W. Liebert, pp.579-582. Schenectady, N.Y.: L. Davis Press.
Da Costa, G.S. (1988). Dwarf Spheroidal Galaxies and Globular Clusters. In the Harlow-Shapley Symposium on Globular Cluster Systems in Galaxies, IAU Symp.126, eds. J.E. Grindlay & A.G.D. Philip, p.217-235. Dordrecht: Reidel.
Da Costa, G.S. & Demarque, P. (1982). Nitrogen variations on the main sequence of 47 Tucanae: implications from stellar structure theory. Ap.J.,259, 193-197.
Da Costa, G.S. & Norris, J. (1988). Variability of ωCentauri blue stragglers: clues to their origin. In the Harlow-Shapley Symposium on Globular Cluster Systems in Galaxies, IAU Symp.126, eds. J.E.Grindlay & A.G.Davis Philip, pp.681-682. Dordrecht: Reidel
Da Costa, G.S., Norris, J. & Villumsen, J.V. (1986). The blue stragglers of Omega Centauri. Ap.J.,308,743-754.
Da Costa, G.S., Ortolani, S., & Mould, J. (1982). Pal 14: an intermediate metal abundance globular cluster in the outer galactic halo. Ap.J.,257, 633-639.
Demarque, P. & Hirschfeld, A.W. (1975). On the nature of the bright variables in dwarf spheroidal galaxies. Ap.J.,202, 346-352.
Demers, S. & Irwin, M.J. (1987). The long period variables of Fornax. I. Search, discovery and periods. M.N.R.A.S.,226, 943
Demers, S. & Irwin, M.J. (1989). The long period variables of Fornax. II. The brighter Cepheids. M.N.R.A.S., in press.
Eggen, O.J. (1952a). A new variable star with the shortest known period. P.A.S.P.,64, 31.
Eggen, O.J. (1952b). The short period variable HD 223065 (=SX Phe). P.A.S.P.,64, 305.
Eggen, O.J. (1970). Ultrashort-period variables and the masses of blue stragglers in the old disk population. P.A.S.P,82, 274-292.
Eggen, O.J. (1971). The nature of the blue stragglers in the old disk population. P.A.S.P,83, 762-767.
Eggen, O.J. (1976). Contact binaries, II. Mem.R.A.S.70, 111-164.
Eggen, O.J. (1979). The classification of intrinsic variables. VIII. Ultrashort period Cepheids. Ap.J.Suppl.,41, 413-434.
Eggen, O.J. & Iben, I. (1988). Starbursts, binary stars, and blue stragglers in local superclusters and groups. I. The very young disk and young disk populations. A.J.,96, 635-669.
Fernley, J.A., Jameson, R.F., Sherrington, M.R. & Skillen, I. (1987). The radii and masses of dwarf Cepheids. M.N.R.A.S.,225, 451-468.
Fitch, W.S. (1970). Pulsation constants and densities for double-mode variables in the Cepheid instability strip. Ap.J.,161, 669-678.
Fitch, W.S. & Szeidl, B. (1976). The three radial modes and evolutionary state of AC Andromedae. Ap.J.,203, 616-624.

Frogel, J. & Twarog, B. (1983). Faint stellar photometry in clusters I. NGC 2204 and E3. Ap.J.,274, 270.

Frolov, M.S. & Irakaev, B.N. (1984). On the SX Phe-type stars. I.B.V.S., No.2462, pp.1-2.

Graham, J. (1975). The RR Lyrae Stars in the Small Magellanic Cloud. P.A.S.P.83, 641-682.

Gratton, R.G. & Ortolani, S. (1984). Deep photometry of globular clusters. II. The remote cluster Pal 3. Astron. & Astrophys.Suppl.,57, 177-188.

Guinan, E.F. & Bradstreet, D.H. (1988). Kinematic clues to the origin and evolution of low mass contact binaries. In Formation and evolution of low mass stars; NATO ASI Meeting at Viana do Castelo, Portugal, Oct. 1987, eds. A.K. Deupree & M.T. Lago, in press. Dordrecht: Reidel.

Halpren, L. & Moon, T.T. (1983). Revised list of pulsating stars with ultra-short periods. Astrophys. & Space Science,91, 43-51.

Harris, H.C. (1985). Population II Cepheids. In Cepheids: Theory and Observations, ed.B.F. Madore, pp.232-245. Cambridge: Cambridge University Press.

Harris, H.C. (1987). Population II variables. In Lecture Notes in Physics, Vol.274, eds. A.N. Cox, W.M. Sparks & S.G. Starrfield, pp. 274-283. Berlin: Springer-Verlag.

Harris, W.E. & Canterna, R. (1980). Color-magnitude photometry to the main sequence for the anomalous globular cluster Palomar 12. Ap.J.,239, 815.

Hesser, J.E., McClure, R.D., Hawarden, T.G., Cannon, R.D., von Rudloff, R., Kruger, B. & Egles, D. (1984). A new color-magnitude diagram for the peculiar star cluster E3=C0921-770. P.A.S.P.,96, 406-418.

Hills, J.G. & Day, C.A. (1976). Stellar collisions in globular clusters. Ap.Letters,17, 87.

Hirshfeld, A.W. (1980). The stellar content of dwarf spheroidal galaxies. Ap.J.,241, 111-124.

Hodge, P.W. & Wright, F.W. (1983). Variable stars in the Leo I dwarf galaxy. A.J.,83, 228.

Jensen, K.S. & Jorgensen, H.E. (1985). CCD based B and V lightcurves for the eclipsing binary NJL 5 in Omega Centauri. Astron. & Astrophys.Suppl.,60, 229-236.

Jorgensen, H.E. (1982). The dwarf Cepheid NJL 79 in Omega Centauri. Astron. & Astrophys.,108, 99-101.

Jorgensen, H.E. & Hansen, L. (1984) The dwarf Cepheids E39 and NJL 220 in Omega Centauri, Astron. & Astrophys.,133, 165-168.

Kaluzny, J. & Shara, M.M. (1987). The discovery of six new short-period variables in the old open cluster NGC 188. Ap.J.,314, 585-593.

Liller, M.H. (1978). An eclipsing binary in the field of Cen. I.B.V.S., No.1527.

Mateo, M., Harris, H.C., Nemec, J.M., Olszewski, E. & Schombert, J. (1988). Blue stragglers as remnants of stellar mergers: the discovery of three contact binaries among the blue stragglers in NGC 5466. In preparation.

Mateo, M., Nemec, J.M., Harris, H.C., Olszewski, E., Schombert, J. & Brent, G. (1989). Variable blue stragglers in NGC 5466. In preparation.
Mathieu, R.D. & Latham, D.W. (1986). The spatial distribution of spectroscopic binaries and blue stragglers in the open cluster M67. A.J.,92, 1364-1371.
Mathieu, R.D. & Mazeh, T. (1988). The circularized binaries in open clusters: a new clock for age determination. Ap.J.,326, 256-264.
McClure, R.D. (1985). The carbon and related stars. Journ.Roy.Astron. Soc.,79, 277.
McClure, R.D., Fletcher, R.D. & Nemec, J.M. (1980). The binary nature of the barium stars. Ap.J.,238, 135-138.
McClure, R.D., Hesser, J.E., Stetson, P.B. & Stryker, L.L. (1985). CCD photometry of the sparse halo cluster E3. P.A.S.P.,97, 665-675.
McClure R.D. & Twarog, B.A. (1977). New photographic data for NGC 188. Ap.J.,214, 111-123.
McCrea, W.H. (1964). Extended main-sequence of some stellar clusters. M.N.R.A.S.,128, 147-155.
McMillan, R.S., Breger, M., Ferland, G.J. & Loumos, G.L. (1976). A Survey for small-amplitude variability among Population II stars. P.A.S.P.,88, 495-509.
McNamara, D.H. & Feltz, K.A.Jr. (1978). GD 428 and the nature of dwarf Cepheids. P.A.S.P.,90, 275-284.
Mochnachi, S. (1981). Contact binary stars. Ap.J.,245, 650-670.
Mould, J. (1983). Star formation history of nearby dwarf galaxies. In Highlights of Astronomy, ed. R.M.West.,6, 179-186. Dordrecht: Reidel.
Mould, J. & Aaronson, M. (1983). The Carina dwarf spheroidal - an intermediate age galaxy. Ap.J.,273, 530-538.
Nemec, J.M. & Cohen, J.G. (1988). Blue straggler stars in the globular cluster NGC 5053. Ap.J. (in press).
Nemec, J.M. & Harris, H.C. (1987). Blue straggler stars in the globular cluster NGC 5466. Ap.J.,316, 172-188.
Nemec, J.M., Mateo, M., Burke, M., Richer, H., Fahlman, G.G., Schombert, J. & Harris, H. (1988). SX Phe variables in the globular cluster NGC 5053. In preparation.
Nemec, J.M. & Rich, R.M. (1985). Ages of galactic halo star clusters. B.A.A.S.,18, 446.
Nemec, J.M., Richer, H. & Fahlman, G.G. (1989). Blue straggler stars in the globular cluster NGC 5897. In preparation.
Nemec, J.M., Wehlau, A., & Mendes de Oliveira, C. (1988). Variable stars in the Ursa Minor dwarf galaxy. A.J.,95, 528-559.
Niss, B., Jorgensen, H.E. & Laustsen, S. (1978). A search for new variables in the globular cluster Omega Centauri. Astron. & Astrophys.Suppl.,32, 387-393.
Norris, J. & Zinn, R. (1975). The Cepheid variables and the stellar populations of the dwarf spheroidal galaxies. Ap.J..202, 335-345.

Olszewski, E. & Aaronson, M. (1985). The Ursa Minor dwarf galaxy: still an old stellar system. A.J.,90, 2221-2238.
Peterson, C. & King, I.R. (1975). The structure of star clusters. VI. Observed radii and structural parameters in globular clusters. A.J.,80, 427-436.
Peterson, R.C., Carney, B.W. & Latham, D.W. (1984). The blue stragglers of M67. Ap.J.,279, 237-251.
Renzini,A., Mengel, J.G. & Sweigart, A.V. (1977). The anomalous Cepheids in dwarf spheroidal galaxies as binary systems. Astron. & Astrophys.,56, 369-376.
Richer, H.B. & Fahlman, G.G. (1987). Deep CCD photometry in globular clusters. VI. White dwarfs, cataclysmic variables, and binary stars in M71. Ap.J.,325, 218-224.
Saha, A., Monet, D.G. & Seitzer, P. (1986). RR Lyrae Stars in the Carina dwarf galaxy. A.J.,92, 302-327.
Saio, H. & Wheeler, J.C. (1980). The evolution of mixed long-lived stars. Ap.J.,242, 1176-1182.
Sandage, A.R. (1953). The color-magnitude diagram for the globular cluster M3. A.J.,58, 61-75.
Sandage, A.R. & Katem, B. (1968). On the intrinsic widths of the subgiant and horizontal branch sequences in the globular cluster M3. A.J.,87, 537-554.
Simon, N. & Lee, A.S. (1981). The structural properties of Cepheid light curves. Ap.J.,248, 291-297.
Smith, G.H. & Dopita, M.A. (1983). The chemical inhomogeneity of the Sculptor dwarf spheroidal galaxy. Ap.J.,271, 113-122.
Smith, G.H., McClure, R.D., Stetson, P.B., Hesser, J.E., & Bell, R.A. (1986). CCD photometry of the globular cluster Palomar 5. A.J.,91, 842-854.
Smith, H.A. & Stryker, L.L. (1986). Anomalous Cepheids in the Sculptor dwarf galaxy. A.J.,92, 328-334.
Stellingwerf, R.F. (1979). Pulsation in the lower Cepheid strip. I. Linear survey. Ap.J.,227, 935-942.
Stetson, P.B. (1984). Spectroscopy of giant stars in the Draco and Ursa Minor dwarf galaxies. P.A.S.P.,96, 128-142.
Swope, H.H. (1968). Thirteen periodic variable stars brighter than the normal RR Lyrae-type variables in four dwarf galaxies. A.J.,73, 204-205.
Teays, T.J. & Simon, N.R. (1985). The unusual pulsating variable XZ Ceti. Ap.J.,290, 683-688.
van Agt, S.L.T.J. (1967). A discussion of the Ursa Minor dwarf galaxy based on plates obtained by Walter Baade. Bull.Astron. Inst.Neth.,19, 275-302.
van Agt, S.L.T.J. (1968). Magnitudes, phases and light-curves of variable stars in the central region of the Ursa Minor dwarf galaxy. Bull.Astron.Inst.Neth.Suppl.,2, 237-258.
van den Bergh, S. (1972a). Search for faint companions to M31. Ap.J.,171, L31-L33.
van den Bergh, S. (1972b). Resolution of one of the companions to M31. Ap.J.,178, L99.

van den Bergh, S., Demers, S. & Kunkel, W.E. (1980). The dying globular cluster E3. Ap.J.,239, 112.
VandenBerg, D.A. & Smith, G.H. (1988). Constraints from stellar models on mixing as a viable explanation of abundance anomalies in globular clusters. P.A.S.P.,100, 314-335.
van den Heuvel, E.P.J. (1968). The origin of Ap and Am stars and other slowly rotating A- and B-type main-sequence stars. Bull.Astron.Inst.Neth.,19, 326-431.
Wallerstein, G. & Cox, A.N. (1984). The Population II Cepheids. P.A.S.P.,96, 677-691.
Walraven, Th. (1955). On the short period variable HD 223065. Bull.Astron.Inst.Neth.,12, 57.
Webbink, R.F. (1976a). The evolution of low-mass close binary systems. I. The evolutionary fate of contact binaries. Ap.J.,209, 829-845.
Webbink, R.F. (1976b). The evolution of low-mass close binary systems. II. $1.50\ M_\odot + 0.75\ M_\odot$: evolution into contact. Ap.J.Suppl.,32, 583-601.
Webbink, R.F. (1979). The evolution of low-mass close binary systems. VI. Population II W Ursae Majoris systems. Ap.J.,227, 178-184.
Webbink, R.F. (1985). Structure parameters of galactic globular clusters. In Dynamics of Star Clusters, eds. J. Goodman & P. Hut, pp. 547-577. Dordrecht: Reidel.
Wheeler, J.C. (1979). Blue stragglers as long lived stars. Ap.J.,234, 569.
Whelan, J.A.J., Worden, S.P., Rucinski, S.M. & Romanishin, W. (1979). AH Cancri: a contact binary in M67. M.N.R.A.S.,186, 729-741.
Zinn, R. (1980). The dwarf spheroidal galaxies. In Globular Clusters, ed.D. Hanes & B. Madore, pp. 191-212. Cambridge: Cambridge University Press.
Zinn, R. (1981). The metal abundance range in the Ursa Minor dwarf galaxy. Ap.J.,251, 52-60.
Zinn, R. (1985). The dwarf spheroidal galaxies and their variable stars. Mem.S.A.It.,56, 223-236.
Zinn, R. & Dahn, C. (1976). Variable 19 in NGC 5466: An anomalous cepheid in a globular cluster. A.J.,81, 527-533.
Zinn, R. & King, C.R. (1982). The Mass of the Anomalous Cepheid in the Globular Cluster NGC 5466. Ap.J.,262, 700-708.
Zinn, R. & Searle, L. (1976). The masses of the anomalous Cepheids in the Draco system. Ap.J.,209, 734-747.

TYPE I INTERMITTENT CHAOS IN HYDRODYNAMIC PULSATION MODELS

Tohsiki Aikawa
Faculty of Engineering
Hohoku-Gakuin University
Tagajo, 985 Japan

Abstract. We investigate the cause of the transition from limit cycles to chaotic oscillations in pulsation of less-massive supergiant stars. For this purpose, we examine one of the models which has a limit cycle but is very close to the transition to the type I intermittency (Aikawa 1988, Astrophys. & Space Sci., 139, 214). It is shown that the pulsational driving by the hydrogen ionization with the M type ionization front, which becomes very effective at amplitudes beyond the limit cycle, is responsible for the existence of another unstable fixed point beyond the limit cycle (see Aikawa, 1988, in press Astrophys. & Space Sci.) A merging process of these two fixed points makes the transition.

DOUBLE MODE PULSATING STARS AND OPACITY CHANGES

G.K. Andreasen
Copenhagen University Observatory

Abstract. The basic physical properties of double mode Cepheids and double mode δ Scuti stars are still highly controversial. Based on results of very recent opacity calculations a detailed consistent, possible solution of the problem of the nature of double mode variable stars has been developed. Extensive studies have been performed to establish the rules governing the morphology of the period ratio diagrams. An ideal reproduction of the period ratio data for both stellar types are obtained assuming an increase of the Cox-Steward opacity by a factor of about 2.5 in the temperature range from 1.5×10^5K to 8.0×10^5K. Consequences for the mass calibration of Petersen diagrams are evaluated.

THE PULSATION OF SOME δ SCUTI STARS WITH UNUSUAL LIGHT CURVES

E. Antonello
Osservatorio Astronomico di Brera
Milano-Merate, Italy

E. Poretti
Osservatorio Astronomico di Brera
Milano-Merate, Italy

Abstract. Among the δ Scuti stars with large amplitude there are a few stars with unusual light curve shape and multiperiodicity. These stars are V1719 Cyg, V798 Cyg, V974 Oph and δ Scuti. The Fourier phase difference for the main pulsation mode is very different from that typical of the other large amplitude δ Scuti, SX Phe and RR Lyrae stars. The period ratio is generally high, ≥ 0.80. Two of the stars (δ Scuti and V1719 Cyg) have known spectral type or Strömgren indices, and these indicate high metallicity which should be related to the δ Delphini type (evolved Am/Fm stars). Some tests with the one-zone model show that the unusual light curve shape could be related to a poor He content in the envelope, while the application of model atmosphere grids shows that the anomalous behavior of the metallicity photometric index of V1719 Cyg is simply due to the high metal abundance of its atmosphere. The conclusion is that diffusion phenomena such as the He settling and the metal enhancement should be able to explain all the observed characteristics of these pulsating stars.

PROBLEMS WITH THE BAADE WESSELINK METHOD

E. Böhm-Vitense
University of Washington, Seattle, Washington

P. Garnavich
University of Washington, Seattle, Washington

M. Lawler
University of Washington, Seattle, Washington

J. Mena-Werth
University of Washington, Seattle, Washington

S. Morgan
University of Washington, Seattle, Washington

E. Peterson
University of Washington, Seattle, Washington

S. Temple
University of Washington, Seattle, Washington

Abstract. It is well known that the Baade-Wesselink method leads to different radii for Cepheids depending on which colors are used to determine the effective temperatures. We try to find the reasons for this discrepancy. We employ yet another version of this method using only maximum and minimum radii, thereby circumventing uncertainties in the phase relations between radial velocities and colors. This has essentially no influence on the derived radii. One major uncertainty is the relation between the photospheric expansion velocity and the measured radial velocity. The main reason for the discrepant results obtained by using different colors appears to be an inconsistency in the difference in the applied temperature-color calibrations. Small changes in the $d(\log T_{eff})/d(color)$ can cause major changes in the derived radii.

MULTIPLE CLOSE FREQUENCES OF THE δ SCUTI STAR θ^2TAU: THE SECOND MULTISITE CAMPAIGN

M. Breger
McDonald Observatory, University of Texas, Austin, Texas,
USA and Institut für Astronomie, Universität Wien, Wien,
Austria

R. Garrido
Instituto de Astrofisica de Andalucia, Granada, Spain

Huang Lin
Beijing Observatory, Academy of Sciences, Beijing, China

Jiang Shi-yang
Beijing Observatory, Academy of Sciences, Beijing, China

Guo Zi-he
Beijing Observatory, Academy of Sciences, Beijing, China

M. Frueh
McDonald Observatory, University of Texas, Austin, Texas,
USA

M. Paparo
Konkoly Observatory, Budapest, Hungary

Abstract. Multisite photoelectric b and V photometry of the δ Scuti variable θ^2(78) Tau has been obtained on three continents. Five close pulsation frequencies (13.22965, 13.48073, 13.69360, 14.31764 and 14.61454 cycles per day) with visual amplitudes between 0.001 and 0.007 mag were found. The four frequencies found in the previous campaign were confirmed. All three main data sets (1982-1986) are in excellent agreement with each other and fit the solution to ±0.0028 mag per single measurement. Over the four years the amplitudes and frequencies of pulsation were constant.

The measured (O-C) values relative to the five-frequency solution are in agreement with the predicted 141-day orbital light-time effects in the binary system and show pulsation of the more massive component. The pulsational Q values of 0.017 to 0.019 days indicate a pulsation in the second or higher overtone ($K \geq 2$). The pattern of frequency spacing cannot be explained by either radial pulsation or rotational splitting alone. The pulsation consists of a complex mixture of modes with different l and m values.

CEPHEID EVOLUTION WITH PULSATIONALLY-DRIVEN MASS LOSS

Wendee M. Brunish
Los Alamos National Laboratory

Lee Anne Willson
Iowa State University

Abstract. We have studied the effects of a pulsationally-driven wind on Cepheid evolution. Mass loss due to the wind, which occurs only when the star is crossing the Cepheid instability strip, is a function of luminosity and radius. We have investigated the evolution of 4, 5, 6, 7 and 8 $M_\odot$ stars using the updated $^{12}C(\alpha,\gamma)^{16}O$ rates.

Our results show that even a small amount of mass loss reduces the rate of period change for all masses. Even a small amount of mass loss moves the blue tip back toward the blue edge of the strip, and since the stars evolve more slowly as they approach the tip, even a slight reddening of the tip can significantly increase the time spent in the Cepheid strip.

Also, since the luminosity, and therefore the period, are not greatly affected, as the star loses mass, its evolutionary mass becomes in better agreement with pulsational masses.

The blue loop tips for the 5 and 6 $M_\odot$ stars lie close to blue edge of the instability strip to begin with, so a small amount of mass loss is sufficient to affect the evolution. A mass loss rate on the order of $10^{-7} M_\odot \cdot yr^{-1}$ will decrease the mass by about 10%, increase the Cepheid lifetime by a factor of two to five and decrease the period rate of change by a similar factor.

THE DISTANCE AND AGE OF THE GLOBULAR CLUSTER M5

Bruce W. Carney
University of North Carolina

David W. Latham
Center for Astrophysics

Rodney V. Jones
University of North Carolina

Judy A. Beck
University of North Carolina

Abstract. Using optical and infrared photometry and echelle spectroscopy of variable V8 in the globular cluster M5, we derive a cluster distance of 6.8 kpc using the Baade-Wesselink method. This agrees with the prediction obtained for the cluster's metallicity using a sample of 19 field stars studied by us and by Liu and Janes (this volume). It also agrees well with estimates for M_V obtained from statistical parallaxes of field stars. It agrees as well with the main sequence fitting procedure where we have used only HD 103095, the field halo dwarf with the most accurate trigonometric parallax (3% error), and which has a metallicity almost identical to that of M5. The star is also cool, hence unevolved, and is not a binary. Using the luminosity of the cluster's main sequence, both Yale and Victoria isochrones yield a cluster age of 18 ± 3 Gyrs.

DISCONTINUITY MODES IN POLYTROPES

Bradley W. Carroll
Department of Physics
Weber State College
Ogden, Utah

Abstract. In the calculation of linear nonradial oscillation modes in composite polytropes with a small density discontinuity, a discontinuity mode may occur. This mode consists of a wave propagating along the discontinuity interface with a large amplitude that declines exponentially away from the interface. The period P of this mode is well-estimated (to within 10%) by

$$P = \frac{2\pi}{\sqrt{\frac{1}{2} g k (\Delta\rho/\langle\rho\rangle)}},$$

where $\Delta\rho/\langle\rho\rangle$ is the fractional density discontinuity and k is the horizontal wavenumber (c.f. Gabriel and Scuflaire 1980, in Nonradial and Nonlinear Stellar Pulsation, eds. H. Hill and W. Dziembowski, Springer-Verlag). For a 12 solar-mass polytrope with a radius of 4.27 $R_\odot$, a 3% density discontinuity at fractional radius 0.15 produces a discontinuity mode with a period of 7.329 hours. As the density discontinuity increases the period P decreases, resulting in avoided crossings with the normal g-mode spectrum. Between these avoided crossings, the discontinuity mode has an unusually large amplitude at the location of the discontinuity.

This research was partially supported by a grant from the American Astronomical Society. Additional support was provided by the Research and Professional Growth Committee at Weber State College.

THE LUMINOSITY-METALLICITY RELATION FOR RR LYRAE STARS AND ITS IMPLICATIONS FOR THE ASTRONOMICAL DISTANCE SCALE

G. Clementini,
Osservatorio Astronomico, Bologna, Italy and
Space Telescope Science Institute, Baltimore, MD, USA

C. Cacciari
Osservatorio Astronomico, Bologna, Italy and
Space Telescope Science Institute, Baltimore, MD, USA

Abstract. The surface brightness version of the Baade-Wesselink method, has been applied to 7 field RR Lyrae stars with metallicity ranging from [Fe/H]= -0.2 to -1.5. V magnitudes, V-R and V-I colors and CORAVEL radial velocities were used, and the analysis was performed over a restricted phase range in order to avoid the complications caused by the pulsating atmospheres. The combination with previous results of the B-W method, which used comparable criteria (Jones, Carney, & Latham, 1988, preprint; Jameson, Fernley, & Longmore 1987, in press M.N.R.A.S; Cohen & Gordon 1987, Ap.J.,318, 215) leads to the following relation between the absolute luminosity and metallicity:

$$M_V = (1.0 \pm 0.05) + (0.17 \pm 0.05) [Fe/H]$$

This relation is in very good agreement with the preliminary results found by Liu and Janes (this volume). The application of the above relation to the RR Lyraes in M31 and the Magellanic Clouds leads to distance moduli of $(m-M)_o$ = 24.21 ± 0.20 for M31, $(m-M)_o$ = 18.26 ± 0.20 for the LMC, and $(m-M)_o$ = 18.85 ± 0.20 for the SMC, and the distance to the galactic center turns out to be approximately 7.2 kpc. From the absolute magnitude of the RR Lyraes and adopting a constant visual magnitude difference between the RR Lyraes and the turn-off ΔV = 3.55 (Buonanno 1986, Mem.S.A.It.,57, 333), we estimate ages of 18.8 and 15.7 Gyr for globular clusters of metallicity [Fe/H] = -2.2 (e.g. M92) and [Fe/H] = -0.8 (e.g. 47 Tuc) respectively, using the age-turnoff luminosity relation derived by Sandage (1982, Ap.J.,252, 553) or 20.9 and 16.9 Gyr using Buonanno's relation.

A POSSIBLE PULSATION MECHANISM FOR B STARS

Arthur N. Cox
Theoretical Division
Los Alamos National Laboratory

Abstract. Numerous possible mechanisms for the cause of B star radial and nonradial pulsations have been investigated in the last 30 years, but none have been clearly shown that they actually work. I briefly review these ideas, and then focus on a new idea about the driving due to time-dependent convection. I suggest that the outward going photons are absorbed by CNONe ions so that they capture the photon momentum. Previous investigations by many have shown that the acceleration of these ions can be two orders of magnitude larger than the local gravity. Levitation and concentration of these elements can occur between temperatures of 100,000K and 600,000K, the former temperature set by the complete ionization of the competing photon capturing helium, and the latter by the existence of simpler and less absorbing excited electron structures of the CNONe elements. Then I suggest that the concentration of these elements gives such a large local photon opacity that a thin convection zone exists down to maybe 10^{-5} of the mass of a typical B star. However, there would be no overlap with the outer surface convection zone that exists down to 50,000K in the outer 10^{-9} of the mass. The long time scale of the convection, or even its periodic turning on and off, gives a source of time-lagged luminosity which can be shown in some cases to drive mechanical pulsations in one or more modes. This envelope pulsation mechanism could apply equally to radial and low degree nonradial B star pulsations. It could also explain why not all B stars pulsate unless this unconventional composition structure occurs, disappears, and maybe reoccurs at some evolution stage.

SOLAR-LIKE OSCILLATIONS IN LATE SPECTRAL CLASS STARS

Arthur N. Cox
Theoretical Division
Los Alamos National Laboratory

Julius H. Cahn
Astronomy Department
University of Illinois

Abstract. There is interest in the minimum mass that can exhibit solar-like nonradial p-mode oscillations. Inspection of models for the solar case shows that there is approximately equal pulsational driving and damping due to the standard radiative effects. It is widely believed that coupling with surface convective elements occasionally tips the balance by strongly driving modes for a few days with the timescale of the convection at the top of the convection zone. Then both the convection and the radiative damping decrease the mode amplitude at the observed decay rate, (which is close to that calculated by Kidman and Cox (1984)) until the next convective reexcitation. Population I composition stellar models at 0.30, 0.45, and 0.75 solar mass have been constructed to see if the near equality of radiative driving and damping exists in them also. If so, then stars down to late spectral class could possibly display solar-like oscillations. We find that the convection timescale decreases to less than 100 seconds at 0.30 solar mass, but the near equality of radiative driving and damping occurs only down to about 0.5 solar mass. Thus there is the possibility that 100 to 300 second oscillations can be observed through most of the K spectral class, if the magnetic activity often seen in these stars does not overwhelm these small amplitude nonradial pulsations.

OSCILLATION FREQUENCIES OF SOLAR MODELS

Arthur N. Cox
Theoretical Division
Los Alamos National Laboratory

Joyce A. Guzik
Astronomy Program, Department of Physics
Iowa State University

Russell B. Kidman
Los Alamos National Laboratory

Abstract. Two precision solar models have been constructed with the Iben evolution program, one with no diffusion of the internal atomic nuclei, and another that includes the effects of gravitational settling, thermal diffusion, and concentration gradient diffusion on the element abundances. The opacity at the bottom of the convection zone was increased 15-20 percent (within its theoretical uncertainty) to allow a few microhertz agreement with the observed solar p-mode frequencies. The original helium mass fractions of the mixtures were 0.291 and 0.289 for the no-diffusion and diffusion models, respectively. The diffusion model evolved to a surface Y=0.256 at the solar age, and the original Z value of 0.0200 decreased to 0.0179. Agreement of asymptotic theory l=0 and 2 p-mode frequency separations and those directly calculated with nonadiabatic theory is good, and there is good agreement with observations also. Calculation of g-mode solutions shows that they do not have equal period spacings until high radial order. Nonadiabatic solutions for these g-modes modes enable us to predict their relative surface visibility. For l=1-5, the lowest order modes seem to be more detectable assuming that they all have the same kinetic energy. Our high helium results in high central temperatures that give over 9 SNUs from the B and 1.5 SNUs from the Be reactions, but a lower more conventional helium abundance would reduce this B neutrino flux by maybe two SNUs. Models with iron condensed-out deeper than the convection zone, and with the presumed WIMPs to cool the central regions and reduce the SNU flux, agree less well with the p-mode frequency separations.

THE EFFECTS OF TIME-DEPENDENT CONVECTION ON WHITE DWARF
RADIAL PULSATIONS

Arthur N. Cox
Theoretical Division
Los Alamos National Laboratory

Sumner G. Starrfield
Physics Department
Arizona State University and
Theoretical Division
Los Alamos National Laboratory

Abstract. After the discovery of pulsations in white dwarfs, predictions were made that these DA and the hotter DB stars should be pulsating in radial modes with periods of a few seconds or less. The mechanisms are the normal kappa and gamma effects that periodically block the flow of radiative luminosity and the blocking effect of the frozen-in convection at the bottom of the convection zone. Blue edges of the instability strips are between 12,000K and 13,000K for the DA and between 32,000K and 33,000K for the DB variables. Extensive observations, however, have shown that these stars pulsate only in the few-hundred-second nonradial modes and not in any few-second radial modes. We have added the time dependent convection model of Cox, Brownlee, and Eilers (1966) to our pulsation analyses to further investigate the white dwarf radial modes. Since the time scale of the convection is usually short compared to the radial pulsation periods, convection is able to carry luminosity rapidly enough to nullify the kappa and gamma effects periodic radiation blocking. We find that most, and maybe all, radial pulsations for 0.6 solar mass carbon-oxygen white dwarfs with thin hydrogen or helium surface layers are stabilized for both these DA and DB classes, now finally in agreement with observations.

S SGE: A CEPHEID TRIPLE SYSTEM

Nancy Remage Evans*
Department of Astronomy, University of Toronto and
Astronomy Programs, Computer Sciences Corp.

Mark H. Slovak
Department. of Astronomy, University of Wisconsin

Douglas L. Welch
Dominion Astrophysical Observatory

Abstract. IUE spectra have been used to investigate the companion of the classical Cepheid S Sge. At the phase of the IUE observations, 45 Dra (F7Ib) is a good match to S Sge, both in $(B-V)_o$ and from 2200 to 3200 Å. However, from 1700 to 1900 Å, the S Sge spectrum has excess flux, as compared to 45 Dra. The flux remaining after subtracting 45 Dra from S Sge is a good match to an A5V to A7V star. The magnitude difference between the Cepheid and the companion is consistent with the absolute magnitude and colors of an A5V star. This corresponds to a mass of 1.9 solar masses (Popper, 1981, Ann. Rev. Astr. Ap., 18, 115). A spectral type range from A3V to A7V corresponds to an uncertainty in the mass of ±0.2 solar masses. However, the mass function from the orbit and an evolutionary mass for the Cepheid require a companion mass of 2.7 solar masses or greater. We infer that the companion is itself a binary.

* IUE Guest Observer

THE BLUE EDGE OF THE HELIUM STAR INSTABILITY STRIP

Yu. A. Fadeyev
Astronomical Council of the USSR Academy of Sciences

Abstract. Hydrodynamic calculations of nonlinear pulsations were done for models of helium stars with mass 1 $M_\odot$, luminosity from 3220 $L_\odot$ to 12820 $L_\odot$ and effective temperature from 6000K to 8000K. The models with L > 8000$L_\odot$ were found to pulsate in the fundamental mode with large amplitude ($\Delta R/R \sim 1$), whereas less luminous models (L < 8000$L_\odot$) revealed small amplitude oscillations ($\Delta R/R \sim 0.2$) in the first overtone. In the luminosity range considered the blue edge of the instability strip corresponds to an effective temperature of 7500K, that is to the upper limit of the effective temperatures of R CrB stars. Application of the period-luminosity relation to the variables R CrB and RY Sgr gives their luminosities to be of 9000$L_\odot$.

ENVELOPE DISTENTION AND MASS LOSS IN W VIR PULSATING
VARIABLES

Yu. A. Fadeyev
Astronomical Council of the USSR Academy of Sciences

H. Muthsam
Institute of Mathematics, Vienna University

Abstract. The hydrodynamic calculations of nonlinear self-excited radial pulsations were done for the models of W Vir stars with mass $0.6\ M_\odot$ and luminosities from $794 L_\odot$ to $3981 L_\odot$. The periods of the models are longer than 10 days. The pulsations are shown to be accompanied by periodic shocks that change the density distribution in the pulsating stellar atmosphere. At radii less than 5 R_{ph}, where R_{ph} is the radius of the photosphere, the mean dynamic scale height is nearly five times the static scale height. In this region of the atmosphere the mean radii of outer mass zones do not change perceptably. On the other hand, at radii larger than 5 R_{ph} the scale height is of the order of R_{ph} so that the outermost layers ultimately expand with a velocity exceeding the local escape velocity. The mass flux in the atmosphere increases with decreasing mass to radius ratio and mass loss rate in W Vir type variables is in the range from $2 \times 10^{-6} M_\odot \cdot yr^{-1}$ to $10^{-5} M_\odot \cdot yr^{-1}$.

FORTHCOMING CEPHEID DATABASE

J. Donald Fernie,
Department of Astronomy
University of Toronto

Nancy Remage Evans
Department of Astronomy
University of Toronto

Abstract. A summary of basic parameters of classical Cepheid variable stars is being prepared. It will contain stars from the fourth edition of the General Catalogue of Variable Stars in the categories CEP, CEP(B), DCEP, and DCEPS, excluding those which belong to Population II according to the criteria of Harris (1985, A.J., 90, 756). V and B intensity means, as well as velocity means are derived from Fourier fits using the N. Simon program MINFIT. In addition to the data contained in the catalogue by Fernie and Hube (1968, A.J., 73, 492), it will provide a comparison of reddenings from a number of sources, binary status, and period change information. We anticipate a hardcopy version, and also a version on floppy disks. The computer version uses dBaseIII, and will include programs, for instance, to calculate phases using changing periods.

We would appreciate being able to include any additional data which will be published shortly.

HYDRODYNAMIC MODELS OF LOW-MASS PULSATING SUPERGIANTS WITH RADIATIVE TRANSFER

A.B. Foken
Astronomical Council of the Academy of Sciences of the USSR

Abstract. A method of calculating nonlinear stellar pulsations including nonstationary radiative transfer in a grey spherical atmosphere is described. With the help of this method eleven type II supergiant radiative models were constructed with masses of $0.6 M_\odot$, luminosities ranging from $128 L_\odot$ to $3123 L_\odot$ and periods in the range from 1.123 to 46 days. A stable limit cycle was found to be accessible only by models with an effective temperature between 5700K and 6165K. The model with $T_e = 6165K$ is stable, whereas the models cooler than 5700K show nonregular behavior. Transition from strictly periodic to nonregular pulsation arises when $M/R \leqslant 0.018$, due to high amplitudes, $\delta r/r \approx 1$, and strong shocks in the atmosphere. The radiative transfer effects lead to some decay in the radial amplitude, as well as to a more significant decrease, about 0.6 magnitudes, in the light variation. A photometric comparison between the light curves of the models calculated with and without transfer and the observed light curve of the variable star No. 154 in M3 shows that the results predicted by the transfer model are in much better agreement with obervational data.

A STUDY OF PERIOD DOUBLING IN A ONE-ZONE PULSATING STELLAR MODEL

A.B. Foken
Astronomical Council of the Academy of Sciences of the USSR

Abstract. A simple one-zone model for nonlinear stellar pulsations is outlined and applied to the study of period doubling observed in some W Virginis and RV Tauri stars. The model reveals a number of period doubling bifurcations as the parameters are varied, similar to those found by Buchler & Kovacs in their series of hydrodynamic models. In the vicinity of a stable limit cycle, despite its large number of degrees of freedom, the model develops an essentially one-dimensional Poincaré's return map, determining the modulation of the amplitude. The analysis of these maps confirmed that period doubling has its origin in a strong nonlinear increase of total energy losses per period as radial amplitude, $\delta r/r$, increases. An additional study of nearly periodic hydrodynamic models with $P > 15$ days, calculated including radiative transfer effects, shows that the rate of energy dissipation per period by shocks in the atmospheres increases rapidly with $\delta r/r$, whereas the excitation rate, δ_o, remains rather stable. This permitted us to construct an analytic return map for maxima of the total kinetic energy which clearly demonstrates the mechanism of successive period doubling as δ_o, the sole bifurcation parameter, grows monotonically.

THE PERIOD-RADIUS RELATION FROM 101 CEPHEID RADII

W.P. Gieren
Universidad Nacional de Colombia

T.G. Barnes III
McDonald Observatory, University of Texas

T.J. Moffett
Purdue University

Abstract. Surface brightness solutions using the (V-R) color index have been carried out for 52 southern Cepheids and 63 northern Cepheids, with 14 stars in common. For the southern stars the data came from the study by Gieren (1986, M.N.R.A.S, 222, 251), with new observations, and for the northern stars, from Moffett and Barnes (1987, Ap.J., 323, 280). All stars were reduced using the same surface brightness relation and the same transformation from radial velocity to pulsational velocity. The 14 stars in common show that there is only a small systematic difference between the southern and northern data sets and reductions. The combined solution for the 101 Cepheids in the period range 3 to 45 days is $\log R = 1.108 + 0.743 \log P$ with uncertainties of ±0.023 in the zero point and ±0.023 in the slope. This result is consistent with the individual results of the above references but has much smaller uncertainty.

IRREGULARITY INTERPRETED AS LOW DIMENSION CHAOS FOR CONVECTIVE MODELS OF W VIR VARIABLES

S. Ami Glasner
Physics Department
University of Florida

J. Robert Buchler
Physics Department
University of Florida

Abstract. Linear and nonlinear pulsational properties of convective stellar envelopes relevant for W Vir and RV Tau stars are surveyed. All models show the same trend to pass from regular to irregular behavior when a control parameter is changed (the effective temperature). The transition to irregular pulsation follows well known systematic routes to chaos (as in the radiative case). Some rich structures were found in special cases; they deserve further research. We show that the chaotic behavior is sustained even when convection is taken into account. The effect of the inclusion of time dependent convection shows up mostly as a shift of Kovacs and Buchler (Ap.J 1988) results in the parameters plane (L, T_{eff}) towards more realistic models.

PERIOD DOUBLING IN VARIABLE STARS: A TENTATIVE INTERPRETATION OF OBERVED LIGHT CURVES OF VARIABLE WHITE DWARFS AND MIRA STARS

M.J. Goupil

A. Baglin

M. Auvergne

Abstract. Irregular pulsations are commonly observed in many groups of pulsating stars. An interpretation in terms of superposition of modes of pulsation is sometimes barely convincing when too many modes have to be considered and when unexpected ratios of frequenceis are observed.

The idea is thus to attribute some of those irregularities to basic purely nonlinear mechanisms, generators of chaos. This is supported by the results of Buchler et al. (1987, Ap.J.,320, L57) who have found series of numerical models of W Virginis stars undergoing a cascade of period doubling bifurcations when the effective temperature is decreased. Such a process is a classical route to chaos encountered in very different, simple or complex, nonlinear dynamical systems. It is therefore likely to occur in variable stars.

The analysis of three variable white dwarfs (Goupil, Auvergne & Baglin 1988, Astron. & Astrophys.,196, L13 and in preparation) reveals the existence of subharmonics in their power spectra which constitutes a first indication that those stars have undergone period doubling bifurcations. A preliminary analysis of a Mira star indicates a similar behavior.

Specific nonlinear methods of analysis (such as return maps and dimension computations) are necesary to confirm these conclusions. However, these require long series of data of very high signal to noise ratios which are not yet available. Programs of observatons devoted to this aim are in progress.

STAR FORMATION HISTORY OF THE LARGE MAGELLANIC CLOUD AND
ASYMPTOTIC GIANT BRANCH EVOLUTION OBTAINED FROM A STUDY
OF LONG-PERIOD VARIABLES.

Shaun M.G. Hughes
Mount Stromlo and Siding Spring Observatories
Australian National University

P.R. Wood
Mount Stromlo and Siding Spring Observatories
Australian National University

Abstract. A survey for Long-Period Variables has been made of the Bar and southern regions of the Large Magellanic Cloud. This has been combined with the results of Reid, Glass and Catchpole (1988) to present a global survey of the LMC. Periods are used as age indicators, giving the star formation history of the LMC from 1 to 10 Gyr, which indicates a major burst of star formation occurred approximately 4 Gyr ago.

Infrared photometry is used to derive bolometric luminosities and shows most are Asymptotic Giant Branch members and that a third are Carbon stars. None of the Carbon stars have a bolometric magnitude brighter than -5.4, and they tend to have lower pulsation amplitudes and longer periods than the M stars. The remaining stars are almost all of spectral type M and their I-band amplitude of pulsation increases with period. Both the Carbon and M stars lie on a well-defined K-(log P) relation, that may be useful as a distance estimator.

ON THE IRREGULAR LIGHT VARIATION OF THE RV TAU STAR R SCUTI

Z. Kollath
Konkoly Observatory, Budapest, Hungary

G. Kovacs
Konkoly Observatory, Budapest, Hungary

Abstract. The nature of irregular light variation in giants and supergiants has long been a matter of dispute. Our lack of understanding has two main sources: (i) a paucity of long, continuous, high quality observations and thorough analysis of the available data and (ii) until very recently, a lack of systematic nonlinear hydrodynamic studies for these stars.

The recent finding of Buchler and Kovacs (1987, Ap.J., 320, L270) that irregular behavior of pulsation models in the RV Tau regime is simple (i.e. low dimensional) chaos and results from period-doubling bifurcations, stimulated our present study.

One hundred and thirty-two years of visual observations of R Sct (compiled by Mora, 1934, Mitt. der Sternwarte Budapest, 3, 1) were studied with various techniques of modern time-series analysis. Power spectra indicate nonstationary behavior with irregularly spaced peaks around $P = 142$ days and ½ P. (O-C)-curves for the moments of light minima also show irregular variation with some phase jumps. Phase-space reconstruction together with the correlation dimension indicate that the corresponding dynamical system can be embedded in low-dimensional (< 5) phase-space. We suggest that the light variation is a result of a chaotic oscillation determined by a low-dimensional dynamic system.

The present analysis will be extended (Kollath, 1988, in preparation) by more data, including the new data published by the AAVSO (Mattei et al. 1988, A.A.V.S.O Monograph 3).

IS DELTA SCUTI STAR SEISMOLOGY POSSIBLE?

G. Kovacs
Konkoly Observatory, Budapest, Hungary

Abstract. The direct fit of theoretical pulsation frequencies to the observations (i.e. stellar seismology) proved to be a very efficient tool in the study of solar oscillations. In the case of other multiperiodic variables, like δ Scuti stars, Ap stars and white dwarfs the method suffers from the disturbing abundance of possible nonradial modes. Colour and/or radial velocity (or line profile) measurements can narrow down the number of possibilities, but these kinds of data are not often available with the desired accuracy and sampling rate. Since pulsational frequencies are the most readily and accurately computed and measured quantities of pulsation, we address the question of the accurate fit of the nonradial pulsation frequencies to the observations in the case of δ Scuti stars.

The parameters we want to infer from the frequency fits are mass, effective temperature, radius and, if rotation is included, rotational frequency.

Available pulsation models are not numerous. The most complete survey is that of Fitch (1981, Ap.J., 249, 218). We took his interpolation formula (7) for the logarithm of the period in the case of no rotation.

Rotation is probably an important factor in limiting the amplitudes and determining the frequency patterns of multi-periodic δ Scuti stars, since they are in general fast rotators (Breger, 1979, P.A.S.P., 91, 5). Therefore, rotation is included in the second order approximation of Saio (1981, Ap.J., 244, 299).

Distribution functions of the goodness of fits were computed in a few examples of four and five frequency fits. In the case of no rotation there were only a few cases of interest even for moderate accuracy (0.1 - 0.5 c/d) of fit. Though in the case of rotation the number of good fits increased, it is suggested that accurate theoretical models (like those available in the solar studies) would be able to decrease this number and make δ Scuti star seismology feasible.

A more detailed discussion will be presented elsewhere (Kovacs, 1989, in preparation).

THE LUMINOSITIES OF 13 FIELD RR LYRAE STARS: THE CORRELATION WITH METALLICITIES

T. Liu
Boston University

K.A. Janes
Boston University

Abstract. We have obtained UBVRI and JK photometry and radial velocities for 13 field RR Lyrae stars with a large range of metallicity. Our sample includes eleven ab-type and two c-type variables. The surface brightness method is applied to these RR Lyrae stars to determine their mean absolute magnitudes $<M_V>$ as well as their distances and radii. We have used the Kurucz's model atmospheres to provide metallicity- and gravity-dependent color-temperature transformations and bolometric corrections. Our results indicate a clear correlation between the luminosities of the RR Lyrae stars and their metallicities such that metal-poor RR Lyrae stars are more luminous.

SUMMARY OF RESULTS

Star	Period	[Fe/H]	E(B-V)	$<R/R_\odot>$	$<\log g>$	$<T_{eff}>$	d(pc)	$<M_V>$
SW And	0.4423	-0.15	0.058	4.32	2.86	6512	514	0.96
TV Boo	0.3126	-2.17	0.000	4.30	2.91	7054	1156	0.67
RR Cet	0.5530	-1.25	0.025	4.95	2.73	6413	586	0.80
SU Dra	0.6604	-1.75	0.046	5.05	2.74	6605	640	0.62
RX Eri	0.5872	-1.32	0.041	5.25	2.68	6284	569	0.78
RR Gem	0.3973	-0.29	0.090	4.00	2.91	6758	1064	0.95
RR Leo	0.4524	-1.35	0.056	4.35	2.84	6761	889	0.81
TT Lyn	0.5974	-1.35	0.014	5.35	2.68	6285	654	0.74
AV Peg	0.3904	0.03	0.055	3.87	2.97	6530	670	1.19
AR Per	0.4255	-0.60	0.350	4.05	2.92	6822	493	0.91
T Sex	0.3247	-1.20	0.041	4.02	2.98	7090	671	0.78
TU Uma	0.5577	-1.30	0.004	4.95	2.73	6352	621	0.85
UU Vir	0.4756	-0.49	0.028	4.30	2.86	6613	816	0.92

GLOBULAR CLUSTER DISTANCES FROM THE RR LYRAE LOG (PERIOD) -
INFRARED MAGNITUDE RELATION.

A.J. Longmore
Royal Observatory, Edinburgh

R. Dixon
University of Edinburgh)

I. Skillen
University of Leicester

R.F. Jameson
University of Leicester

J.A. Fernley
University College, London

Abstract. The log (Period) - infrared (2.2 micron) magnitude relationship has been measured for 7 clusters (M3, M4, M5, M15, M107, ω Cen and NGC 5466). For the clusters where there are several observations for each of the variables, enabling a good mean magnitude to be derived, there is no evidence for scatter in the relation outside observational error. This conclusion applies also to ω Cen, even though the variables observed covered a range of metallicity. It is argued that very accurate relative distances can be obtained which are insensitive to reddening errors and the effects of metallicity. Any mass difference between variables in different clusters may still introduce uncertainties in the relative distances. Recent Baade-Wesselink results by Fernley et al. have been used as an absolute calibration, including a small residual correction for metallicity effects. On this scale, the mean distance modulus for 6 of the clusters (excluding NGC5466) is essentially the same as that derived by Buonanno, Corsi and Fusi-Pecci (preprint) using a combination of horizontal branch magnitudes and main sequence fitting. However, individual distance moduli differ by typically 0.09 magnitudes. There is no clear correlation of this residual with metallicity.

RR LYRAE STARS AND THE SANDAGE PERIOD-SHIFT EFFECT EXAMINED USING IR-DERIVED TEMPERATURES

A.J. Longmore
Royal Observatory, Edinburgh

R. Dixon
University of Edinburgh

I. Skillen
University of Leicester

R.F. Jameson
University of Leicester

J.A. Fernley
University College, London

Abstract. Mean temperatures for RR Lyrae stars in 7 globular clusters (M3, M4, M5, M15, M107, ω Cen and NGC 5466) have been determined using the optical-infrared colour <V>-<K> as a temperature indicator. Where <K> has been relatively well determined, from means of 3 or more observations, the scatter in relationships such as Log P' vs log (temperature) and log (temperature) vs (blue amplitude) is significantly reduced when IR-derived temperatures are used instead of those derived from (B-V). Within the observational errors, the gradient in the log P' vs log (temperature) diagram is the same for each cluster. Temperatures derived from <V>-<K> should also be less sensitive to metallicity differences than their optically derived counterparts. The Sandage Period-Shift Effect has therefore been re-examined using 6 of the 7 clusters (NGC 5466 was excluded because of too few data). A strong correlation between period-shift and metallicity is found; a smaller shift (but in the same sense) is also found for the temperature - amplitude relationship.

Although the absolute <V>-<K> to effective temperature scale is not yet well established (particularly for HB stars) the present results give consistently lower temperatures for the RR Lyrae stars than have previously been derived from the optical measurements. This shift in the temperature of the instability strip has interesting consequences for pulsation theory.

SECULAR CHANGES IN THE LIGHT CURVE OF THE SHORT-PERIOD CEPHEID EU TAU

J.M. Matthews
Department of Astronomy, University of Western Ontario

W.P. Gieren
Observatorio Astronomico Nacional, Bogota, Colombia

Abstract. UBVRI photometry and radial velocity measurements of EU Tau obtained during three weeks in November/December 1987 have been used to (1) confirm the period, P = 2.1025 days, and epoch established by Fernie (1987, P.A.S.P., 99, 1093) and earlier observers, and (2) investigate apparent phase and/or amplitude discrepancies in the light curve first reported by Gieren and Matthews (1987, A.J., 94, 431). Comparison of these new data with photometry collected by different observers covering over two decades suggests that the light curve of EU Tau undergoes short-lived (less than about 2 pulsation cycles) changes which appear to occur with a period of roughly 19 - 21 days. Various possible sources of modulation - including binarity, double-mode beating and nonradial pulsations - are considered and rejected as unsatisfactory.

The general behaviour of the light curve can be simulated by a frequency spectrum with a central component at the observed pulsation frequency and 10 sidelobes of low amplitude (< 0.01 mag) and equal spacing (1/21 c/d). Such a pattern could be produced by weak rotational modulation of a single radial mode, but this requires a rotation period no greater than 21 days and a low inclination which would tend to lessen rotational effects. We argue that the most likely explanation of the proposed frequency spectrum is a variation in the star's pulsational limit cycle, caused by instability or resonant mode coupling. Auvergne (1986, Astron. & Astrophys.,159, 197) has already invoked a similar effect to account for the amplitude modulation of another short-period Cepheid, HR 7308. If such processes are occuring in EU Tau and other Cepheids, they will provide new observational constraints on models of Cepheid structure.

PRELIMINARY RESULTS OF A WORLD WIDE PHOTOELECTRIC CAMPAIGN ON δ SCUTI

B. McNamara
S. Baggett
J. Pena
K. Thompson
G. Moore
L. Mantegazza
K. Sekiguchi
M. Candy

Abstract. During June 17-27, 1987 observers in Mexico, Chile, Italy, South Africa, Australia and the United States monitored the light output of the prototype variable star, δ Scuti. The goal of the network was to obtain this star's frequency spectrum. During the campaign, over 17,000 observations were obtained in the Strömgren b and y filters. A preliminary analysis has been performed and the frequencies listed in Table 1 have been tentatively identified.

A second campaign is planned for September 7 - 18, 1988.

Table 1

Frequency (c/d)	Amplitude
5.1608 ± 0.0001	0.0638 ± 0.0002
5.358 ± 0.001	0.0126 ± 0.0002
10.323 ± 0.001	0.0073 ± 0.0002
8.317 ± 0.002	0.0040 ± 0.0002
8.597 ± 0.003	0.0036 ± 0.0002
10.529 ± 0.003	0.0030 ± 0.0002

OBSERVATIONS OF VARIABILITY IN THE RADIAL VELOCITY OF α BOO

W.J. Merline
Lunar and Planetary Laboratory, University of Arizona

Abstract. We have made precise radial velocity observations of α Boo (K1 III) which raise the possibility that giant stars redward of the Cepheid instability strip can pulsate. Our first measurements, made in early 1986 (Smith, McMillan and Merline 1987, Ap.J.,317, L79) revealed a systematic, apparently periodic, variation in radial velocity having a "sawtooth" waveform with a full amplitude of 160 m·sec^{-1} and period of 1.84 or 2.18 days. This discovery was qualitatively confirmed by Cochran (1987, B.A.A.S.,19, 1052). Additional data show that large night-to-night velocity variations of magnitude about 80 m·sec^{-1} persist over a 192 day baseline. A sharp velocity transition has been monitored on at least one night, which covers a range of 100 m·sec^{-1} with an acceleration of 19m·sec^{-1}·hr^{-1}. The positive slope during this transition argues for a 2.18 day rather than the 1.84 day period. Furthermore, the amplitude and shape of the variation may be changing, perhaps suggesting multiple periods. To test the likelihood that radial pulsation could be responsible for the observed variations, stellar pulsation models are being conducted with Dr. Arthur Cox at Los Alamos National Laboratory. Using a structure model based on our best estimates for the stellar parameters, we have found that the 2 day period would require the third or fourth radial overtone, which are driven (unstable) with relatively high growth rates.

This work is supported by NASA grants NAG2-52 and NAGW-1283 and NSF grants AST-8403285 and AST-8714817, Tom Gehrels' Spacewatch Project, University of Arizona Observatories and the T-6 group at Los Alamos National Laboratory.

THE EFFECTS OF METAL ENHANCEMENT ON THE PERIOD RATIO OF DOUBLE MODE CEPHEIDS

Siobahn Morgan
University of Washington, Seattle WA, USA

Arthur N. Cox
Los Alamos National Laboratory, Los Alamos NM, USA

Abstract. We increase the metal abundance in the envelope of Double Mode Cepheids (DMCs) in order to remove the period ratio mass discrepancy. By increasing the value of Z, we are able to increase the opacity by a factor of 2 to 2.5. We find that for a star of 5 solar masses, with a temperature of 5957 K, and a luminosity of 1050 times that of the sun, an increase in the metal abundance starting at the Log (T) value of 4.79 and reaching a maximum Z value of 0.3 in the Log (T) range of 4.8 to 5.3 produces a period ratio $P_1/P_0 = 0.7089$, and a value of $P_0 = 3.001$ days. The value of Z returns to 0.02 at Log (T) = 5.62. The enhanced zone (part of which convects) is located well below the H and He convective regions. The enhanced region is also stable against downward mixing since the radiative gradient is so sub-adiabatic.

Further calculations are being pursued to investigate the formation of a layer of enhanced metals by radiative acceleration, which will levitate higher Z material upward to regions of lower density and thereby increase the heavy element abundance. We find that the radiative acceleration is greatest in a region of Log (T) = 5.8 to 6.0, about 15% of the mass into the star. This could lead to an increase in the Z abundance at lower temperatures where the acceleration is not as great, and where CNONe could be accumulated.

THE IMPORTANCE OF 3:1 RESONANCES IN STELLAR PULSATIONS

Pawel Moskalik
Physics Department
University of Florida
Gainesville

J. Robert Buchler
Physics Department
University of Florida
Gainesville

Abstract. The 2:1 resonance between the fundamental and the second overtone modes has received a great deal of attention in the context of Cepheids. It was clearly shown that it causes the Hertzsprung bump progression and brings about the very characteristic observed variation of the Fourier phases with period (Buchler & Goupil, 1984, Ap.J., 279, 394; Klapp, Goupil & Buchler, 1985, Ap.J., 296, 514; Buchler & Kovacs, 1986, Ap.J., 303, 749).

Again the amplitude equation formalism again here we show that the occurrence of a 3:1 resonances can have a very similar effect on the pulsation and, in particular, on the behavior of the Fourier phases as the resonance region is crossed. There does not seem to be any simple way to discriminate in observational data between the two types of resonances. The astrophysical implications of 3:1 resonances are being explored.

This work was supported in part by NSF grant AST86-10097.

STATISTICS OF PULSATING VARIABLES

M.F. Novikova
Astronomical Council of the Academy of Sciences of the USSR

Yu.A. Fadeyev
Astronomical Council of the Academy of Sciences of the USSR

Abstract. Classification of pulsating variables is based on a number of properties but the most important of them are the period and amplitude of the light variation. These quantities also play an important role in understanding the evolutionary status of the pulsating stars since the pulsation period is related to the mass and radius through the period--mean density relation while the amplitude of the light curve characterizes the efficiency of the mechanism responsible for the pulsational instability. In the present study we considered the period--frequency and light amplitude--frequency distributions for ten types of the most numerous pulsating variables from the fourth edition of the General Catalogue of Variable Stars. The following types of pulsating variables were considered: DSCT, DSCTC, DCEP, RR, RRAB, RRC, CWA, CWB, RV and M. Using these distributions we estimated the upper and lower limits of the period and light amplitude within which 95% and 99% of the pulsating stars of a given type are contained.

NONLINEAR RR LYRAE MODELS AND DOUBLE MODE PULSATION

Dale A. Ostlie
Physics Department
Weber State College
Ogden, UT

Abstract. Nonlinear models of RR Lyrae variables were calculated in the region of the HR diagram containing double mode pulsation. In all models T_{eff} = 7000 K, the luminosity was 60 times solar, X = 0.70, Y = 0.299, and P_1/P_0 = 0.746, characteristic of double mode RR Lyrae variables in M15 (Cox, Hodson, and Clancy, 1983). Models of 0.65 solar masses were calculated using standard opacities and models of 0.75 solar masses were calculated with opacities artificially enhanced by a factor of 1.2 between log(T) = 5.2 and log(T) = 5.9 (Andreasen and Petersen, 1988).

Incorporating time dependent convection generally decreased fundamental mode growth rates and increased first overtone growth rates. However, the first overtone enhanced opacity model appears to be exceptional. Although higher opacities lowered linear growth rates for both the fundamental mode and the first overtone, the growth rates were found to increase in the nonlinear regime. In particular, preliminary calculations indicate that the enhanced opacity model exhibits a positive fundamental mode growth rate near the first overtone limit cycle and a positive first overtone growth rate near the fundamental mode limit cycle. Apparently these models show double mode behavior.

PULSATIONS OF EÖTVÖS SPHERES

W. Dean Pesnell
Department of Astronomy
New Mexico State University
Las Cruces, NM

Abstract. A derivation of Eötvös spheres and their pulsations is presented. These spheres incorporate a short range force in their gravitational potential. We used a Yukawa formulation to agree with other workers in this field, although a slightly different form would guarantee a gravitational force that is always attractive. Various relationships are obeyed by these obejcts, much like the polytropes to which they are similar. An attempt is made to test whether a short range force can resolve two outstanding problems in astrophysics. These are the Cepheid mass discrepancy and the frequencies of high-order solar p-modes. The disagreement in predicted and observed Cepheid masses cannot be explained by a repulsive force, but the solar p-modes have their asympotoic frequency decreased. Linear fits to sequences of models are presented for the radial eigenfrequencies and asymptotic frequencies for nonradial pulsations.

A SEARCH FOR RR LYRAES IN WIDE BINARY SYSTEMS

Charles F. Prosser
Lick Observatory, Board of Studies in Astronomy and
Astrophysics
University of California, Santa Cruz

Abstract. RR Lyraes that have nearby visual companions may be examined to decide if they do indeed belong to a real binary system. Identification of RR Lyraes in binary systems could lead to the determination of physical characteristics of the variable and to an understanding of the history and evolution of such systems in the Galaxy. Where possible, proper motion analysis, using astrograph plates from the Lick Proper Motion Program, for some visual binaries has been done and has enabled the rejection of some variables from consideration. The variables which have been shown not to be physically associated with their visual companions are YZ Cap, V445 Oph, BH Peg, and TX Vir. EZ Lyr, RW Ari, and V494 Sco still remain possible cases requiring relative radial velocities between the variable and companion. Limited plate material has meant that one must resort to radial velocity measurements for the RR Lyrae and close companion in order to determine the nature of the pair.

Partial funding to attend this meeting was received from NSF grant no. AST 84-14142 (to B. Jones) and from the University of California.

THE BINARY CEPHEID S SAGITTAE REVISITED

Mark H. Slovak
University of Wisconsin
Washburn Observatory

Thomas G. Barnes, III
University of Texas
McDonald Observatory

Nancy R. Evans
University of Toronto
Department of Astronomy and
Computer Science Corporation
Astronomy Programs

Douglas L. Welch
Dominion Astrophysical Observatory
Herzberg Institute of Astrophysics

Thomas J. Moffett
Purdue University
Department of Physics

Abstract. A new orbit has been derived for the binary Cepheid S Sagittae = HD 188727 = BD+16°4067, using both extant data and new radial velocities derived from radial velocity spectrometer observations at McDonald Observatory and the Dominion Astrophysical Observatory. We deconvoled both the pulsational and orbital components from the observed velocities in order to fit a truncated Fourier series to the pulsation velocities and to derive orbital elements respectively. We confirm the orbital elements given in Herbig and Moore (1952) and speculate that the secular trend evident in the orbital (O-C) diagram may be interpreted as evidence for a third component. (See Evans, Slovak and Welch, this volume, for IUE observations supporting this interpretation.)

CCD OBSERVATIONS OF VARIABLE STARS IN GLOBULAR CLUSTERS

H.A. Smith
Michigan State University

J.R. Kuhn
Michigan State University

J. Curtis
Michigan State University

Abstract. BVR observations of the relatively metal-rich globular cluster NGC 6388 have been obtained with a CCD on the CTIO 0.9 m telescope. Eighteen possible short period variable stars have been discovered in or near the cluster. At least 10 of these are probable RR Lyrae members of NGC 6388. We confirm the finding of Hazen and Hesser that this cluster is one of the most metal-rich to contain a significant number of RR Lyraes. A program of CCD photometry of field and cluster variable stars has been initiated on the 0.6 m telescope of the Michigan State University Observatory.

THE CHROMOSPHERE OF β CASSIOPEIAE

Terry J. Teays
Behlen Observatory
Department of Physics and Astronomy
University of Nebraska-Lincoln and
Astronomy Programs
Computer Sciences Corporation

Edward G. Schmidt
Behlen Observatory
Department of Physics and Astronomy
University of Nebraska-Lincoln

Massimo Fracassini
Dipartimento di Fisica
Università degli Studi di Milano

Laura E. Pasinetti Fracassini
Dipartimento di Fisica
Università degli Studi di Milano

Abstract. We have carried out high dispersion, long wavelength IUE observations and ground based photometry of the δ Scuti star, β Cas. Our ground based observations were used, together with the previous results of Antonello et al. (1986, I.B.V.S. 2958, 1986), to ensure that the IUE observations were correctly phased relative to the photometric variation. Fluxes for the emission core of the Mg II k line (2796 Å) were obtained from 23 ultraviolet spectra taken over several cycles in 1986 and 1987. The emission flux, if present, was measured with respect to the mean line profile of all of the spectra. Emission was present at phases between 0.4 and 0.5. This is in contrast to what has been observed for another δ Scuti star, ρ Pup, and for the classical Cepheids, where the emission appeared at maximum light.

THE UNUSUAL PERIOD DISTRIBUTION OF THE RR LYRAE VARIABLES IN NGC 5897

Amelia Wehlau
Department of Astronomy
The University of Western Ontario
London, Ontario, Canada N6A 3K7

Abstract. Attention is called to the rather unusual distribution of the periods of the RR Lyrae variables in NGC 5897, a metal-poor halo globular cluster with a very low central concentration. Of the seven RR Lyrae stars known in the cluster, three have periods between 0.797 and 0.856 day and two have periods of 0.45 and 0.42 day. The other two have periods of 0.34 and 0.35 day with much lower amplitudes of variation. Due to the lack of crowding in this cluster photoelectric observations and Fourier decompositions of the resulting light curves should be possible for at least six of the RR Lyrae variables. In addition, the cluster appears to contain a non-variable horizontal branch star, SK 120, lying within the instability strip. As this is the only well documented case of such a star, photoelectric observations of this star would also be desirable.

PULSATION OF α CIR (HD 128898)

Werner W. Weiss
Institut for Astronomy, University of Vienna, Vienna, Austria

Hartmut Schneider
Universitaets-Sternwarte
Goettingen, Federal Republic of Germany

Abstract The group of pulsating CP2-stars (also called "rapidly oscillating Ap stars" provides asteroseismology with oscillation spectra of high accuracy. The potential as a diagnostic tool for modelling stellar interiors is widely appreciated. The identification of pulsation modes is important for such an analysis. However, this is rarely possible in an unambiguous manner. To improve the situation and to make use of additional information, we observed HD 128898 simultaneously spectroscopically and photometrically at ESO. For each of our individual CAT-CES spectra (1 minute integration time) it was thus possible to determine the pulsation phase at mid-exposure. A total of 887 spectra (R = 50000) were binned according to their pulsation phase and coadded to improve significantly the signal to noise ratio.

Although a full amplitude of about 6 millimagnitudes in Strömgren v was observed, we were unable to detect significant variations in radial velocity or spectral line profiles with the pulsation period of 6.3 minutes. An upper limit for variations in radial velocity can be estimated to be about 100 m·sec^{-1}. This value gives an upper limit for the radial velocity to light amplitude ratio in B of about 20 km·sec^{-1}·magnitude^{-1}, which is considerably smaller than the value of 60 km·sec^{-1}·magnitude^{-1} as is published for HR 1217 by Matthews et al. (1988, Ap.J.,324, 1099).

For the full paper consult Astronomy & Astrophysics, 1988.

INDEX OF TOPICS

A stars	65, 66, 79, 288
Abundances, Helium	164
Abundances, metals	7, 12, 54-55, 89-94, 137-139, 142-150, 171, 278
Angular diameters, stellar	193-194
Anomalous Cepheids	see Cepheids, anomalous
Asymmetric drift	86-89, 205-206
Asymptotic giant branch	30, 38, 67-79, 172, 269
B star pulsation	256
Baade-Wesselink method	132, 170, 173, 191-203, 250, 253, 255, 266
Baade-Wesselink method, inverted	199
Be stars	65
Beta Cephei stars	14
Binaries	
contact	236-237
detached	237-238
Blue band	28, 53
Blue edge	23
Blue loop	27, 32, 53, 172
Blue stragglers	66, 222-239
in color magnitude diagrams	224-225
in globular clusters	222-223
luminosities	223
masses	225-226
variable	226-238
Boltzman equation, collisionless	85
Bump mechanism	22
Carbon stars	38, 68, 74, 269
Cepheids, anomalous	216-221
light curves	219-220
masses	216-217, 220-221
periods	217
pulsation modes	217
Cepheids, classical	19, 64, 65, 67, 79, 178, 252, 263, 280
binary	3, 260, 284
bump	5, 279
as distance indicators	169-175, 177-178
double mode	6, 14, 248, 278
extragalactic	171-175, 177-188
instability strip	23, 27
luminosities	43, 55
masses	3, 5, 7-14, 24, 42-44, 252, 282

Index of topics

Cepheids, classical (continued)
    period-frequency
        distribution   45-50, 178
    period-luminosity
        relation   169-170, 171
    pulsation modes   4-6, 275
    radii   2, 5, 14, 266
Cepheids, type II   172, 178, 262, 265, 267, 268, 280
Chandrasekhar limit   24
Chaos   247, 267, 268, 270
Charge coupled devices   179
Chemical evolution   92
Chromospheres   286
Clusters, globular   35, 36
Clusters, galactic   35
Collapse, dissipative   90
Color magnitude diagram   40, 142, 147-150
Convective overshoot   7, 15, 26, 33, 34-38, 42, 44, 103, 108
Cooling, gas cloud   94
CORAVEL   199, 255
Core mass   104
CORS method   195, 201

Delta Delphini stars   249
Delta Scuti stars   3, 12, 64, 66, 228, 235-236, 248, 249, 251, 271, 276, 280, 286
Distance scale   159-161, 169-175
Drift, asymmetric   86-89, 205-206
Duplicity   34

Element levitation   3
Eötvös spheres   282
Equation of state   60

FK Com stars   237
Fourier decomposition   6, 178-179, 180, 200

Galactic bulge   208, 209
Galactic spheroid   95-98
Galactic thick disk   98-101
Galaxies, distances   173
Gamma mechanism   3, 21, 259
Globular clusters,
    ages   135-139, 253, 255
    blue stragglers   222
    distances of   273
    as distance indicators   175
    main sequence fitting   122-132
    mass segregation   225-226
    metallicity   125-126, 285
    Miras in   208
    Second parameter
        problem   112

Index of topics

Hayashi limit	27, 42
Helium burning, core	26-28, 36, 103
Helium flash	28, 78, 104
Helium stars	261
Horizontal branch	24, 36, 103-119, 122-126, 162, 170, 172
Hubble's constant	172-175
Hydrogen burning, core	24-26, 36
Jeans' equations	85
K giants, metal poor	95
Kappa mechanism	3, 22, 259
Kelvin-Helmholtz time	75-79
Local Group galaxies	216
Long period variables	171-172, 269
Luminosity function	41
Magellanic Clouds	6, 15, 35, 38, 39
Magnetic fields	14
Main sequence fitting	159
Main sequence stars	66, 128-132
Mass loss	2, 26, 30, 32-33, 44, 63-80, 103, 104, 112, 252, 262
Reimers relation	64, 74, 76
Mass function	42
Maximum Likelihood method	196, 202
Miras	64, 65, 67-74, 79, 172, 205-212, 268
Kinematics	205-212
Masses	206-209
Period-luminosity relation	206-208
Progenitors	211
Pulsation mode	68, 70, 208
Novae	171
OB stars	65, 66
Oosterhoff effect	see Period shift, Sandage
Opacities	8, 10, 11, 26, 31, 50-54, 59-62, 248, 278
Period doubling	265, 268, 270
Period shift, Sandage	103-104, 108-112, 153, 158, 159, 161, 274
Planetary nebulae	30, 65, 68, 78-79, 174, 210
Polytropes	254
Population II stars	94, 99
Pre-main sequence stars	65
Pulsation	
instability	20-23
levitation	65
R Coronae Borealis stars	65, 261

Index of topics

R mechanism	22
Reactions, nuclear	27, 30, 31, 35
Red giants	26, 28, 40, 65, 66, 171, 172, 277
Resonances	279
Rotation	34
RR Lyrae stars	24, 28, 64, 65, 67, 89, 108, 121-139, 142, 151-158, 178, 217, 218, 236, 249, 280, 287
binary	283
as distance indicators	159, 171, 208, 217, 255, 273
double mode	281
and galactic structure	95, 97, 99, 206
Helium abundances	164
luminosities	121-139, 159-161, 163, 164, 170, 272
masses	163, 164
metallicities	122-139
periods	111
pulsation modes	217
pulsation properties	126-127, 281
radii	200, 272
and stellar evolution	161-163
temperatures	111, 157, 274
RV Tauri stars	67, 265, 267, 280
Semiconvection	25, 27, 103, 108
Semiregular variables	207
Solar oscillations	257, 258, 271
Standard candles	173-174
Star counts	96
Stellar winds	32, 104
Subdwarfs	95
Supergiants, pulsation	264
Supernovae	2, 24, 77, 83, 92-94, 174
Superwind	74-78
Surface brightnesses, stellar	193-200
SX Phoenicis stars	227, 236, 249
amplitudes	233-236
in clusters	227, 229-232
extragalactic	232-233
period-luminosity relation	233
pulsation modes	235
Thermal pulses	30
Tully-Fisher relation	172-175
Universe, age	66
W Ursae Majoris stars	236-237
W Virginis stars	see Cepheids, type II
Wesselink method	see Baade-Wesselink method
White dwarfs	74, 210, 259, 268, 271
Wolf-Rayet stars	65

INDEX OF OBJECTS

Andromeda
    Andromeda I   221
    Andromeda II   221
    Andromeda III   221
    CO And   6, 15
    Eta Aql   4, 197
    GP And   235
    SW And   272
Aquarius
    CY Aqr   227, 228, 233, 235
Aquila
    TT Aql   199
    U Aql   3
Aries
    RW Ari   283
    VW Ari   226
    X Ari   200

BD −12°2669   226
BD +25°1981   226
Bootes
    Alpha Boo   277
    TV Boo   272

Camelopardalis
    BL Cam   227, 228, 233
    RU Cam   3
Cancer
    AH Cnc   236
Capricornus
    YZ Cap   283
Carina
    anomalous Cepheids 217, 221
    dwarf galaxy   222, 232
Cassiopeia
    Beta Cas   286
    BM Cas   14
    SU Cas   4, 15
Centaurus
    Omega Cen   109, 141-167, 208, 227, 229-230, 233, 273, 274
    Omega Cen-NJL5   237
Cetus
    RR Cet   272
    XZ Cet   6, 221
Circinus
    Alpha Cir   288

Index of objects

Corona Borealis
    R CrB   261
Crater
    SU Crt   226, 235
Cygnus
    DT Cyg   5, 15
    SU Cyg   3, 7, 9, 14
    V798 Cyg   249
    V1719 Cyg   249

Draco
    anomalous Cepheids   217, 220
    dwarf galaxy   222, 225, 232
    SU Dra   272
    45 Dra   260
DV14   6

E 3   222, 223, 225
Eridanus
    RX Eri   272

Fornax
    cluster   175
    dwarf galaxy   217

Gemini
    RR Gem   272

HD 103095   253
HD 144972   5, 15
HR 7308   see V 473 Lyr
HV 2345   6
Hydra
    KZ Hya   227, 228, 233, 235

Index Catalogue objects
    IC 1613   170, 171, 177, 181, 182
    IC 4182   174
    IC 4499   127

Leo
    group   174
    Leo I,
      Anomalous Cepheids 217, 221
    RR Leo   272
Lynx
    TT Lyn   272
Lyra
    EZ Lyr   283
    V 473 Lyr   3, 15, 275
Large Magellanic
        Cloud   171, 172, 179, 206, 208, 217, 255, 269

Index of objects

Magellanic Clouds	177, 178, 179
Messier objects	
M3	105-108, 113, 115, 116, 122-124, 127, 163, 165, 222, 225, 226, 227, 232, 233, 273, 274
M3-V154	264
M4	123-126, 273, 274
M5	200, 273, 274
M5-V8	253
M15	108, 122-124, 127, 161, 163, 223, 273, 274, 281
M31	170, 171-172, 173, 181, 217, 255
M33	171-172, 173, 177, 178, 180, 181-182
M67	226, 233, 236, 237-238
M67-V5	237
M67-V8	237
M71	222
M80	223
M92	107, 108, 113, 114, 165
M101	169
M107	273, 274
Milky Way	83-101, 208, 255
Monoceros	
T Mon	4
Musca	
S Mus	3, 7
New General Catalogue objects	
NGC 188	225, 233, 236, 237
NGC 288	222
NGC 1866	35, 39-41
NGC 2419	222
NGC 2808	222, 225
NGC 3109	170
NGC 5053	215, 222-225, 227, 231, 233, 238
NGC 5466	215, 222, 223, 225, 227, 230, 231, 233, 236, 273, 274
NGC 5466-V19	216, 217, 219, 236
NGC 5466-V27	236
NGC 5466-V30	236
NGC 5466-V35	236
NGC 5897	222, 287
NGC 5927	206
NGC 6356	206
NGC 6388	285
NGC 6522	208, 212
NGC 6553	206
NGC 6637	206
NGC 6712	206
NGC 6752	112-113
NGC 6822	170, 177, 178, 180, 181-187
NGC 7492	222, 225

Index of objects

Norma
    S Nor   43, 44

Omega Centauri   see Centaurus, Omega Cen
Ophiuchus
    V445 Oph   283
    V974 Oph   249

Palomar clusters
    Pal 3   222
    Pal 4   222
    Pal 5   222
    Pal 12   222, 226
    Pal 13   222
    Pal 14   222
    Pal 15   222
Pegasus
    AV Peg   272
    BH Peg   283
    DY Peg   227, 228, 233, 235
Perseus
    AR Per   272
Phoenix
    SX Phe   227, 228, 233
Polaris   see Ursa Minor, Alpha UMi

Sagitta
    S Sge   260, 284
Sagittarius
    RY Sgr   261
    U Sgr   9, 43, 44
Scorpius
    V494 Sco   283
    V636 Sco   3, 7
Sculptor
    dwarf galaxy   222, 232
Scutum
    Delta Scuti   276
    EW Sct   6
    R Sct   270
Sextans
    Sextans A   170
    Sextans B   170
    T Sex   272
Small Magellanic
        Cloud   179, 208, 217, 221, 255
Supernova 1987A   2, 25

Index of objects

Taurus
    EU Tau   5, 15, 275
    Theta 2 Tau   251
Tucana
    BS Tuc   226
    47 Tuc   206

Ursa Major
    TU UMa   272
Ursa Minor
    Alpha UMi   4
    anomalous Cepheids   219
    dwarf galaxy   217, 219, 222, 232

Virgo
    cluster   169, 174, 175
    TX Vir   283
    UU Vir   272

WLM   170

MAY 2 5 1990